ERRATA: Diagram on p. 19

Entries of line 2 from above

Tibetan	Burmese-Lolo	Chin	Karen	Others	

should read:

Burman	Old Burman	Chin	Kachin	Naga	Lolo

IN SEARCH OF CHIN IDENTITY

NORDIC INSTITUTE OF ASIAN STUDIES
Recent Monographs

67. Asta Olesen: *Islam and Politics in Afghanistan*
68. Hans Antlöv: *Exemplary Centre, Administrative Periphery*
69. Arne Kalland: *Fishing Villages in Tokugawa Japan*
70. Weng Eang Cheong: *The Hong Merchants of Canton*
71. Christine Dobb*in: Asian Entrepreneurial Minorities*
72. Eldrid Mageli: *Organising Women's Protest*
73. Vibeke Børdahl: *The Oral Tradition of Yangzhou Storytelling*
74. Cecilia Nathansen Milwertz: *Accepting Population Control*
75. Sharifah Zaleha Syed Hassan and Sven Cederroth: *Managing Marital Disputes in Malaysia*
76. Antoon Geels: *Subud and the Javanese Mystical Tradition*
77. Kristina Lindell, Jan-Öjvind Swahn and Damrong Tayanin: *Folk Tales from Kammu – VI: A Story-Teller's Last Tales*
78. Alain Lefebvre: *Kinship, Honour and Money in Rural Pakistan*
79. Christopher E. Goscha: *Thailand and the Southeast Asian Networks of the Vietnamese Revolution, 1885–1954*
80. Helle Bundgaard: *Indian Art Worlds in Contention*
81. Niels Brimnes: *Constructing the Colonial Encounter*
82. Ian Reader: *Religious Violence in Contemporary Japan*
83. Bat-Ochir Bold: *Mongolian Nomadic Society*
84. Shaheen Sardar Ali and Javaid Rehman: *Indigenous Peoples and Ethnic Minorities of Pakistan*
85. Michael D. Barr: *Lee Kuan Yew: The Beliefs Behind the Man*
86. Tessa Carroll: *Language Planning and Language Change in Japan*
87. Minna Säävälä: *Fertility and Familial Power*
88. Mario Rutten: *Rural Capitalists in Asia*
89. Jörgen Hellman: *Performing the Nation*
90. Olof G. Lidin: *Tanegashima – The Arrival of Europe in Japan*
91. Lian H. Sakhong: *In Search of Chin Identity*
92. Margaret Mehl: *Private Academies of Chinese Learning in Meiji Japan*
93. Andrew Hardy: *Red Hills*

IN SEARCH OF CHIN IDENTITY

A Study in Religion, Politics and Ethnic Identity in Burma

Lian H. Sakhong

NIAS
Press

Nordic Institute of Asian Studies
Monograph Series, No. 91

First published in 2003 by NIAS Press
Nordic Institute of Asian Studies
Leifsgade 33, DK–2300 Copenhagen S, Denmark
tel: (+45) 3254 8844 • fax: (+45) 3296 2530
E–mail: books@nias.ku.dk • Website: http://www.niaspress.dk/

Typesetting by NIAS Press
Printed and bound in Thailand

Publication of this book was assisted with
economic support from the Swedish Research Council

British Library Catalogue in Publication Data

Sakhong, Lian H.

In search of Chin identity : a study in religion, politics and ethnic identity in Burma. - (NIAS monograph ; 91)
1. Chin (Southeast Asian people) - Burma - Race identity
2. Chin (Southeast Asian people) - Burma - Religion
3. Ethnology - Burma 4. Nationalism - Burma - Religious aspects - Christianity 5. Christianity and politics - Burma
I. Title II. Nordic Institute of Asian Studies
306'.08'09591

ISBN 0-7007-1764-1 (European edition)
ISBN 87-91114-15-2 (North American edition)

CONTENTS

MAPS

Acknowledgements

This book is a revised and expanded version of my doctoral dissertation, submitted to Uppsala University in 2000. It would never have been undertaken, completed or published were it not for the support of numerous extraordinary people and research institutions.

In order to obtain my materials, I have had the privilege to consult with libraries in various parts of the world. In Burma I used the libraries of Rangoon University, Mandalay University and the Burma Institute of Theology. In Singapore I explored the National Library and the Library of Trinity Theological College. In Thailand I visited the Library of Chiangmai University twice, and in India I depended on the Library of United Theological College, Bangalore, where many research papers and theses on Christian movements in the Mizoram State of West Chinram have been written and preserved. The British Library in London and the Library of the University of London preserve the most outstanding, indispensable materials concerning the British colonial period in India and Burma. In the United States I used the Library of Congress in Washington DC. In Scandinavia I consulted with the Library of the Nordic Institute of Asian Studies in Copenhagen, Denmark, which I visited twice, and the Library of Uppsala University, called Carolina Rediviva, where a number of books on contemporary Burmese study are preserved.

Moreover, I draw heavily on archival materials from many different countries, including the National Archives in Rangoon, Burma, and the Archive of the Tribal Research Institute in Aizawl, Mizoram, India. In order to obtain the missionary sources, I did archival research at the Official Archives of the American Baptist Churches at Valley Forge, Pa, USA, during the summer of 1997. I also visited the Official Archives of the American Baptist Foreign Missionary Society at Rochester, New York, where I bought eighteen reels of microfilm that contain the originals of all the missionaries' correspondence during the entire history of the American Baptist Chin Mission.

I would specially like to acknowledge the steadfast support from my professors and colleagues at the Department of Theology, Uppsala University. My deepest gratitude goes to the late Professor Carl F. Hallencreutz, whose theoretical insight and unfailing support throughout my study enabled me to keep going. Professor Alf Tergel, Dr Sigbert Axelsson and Dr Axel-Ivar Berglund also deserve my deepest appreciation for their dedicated and gentle guidance, both academic and personal.

Thanks are also due to the Swedish Research Council for the generous grant to facilitate the publication of the present book.

I would like to thank again the Nordic Institute of Asian Studies (NIAS), for the publication of this book. I am indebted to the staff of NIAS Press, especially my editor Leena Höskuldsson, for their patient contribution.

Finally, I would like to extend my appreciation and gratitude to my close friends and family. My good friend Frederic Love deserves a special word of heartfelt thanks for his editing work at the final stage of my dissertation; thanks also to Bob Delaplane and Alice Rinell Harmansson for their unfailing support and help throughout my study. The love of my dearest family has always been the greatest source of my strength: Per and Margareta Ehnfors, who always stand beside me as my godparents in Sweden, and my wife Aapen and our beloved children Van and Laura deserve my most heartfelt thanks. All have been instrumental in the completion of this book, and I humbly dedicate this work to them.

Uppsala, January 2003
Lian H. Sakhong

Abbreviations and Acronyms

ABCM	American Baptist Chin Mission
ABFMS	American Baptist Foreign Mission Society
ABM	American Baptist Mission (mission in Burma)
ABMU	American Baptist Missionary Union (earlier name of ABFMS)
AFO	Anti-Fascist Organization
AFPFL	Anti-Fascist People's Freedom League
BIA	Burma Independent Army
BMSL	Baptist Missionary Society of London
BNA	Burma National Army
CHBA	Chin Hills Baptist Association
DFA	Director of Frontier Areas
FA	Frontier Areas
FAA	Frontier Areas Administration
FACE	Frontier Areas Committee of Enquiry
GCBA	General Council of Buddhist Associations
HMG	His/Her Majesty's Government
KIH	Kaiser-I-Hind
KNU	Karen National Union
KYO	Karen Youth Organization
LPM	Lakher Pioneer Mission
PVO	People's Volunteer Organization
SCOUHP	Supreme Council of the United Hills People
SPFL	Shan People's Freedom League
UNLD	United Nationalities League for Democracy
YMBA	Young Men's Buddhist Association
ZBC	Zomi Baptist Convention

Glossary

Arnak-hman	sacrificial ceremony for marriage couple; literally, "black-hen sacrificial ceremony". (Black is religious symbol for 'invisible force' which refers to traditional religious concept of '*Zing*'.)
Bawi-phun	nobility
Biaknak	religion
Biak-inn	church, Christian religious institute
Chia-phun	ordinary class, commoners
Chung-um	household god, which protected family and clan; literally, the guardian god who is residing inside the house
Ding-thlu	the attribute of the righteousness of Supreme God, *Khua-zing*, in the concept of Chin traditional religion; literally, the one who paves the way of righteousness
Do-dang-tu	religious title of Chin traditional chief (*Ram-uk*) and priest (*Tlang-bawi*); literally, the one who offers sacrificial ceremonies for family, clan and community; the one who can communicate with guardian god called *Khua-hrum*
Khua-bawi	headman; the one who rules the village
Khua-bawi system	Chin traditional administrative system where the *Khua-bawi* ruled his village and community independently
Khua-chia	evil spirits, which, according to Chin traditional religious belief, caused all kind of sickness and harmed the soul and body of human being
Khua-hrum	Guardian god of village and community; the basic foundation for Chin traditional religion, for it was known as *Khua-hrum Biaknak*

Khua-hrum biaknak	Chin traditional religion, the religion of *Khua-hrum*
Khua-man	religious tax imposed on village, not on individual person
Khua-vang	shaman; literally, visible power of darkness, mysterious power of light, the one who can see in darkness
Khua-zing	Supreme God; the creator of the entire universe, including spiritual beings and human beings, the giver of *Zing* – the vital force of life, the Father of all, the meaning of everything and the source of life
Krifa biaknak	Christianity; literally, the sons of Christ or the followers of Jesus Christ
Krifa phung	Christian ways of life; Christians who live Christian ways of life, being a Christian
Lai-phung	Chin traditional ways of life which centred upon the worship of *Khua-hrum*
Lai-rel	the attribute of Supreme God, *Khua-zing*; literally, the one who judges fairly and correctly
Lulpi	celestial being who abode in heaven but not the place where *Khua-zing* lives, and who is the helping spirit of *Khua-vang*
Mithi-khua	the world of death; the place where the departed human soul lives life after death
Phun	social class
Phung	culture
Phunglam	ways of life, including religious system, political system and almost every aspect of life
Pulthi-khua	a place for who died a normal death at home
Rai-thawi	sacrificial ceremonies connected with sickness; literally, appeasement to the moving evil spirits
Ram	nation-state, country, homeland, territory independently ruled by Chin traditional Chieftain called *Ram-uk*
Ram-uk	chief, the ruler of the land and the people, the lord of the soils

Ram-uk system	Chin traditional administrative system where tribal or local chief called *Ram-uk* ruled at least two villages, but usually the entire tribe or several villages and communities; feudal system
Sa-khua	religious tax, particularly for community sacrificial ceremonies such as *Tual Khua-hrum Dangh* and *Tlang Khua-hrum Dangh.* This also can be referred to other religious taxes collected by the Chief (*Ram-uk*) and the Priest (*Tlang-bawi*).
Sal	slave
S*arthi-khua*	g place for who died accidental death
Thim-zing	great darkness, related to Chin traditional religious belief in the beginning and the end of the world; great flood (*Chun-mui* in Lia dialect, *Khazanghra* in Mara dialect)
Thla-rau	human soul
Tlang-bawi	priest
Zing	invisible force of life that comes from God, *Khua-zing*; the source of life
Zing-dangh	sacrificial ceremonies related to individual *Zing*, but it can also be referred to all kind of sacrificial ceremonies for the Chin traditional religious was fundamentally based upon the belief in *Zing* and its giver: *Khua-zing.*

Introduction

In many Asian countries today, potential conflicts exist between the majority religion and culture and ethnic minorities who practise another religion. Problems are easily aggravated if the government applies a confessional policy on religion, which favours the majority religion, while minority religions are marginalized or even suppressed. Contemporary Burma, or Myanmar, is one example. Actually, the very name 'Myanmar' implies confessional claims and ethnic exclusiveness, even if the present military junta would prefer to be characterized as having opted for a secular policy on religion.[1]

In Burma, the current situation of the Chin, Kachin and Karen, for instance, can be seriously questioned from the human rights point of view. As far as religious conditions are concerned, there is complex interaction between Christianity, the religion of these ethnic minorities, and Buddhism, the majority religion. Such problems make Burma an obvious case for the study of religious and ethnic identity within the framework of a multi-ethnic, multi-religious plural society of a modern nation-state. In this study, I shall investigate the complex interaction between religious and ethnic identity among the Chin people in what is now called the Union of Burma.

STUDY IN ETHNICITY

Studies in ethnic identity, or what is now generally called ethnicity, have advanced considerably since the mid-1950s, and a voluminous literature has sprung up around this concept, much of it concerned with the problems of how ethnicity should be defined in relation to concepts such as gender, class and state. The term ethnicity seems to be rather new, first appearing in English as late as the 1950s. The meaning of the term still is subject to discussion.[2] Pioneer theories will be looked at closer in Appendix I (p. 245 below).

In this study, ethnicity will be used synonymously with nationality and the terms ethnic identity and national identity will be used synonymously. I shall come back to my definition of national identity, nation and state later.

I shall not lean on one particular school of thought, but utilize different theories when they seem appropriate with the context of the study. However, I shall follow, at least for the sake of convenience, Anthony

Smith's and John Armstrong's approaches of the 'ethno-symbolic', when defining terms such as 'ethnicity' 'nation', 'state', etc. below.

Anthony Smith's definition of 'ethnic group', or *ethnie,* includes:

1. a common *proper name,* to identify and express the 'essence' of the community;
2. a myth of *common ancestry,* a myth rather than a fact, a myth that includes the idea of a common origin in time and place and which gives an *ethnie* a sense of fictive kinship, a super-family;
3. shared *historical memories,* or better, shared memories of a common past or pasts, including heroes, events and their commemoration;
4. one or more *elements of common culture,* which need not be specified but normally include religion, customs and/or language;
5. a *link* with a *homeland,* not necessarily its physical occupation by the *ethnie,* only its symbolic attachment to the ancestral land, as with Diaspora peoples; and
6. a *sense of solidarity* on the part of at least some sections of the *ethnie* population (A. Smith 1986: 14; Hutchinson and Smith 1996: 6–7).

In addition to his definition of ethnic groups, Anthony Smith speaks of 'nation' as 'a named human population sharing territory, common myths and historical memories, a mass, public culture, a common economy and common legal rights and duties for all members' (A. Smith 1986: 14). According to *The Blackwell Dictionary of Political Science,* the term 'nation' is defined as follows:

> [Nation is] a body of people who possess the consciousness of a common identity, giving them a distinctiveness from other peoples. Hence 'togetherness' and 'separateness' are important parts of national consciousness. The consciousness will be based upon common historical experiences (which may be partly based on myth), and other shared features such as geographical propinquity and a common culture including literature and language. As with all feelings, that of belonging to a nation is bound to be relative. Different nations may be encompassed within the boundaries of states (Beady 1999: 219).

While 'nation' here is defined in historical and cultural terms, the term 'state' is defined by 'territory and coercive power' (ibid.: 219). The political sovereignty therefore is the indispensable feature of a 'state'. However, according to the Constitution of the Union of Burma in 1947, which was a federal union at least in terms of theory, and even the Union Constitution of 1974, the States (with a capital 'S') were formed within the Union, such as the Chin State, Kachin State, Karen State, etc., but without political sovereignty. According to constitutional federal theory, this kind of State (capital 'S', within the Union) must be invested with political power of legislative, judiciary and administrative authority, which are the basics for self-determination, for a people and a nation. Self-determination does not necessarily mean political sovereignty. In order to avoid terminology confusion, however, I shall use in this study the term 'nation-state' for a 'state' which is politically independent and recognized by the world as a sovereign country.

I therefore define the Chin people as a 'nationality' or 'ethnic nationality', and Chinland or Chinram[3] as a 'nation', but not as a nation-state, based on all the well-recognized theories that I have just mentioned and also based on the traditional Chin concepts of *Miphun, Ram,* and *Phunglam.* The meaning and concept of *Miphun* is an *ethnie* or a 'race' or a 'people' who believe that they come from a common descent or ancestor. *Ram* is a homeland, a country or a nation with well-defined territory and claimed by a certain people who have belonged to it historically; and the broad concept of *Phunglam* is 'ways of life', which includes almost all cultural and social aspects of life, religious practices, belief and value systems, customary law and political structure and the many aesthetic aspects of life such as dance, song, and even the customs of feasts and festivals, all the elements in life that 'bind successive generations of members together' as a people and a nationality, and at the same time separate them from others.

OBJECTIVES OF THE STUDY

Prior to British annexation in 1896, the Chin were independent people ruled by their own traditional tribal and local chiefs, called *Ram-uk* and *Khua-bawi,* respectively. Although all the tribes and villages followed the same pattern of belief systems, the ritual practices in traditional Chin religion – called *Khua-hrum* worship – were very much mutually exclusive and could not serve to unite the entire Chin people under a single religious institution. Thus, until the British occupation, Chin society remained tribal: people's identification with each other was tribally exclusive and their common national identity remained elusive.

By the turn of twentieth century, however, Chin society was abruptly transformed by powerful outside forces. The British conquered Chinram, and Christian missionaries followed the colonial powers and converted the people. Within this process of change, the Chin people found themselves in the midst of multi-ethnic and multi-religious environments, which they did not welcome. They also realized that their country was not the centre of the universe but a very small part of a very big British Empire. After the colonial period, they found themselves again being separated into three different countries – India, Burma and Bangladesh – without their consent. While West Chinram (the present Mizoram State) became part of India, East Chinram (present Chin State) joined the Union of Burma according to the Panglong Agreement signed in 1947. The smaller part of Chinram became part of East Pakistan, present-day Bangladesh.

Primary agent of change, in my hypothesis, is modern political systems represented by British colonial power and its successors – independent India and Burma. Political development, of course, is the only agent with necessary power to force change. In tribal society, 'distinction cannot easily be made between religious, social, cultural and political elements' (Downs 1994: 28). Anything that effects one aspect of life can strongly effect every aspect of life. In fact, 'tribal society can only be maintained through traditional instruments of integration, if they remain in fundamental isolation from other societies' (ibid.). When centuries-old isolationism in Chinram was broken up by British colonial power, the traditional way of maintaining the tribal group's identity was no longer effective, and the process of de-tribalization had begun.

The process of de-tribalization can be a dangerous moment because that process could either become what Frederick Downs called the process of 'dehumanization', or a process which Swedish scholar Eric Ringmar called a 'formative moment' (Ringmar 1996: 145). If the process becomes that of dehumanization, that is, 'to rob them of their essential life' of 'people's soul', the existence of tribal peoples could be in danger. There are many examples, according to Downs, in the Americas, Africa, India and other parts of Asia where many tribal peoples have become extinct (Downs 1994: 24). On the other hand, the process of de-tribalization can become a 'formative moment' if the people find other alternatives, instead of seeking 'to revitalize the old culture'. In my hypothesis, the process of de-tribalization in Chin society became a process of 'formative moment', a time in which new meaning became available and people were suddenly able to identify themselves with something meaningful. Christianity provided the Chin people the new meanings and symbols within this process of 'formative moment', but without 'a complete break with the past'. I shall therefore focus in this study on how Christianity helped the Chin people – no longer as divided tribal groups, but as the entire nationality of Chin ethnicity – to maintain their identity, and how Christianity itself became a new creative force of national identity for the Chin within this 'formative' process of powerful changes.

In order to highlight the Chin response to the new religious challenge and how they became Christians, I shall approach this study from the Chin local perspective. Thus, instead of investigating purely institutional development of the Chin churches, I shall try to investigate the gradual shift from traditional Chin religion to Christianity as an integrating factor in the development of Chin self-awareness. In this way, I shall analyse the local stories that people tell about their society and about the past, especially events personified in ancestors and other historic figures. Through such stories, small and large, personal and collective, the Chin people do much of their identity work together. In other words, such stories hold 'history and identity together' (White 1995: 3).

The most prominent and frequently repeated local stories are about the moment of first confrontation with colonial power and the Christian mission, and subsequent conversion to Christianity. The stories of conversion are repeatedly told and retold, often in narrative accounts as writings, songs, sermons and speeches passed on during such occasions as religious feasts, celebrations and worship services. These are times when people engage in exchange practices that define social and political relations. Although the wars against British annexation (1872–96), the Anglo–Chin War (1917–19), World War II and Japanese invasion (1939-45), and the Independence of Burma (1948) are significant junctures in temporal consciousness, the events of Christian conversion are uniquely important in the organization of a socio-historical memory.

In present Chin society, telling dramatic versions of the conversion stories has become almost a ritual practice during Sunday worship services and the annual Local and Association Meetings, called *Civui,* where villages and communities commonly gather to recall the past. Narratives of shared experience and history do not simply represent identity and emotion, they even constitute them. In other words, histories told and remembered by those who inherit them are discourses of identity, just as identity is inevitably a discourse of history. Thus 'history teaching', as Appleby claims, 'is identity formation' (Appleby 1998: 14). Especially for the people who live in communities transformed by powerful outside forces, the common perception of a threat to their existence, as well as the narrative accounts of socio-religio-cultural contact with the outside world, created identity through the idiom of shared history. However, just as 'history is never finished, neither is identity' (White 1995: 3). It is continually refashioned as people make cultural meaning out of shifting social and political circumstances. I shall therefore analyse in this study how the old tribal and clan identities were gradually replaced; how Christianity provided a means of preserving and promoting the self-awareness of Chin identity through its theological concepts and ideology and its ecclesiastical structure; and how the Chin people gradually adjusted to Christianity through an accelerated religious change in their society.

When the Chin of the Chin State in Burma celebrated the centennial anniversary of the arrival of Christianity on 15 March 1999, more than 80 per cent of the Chin population professed Christian faith.[4]

In West Chinram (Mizoram State) in India, church growth has been even faster and the life of Christianity more vigorous than in the Chin State of Burma. This means that in a mere century, almost the entire Chin population has accepted the Christian faith. Such unusually fast growth of Christianity on both sides of Chinram and the vitality of its practice are the factors that I intend to explore in this study. However, I shall limit myself within the framework of the first fifty years of Christian activity in East Chinram (Chin State in Burma).

Christianity in East Chinram is a fruit of the work of the American Baptist Mission. There have also been significant indigenous Chin missions, notably the Chins for Christ in One Century (CCOC).[5] During the first five decades of the twentieth century, a great number of Chin became Christians, and the church became deeply rooted in the socio-cultural tradition of the people.[6] In my hypothesis, the growth of Christianity in East Chinram during the first five decades of the twentieth century was based on the contribution of at least three factors: 1) the socio-political change represented by British colonial power, 2) the missionary factor (both foreign and indigenous) and 3) the theological similarity between Christianity and traditional Chin religion, which meant that conversion was not a radical change but a religious transformation from *Khua-hrum* oriented ritual practices to a *Khua-zing* oriented worship service within the same conceptual pattern of belief system.

In addition to the rapid growth of Christianity among the Chin, another problem that requires explanation is the indigenous character of Christianity in Chinram, or what Chin theologian Mangkhosat Kipgen called 'a uniquely *Chin* form of Christianity' and that 'it was this indigenization of Christianity which made it possible for the rapid growth to take place' (Kipgen 1996: 4). One of the main purposes of this study, therefore, is to explore the theological continuity between the traditional Chin religion and Christianity, which serves as a link between indigenization and the growth of Christianity in Chinram. This also is a reason why Christianity could provide the Chin people a means of preserving their identity and promoting their interests in the face of powerful change.

In order to understand the nature of change and continuity in Chin society, I shall first explore the Chin social structure and traditional ways of life – including a religious system, ritual practices and political and economic systems. Second, from the Chin perspective of resistence, I shall investigate the nature of change, which was thrust upon Chin society by outside forces. In this way, I shall highlight how the Chin adjusted to their new faith within this process of powerful change, and how Christianity provided a means of preserving their identity and self-awareness in the new contexts of multi-racial and multi-religious environments.

STRUCTURE OF THE BOOK

This study is arranged into three parts. Part One deals with pre-Christian Chin social structure, traditional religion and historical developments. After this general introduction, I shall describe the historical development of the Chin people – such as the origin of the Chin, migration patterns and Chin settlements in the Chindwin Valley and present Chinram; and some conceptual structures of Chin traditional religion. It is important to discover the significance of the traditional Chin belief system, the uniqueness of its culture and the relationship between religious, cultural, political and social elements in Chin society, so that we shall be able to lay the groundwork for more systematic and rigorous testing of my hypothesis in Parts Two and Three.

The sense of belonging to a 'national group' or 'ethnicity' was consolidated among various Chin clans and tribes by their possession of common characteristics, including traditional religion and cultural, historical and linguistic traits. Without a serious investigation of the wider historical circumstances in which ethnic phenomena take place, any understanding of Chin ethnicity will be blurred by our immersion in the intricate details of data, their locality and tribal contexts. Chapter 1 will examine the context in which the Chin construct themselves as a people and a nationality with a distinct common descent, homeland, religion, language or related dialects, culture and collective historical memory. Chapters 2 and 3 will examine the Chin traditional ways of life, *Phunglam,* in which the ritual system can be viewed as the confirmation of power and status in the society. Moreover, I shall analyse in Chapter 3 how the Chin traditionally viewed life here on earth as a preparation period for future life in *Mithi-khua.*

Part Two covers historical developments from an early British attempt to annex Chinram until the promulgation of the Burma Act of 1935. In Chapter 4, I shall focus on the British policy of expansion and illustrate how the Chin in their attempt to resist British policy experienced a new sense of common national solidarity. The main objective in Part Two, therefore, is to analyse the development of these changes. It is important to see how these developments of change brought about a crisis in Chin social life, mainly as a result of the changing traditional political system of chieftainship, and also because of economic crises caused by the prolonged war of resistance. The crisis in life, in my hypothesis, is closely linked to the conversion to Christianity. For the people who lived in communities transformed by powerful outside forces, the narrative account of socio-religio-cultural contact with the outside world was marked by religious conversion and the common perception of a threat to their existence, both of which became new identity-creating sources through the idiom of common memory and shared history.

Chapters 5 and 6 deal with the coming of Christianity to East Chinram, and the Chin response to it. Attention will be given to the formative process of the Chin Baptist Churches since the coming of the pioneer American Baptist missionaries in 1899. For various reasons, the Anglo–Chin War (1917–19) provided an important turning point. The confrontation between traditional Chin religion and Christianity, as noted already, produced a uniquely Chin form of indigenous Christianity. Not only did this Christianity become their new national religion within a short period; it also provided a means of preserving and promoting Chin self-consciousness of national identity through its theological concepts, ideology and ecclesiastical structure. Thus, I shall analyse how the Chin gradually adjusted to Christian ways of life – *Krifa Phung* – by finding the common ground between Christian teaching and traditional Chin religious concepts and belief systems.

In Part Three, my primary focus will be on how the Chin increasingly related to the Burmese attempt to form an independent Federal Union. Since the Burma Act of 1935 was promulgated in 1937, the Chin Christian and political movements in East Chinram became integral parts of the Burmese independence movement. Part Three therefore covers a momentous period in the history of the Chin people within the context of Burmese history as a whole – from the beginning of Burmese nationalist movements to the outbreak of World War II, the formation of the Burmese Independence Army (BIA) and the Japanese invasion of Chinram, from the negotiations between the Chin and the Burman nationalist leader Aung San at the Panglong Conference to the joining of Independent Burma. All these political events had a great impact on the Christian movement in East Chinram, which eventually became a member state of the Union of Burma. Chapter 7 is particularly focused on the repercussions of the Burma Act of 1935 and World War II, and Chapters 8 and 9 analyse the political reasons why the Chin and other ethnic groups voluntarily joined the Union of Burma at the Panglong Conference. Chapter 10 evaluates Chin spirituality – which is indeed the creative source of their new national identity – in the new context of the Union of Burma. In this way, I shall illustrate how the Chin increasingly articulated their own democratic values and Christian identity within the process of change that was thrust upon them.

After the Conclusion, some theoretical considerations and earlier studies will be presented in Appendices I and II.

NOTES

1 Ever since the first Myanmar kingdom of the Pagan dynasty, founded by King Annawrattha in 1044, the term ' Myanmar' has been used to denote the ethnicity of Myanmar, which in turn is inseparably intertwined with their state religion of Buddhism. The term Myanmar is exclusive, and does not include the Chin and other ethnic groups who joined the Union of Burma in 1947 on the principle of equality. Though the present military junta changed the country name from Burma to Myanmar after the unlawful military coup in 1989, almost all ethnic groups and democratic forces of Burma, led by Daw Aung San Suu Kyi, do not recognize the name, for it was changed by an illegitimate *de facto* government. I shall therefore use the term Burma to denote the country, and the term Myanmar will be used to denote the ethnic group of Myanmar interchangeable with the word Burman.

2 The term 'ethnicity' is first recorded in the *Oxford English Dictionary* in 1953. It is derived from the much older and commonly used adjective 'ethnic', which in turn derives from the ancient Greek term *etnos;* it was used as a synonym of *gentile*, that is, non-Jews or pagans in the New Testament Greek. In French, for example, the Greek noun survives as *ethnie*, which is associated with the adjective *ethnique*. As the English language has no concrete noun for *ethnos* or *ethnie,* the French term is generally

used to denote an 'ethnic community' or 'ethnic group'. See Hutchinson and A. Smith (eds) 1996, 4.

3 I use Chinland and Chinram interchangeably. At the 'Chin Seminar', held in Ottawa on 29 April–2 May 1998, Dr Za Hlei Thang, one of the most outstanding politicians and scholars among the Chin, proposed that the word *ram* in Chin should be used instead of the English word *land,* as *Chinram* instead of *Chinland.* It was widely accepted by those who attended the seminar.

4 'The Chin Christians Centenary Jubilee', a report in Chin written by Rev. Thawng Ling Mualhlun, on the Miavan Thawngttha Canhmannak (Internet). According to the report, the population of Chin State in Burma is 368,949 (1983 census).

5 The Chin indigenous mission program called 'Chins for Christ in One Century' (CCOC), began as 'Chin Hills for Christ' in 1964. Since I limit myself within the scope of 1896–1949, I shall not discuss in detail the CCOC program in this study. For further reading, I refer to Pum Suan Pau's doctoral dissertation: *Growth of Baptist Churches in Chin State: The Chins for Christ in One Century Experiences,* 1998.

6 During the first three decades of the twentieth century, 'the great majority of' the Chin people in Mizoram State in India had become Christian already. See Kipgen 1996, 1.

PART ONE

THE CHIN BACKGROUND

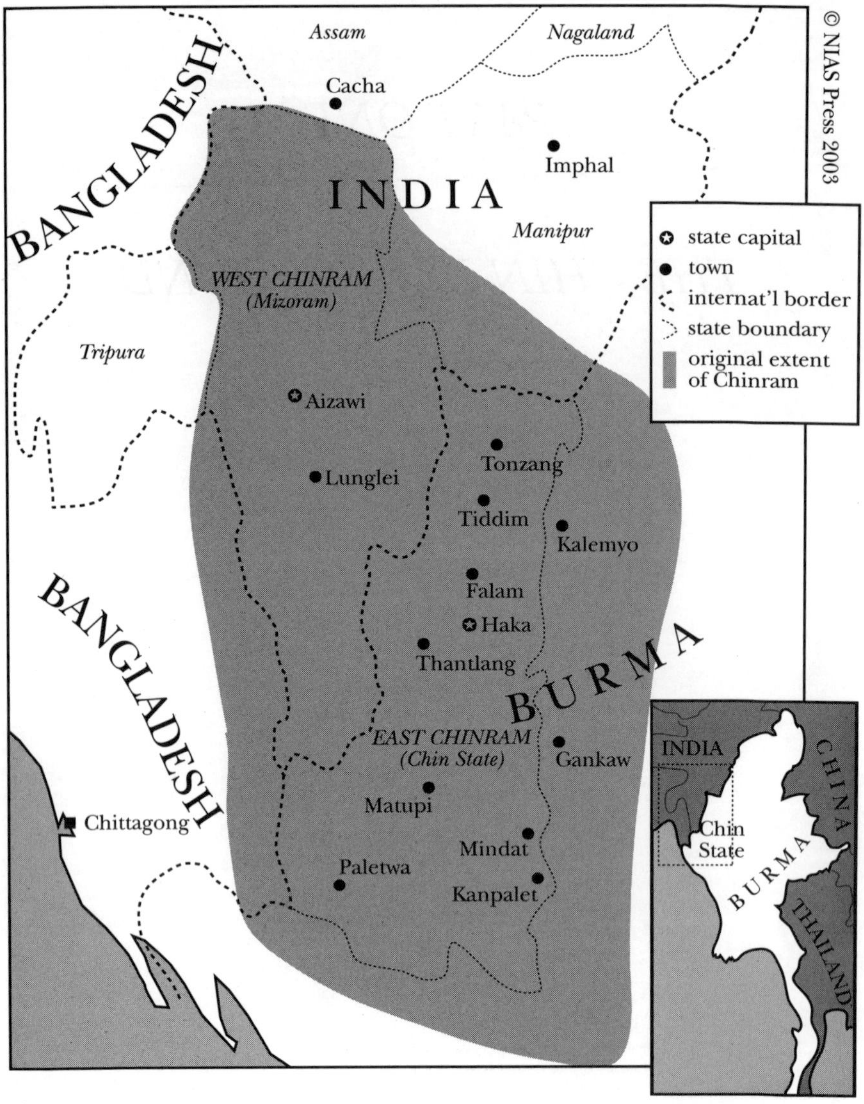

Map 1: Chinram

ONE

Chin Ethnicity

The introduction to this study characterized the Chin people as an 'ethnic nationality' by applying ethno-symbolic theory. In this chapter, I shall investigate who the Chins are, and why they can be described as an ethnic group; in other words, 'What makes a Chin a Chin?', 'What are the chief features that distinguish the Chin people as an ethnic nationality from other human collectives or ethnic groups?' and 'Which criteria make it possible for them to be recognized as a distinctive people and nationality?'. As already noted, ethno-symbolic theorists such as A. D. Smith suggest six main features which serve to define 'ethnic nationality': (i) a common proper name, (ii) a myth of common descent, (iii) a link with a homeland, (iv) collective historical memories, (v) one or more elements of common culture and (vi) a sense of solidarity (A. Smith 1986: 21–24).

In this chapter I shall explore the first four features, which correspond to traditional Chin concepts of *Miphun* and *Ram*. *Miphun* involves the common name and the myth of common descent, and *Ram* covers the concept of a common homeland and collective historical memories.

THE CHIN CONCEPT OF *MIPHUN*

A Collective Name

The common proper name of the 'Chin' is inseparably intertwined with 'the myth of common descent' and the 'myth of the origin' of the Chin. According to the origination myth, the Chin people emerged into this world from the bowels of the earth or a cave or a rock called 'Chinlung',[1] which, as we will see below, is spelled slightly differently by different scholars, based on various Chin dialects and local traditions: 'Chhinlung', 'Chinn-lung', 'Chie'nlung', 'Chinglung', 'Ciinlung', 'Jinlung', 'Sinlung', 'Shinlung', 'Tsinlung' and so on.

The tradition of 'Chinlung' as the origin of the Chin has been kept by all tribes of the Chin in various ways, such as folksongs, folklore and legends known as *Tuanbia*. For people with no writing system, a rich oral tradition consisting of folksong and folklore was the most reliable means of transmitting past events and collective memories through time. The songs were sung repeatedly during feasts and festivals, and the tales that made up Chin folklore were told and retold over the generations. In this way, such collective memories as the origin myth and the myth of common ancestors were handed down. Different tribes and groups of Chin kept the tradition of 'Chinlung' in several versions; the Hmar group of the Mizo tribe, who now live in Mizoram State of India, which I refer in this study as West Chinram, have a traditional folk song:

Kan Seingna Sinlung [Chinlung] ram hmingthang
Ka nu ram ka pa ram ngai
Chawngzil ang Kokir thei changsien
Ka nu ram ka pa ngai.

In English it translates as: 'Famous Sinlung [Chinlung] is my motherland and the home of my ancestors. It could be called back like *chawngzil*, the home of my ancestors' (Chaterjee 1990: 328). This folksong also describes that the Chins were driven out of their original homeland, called 'Chinlung'. Another folksong, traditionally sung at the *Khuahrum* sacrificial ceremony and other important occasions, reads as follows:

My Chinland of old,
My grandfather's land Himalei,
My grandfather's way excels,
Chinlung's way excels. (Kipgen 1996: 36)

Modern scholars generally agree with the traditional account of the origin of the name 'Chin': the word comes from 'Chinlung'. Hrang Nawl, a prominent scholar and politician among the Chin, confirms that the term 'Chin ... come(s) from Ciinlung, Chhinlung or Tsinlung, the cave or the rock where, according to legend, the *Chin* people emerged into this world as humans' (quoted by Vumson 1986: 3). Even Vumson could not dispute the tradition that the Chin 'were originally from a cave called Chinnlung, which is given different locations by different clans' (ibid.: 26).

In addition to individual scholars and researchers, many political and other organizations of the Chin accepted the Chinlung tradition not only as a myth but as a historical fact. The Paite National Council, formed by the Chin people of Manipur and Mizoram States, claimed Chinlung as the origin of the Chin people in a memorandum submitted to the Prime Minister of India. The memorandum stated: 'The traditional memory claimed that their remote original place was a cave in China where, for fear of enemies, they hid themselves, which is interpreted in different dialects as "Sinlung" [Chinlung] in Hmar and Khul in Paite and others.'[2] In this memorandum, they suggested that the Government of India take initiative to group all Chin people inhabiting the Indo-Burma border areas within one country as specified and justified for the safeguard of their economic, social and political rights.

The literal meaning of *Chin-lung* is 'the cave or the hole of the Chin', the same meaning as the Burmese word for *Chindwin,* as in 'Chindwin River', also 'the hole of the Chin' or 'the river of the Chin' (Lehman 1963: 20). However, the word *Chin-lung* can also be translated as 'the cave or the hole where our people originally lived' or 'the place from which our ancestors originated' (Z. Sakhong 1983: 7). Thus, the word *Chin* without the suffix *lung* is translated simply as 'people' or 'a community of people' (Lehman 1999: 92–97). A Chin scholar, Lian Uk, defines the term *Chin* as follows:

> The Chin and several of its synonymous names generally means 'People' and the name Chinland is generally translated as 'Our Land' reflecting the strong fundamental relationship they maintain with their land (Lian Uk 1968: 2).

Similarly, Carey and Tuck, who were the first to bring the Chin under the system of British administration, defined the word *Chin* as 'man or people'. They recorded that the term *Chin* is 'the Burmese corruption of the Chinese "Jin" or "Jen" meaning "man or people"' (Carey and Tuck 1976: 3).

Evidently, the word 'Chin' had been used from the very beginning not only by the Chin themselves but also by neighboring peoples, such as the Kachin, Shan and Burman, to denote the people who occupied the valley of the Chindwin River. While the Kachin and Shan still called the Chin as 'Khyan' or 'Khiang' or 'Chiang', the Burmese usage seems to have changed dramatically from 'Khyan' to 'Chin'.[3]

In stone inscriptions, erected by King Kyanzittha (1084–1113), the name Chin is spelled as 'Khyan' (Luce 1959b: 75–109). These stone inscriptions are the strongest evidence indicating that the name *Chin* was in use before the eleventh century.

Prior to British annexation in 1896, at least seventeen written records existed in English regarding research on what was then called the 'Chin-Kuki linguistic people'. These early writings variously referred to what is now called and spelled 'Chin' as 'Khyeng', 'Khang', 'Khlang', 'Khyang', 'Khyan', 'Kiayn', 'Chiang', 'Chi'en', 'Chien', and so on. Father Sangermono, an early Western writer, to note the existence of the hill tribes of Chin in the western mountains of Burma, lived in Burma as a Catholic missionary from 1783 to 1796. His book *The Burmese Empire,* published in 1893, almost one hundred years after his death, spells the name Chin as 'Chien' and the Chin Hills as the 'Chein Mountains'. He thus recorded:

> To the east of Chein Mountain between 20'30' and 21'30' latitude is a petty nation called 'Jo' (Yaw). They are supposed to have been Chien, who in the progress of time, have become Burmanized, speaking their language, although corruptly, and adopting their customs.[4]

In Assam and Bengal, the Chin tribes – particularly the Zomi tribe who live close to that area – were known as 'Kuki'. The term *Kuki* is Bengali word, meaning 'hill-people or highlanders', which was, as Reid described in 1893:

> [O]riginally applied to the tribe or tribes occupying the tracks immediately to the south of Cachar. It is now employed in a comprehensive sense, to indicate those living to the west of the Kaladyne River, while to the west they are designated as Shendus. On the other hand, to anyone approaching them from Burma side, the Shendus would be known as Chiang, synonymous with Khyen, and pronounced as 'Chin' (Reid 1893: 238).

The designation of Kuki was seldom used by the Chin people themselves, not even by the Zomi, for whom the word is intended. Soppit, who was Assistant Commissioner of Burma and later Sub-Divisional Officer in the North Cacher Hills, Assam, remarked in 1893 in his study of *Lushai-Kuki*:

> The designation of Kuki is never used by the tribes themselves, though many of them answer to it when addressed, knowing it to be the Bengali term for their people (Soppit 1893: 2).

Shakespear, an authority on the Chin, said in 1912:

> The term Kuki has come to have a fairly definite meaning, and we now understand by it certain ... clans, with well marked characteristics, belonging to the Tibeto-Burman stock. On the Chittagong border, the term is loosely applied to most of the inhabitants of the interior hills beyond the Chittagong Hills Tracts; in the Cachar it generally means some families of the Thado and Khuathlang clans, locally distinguished as new Kuki and old Kuki. Now-a-days, the term is hardly employed, having been superseded by Lushai in the Chin Hills, and generally on the Burma border all these clans are called Chin. These Kuki are more closely allied to the Chakmas, and the Lushai are more closely to their eastern neighbours who are known as Chin.

He concluded by writing: 'Nevertheless, there is no doubt that the Kukis, Lushais and Chins are all of the same race' (Shakespear 1912: 8).

In 1826, almost one hundred years before Shakespear published his book, Major Snodgrass, who contacted the Chin people from the Burma side, had already confirmed that Kukis and Lushai were of the Chin nation, but he spelled Chin as Kiayn. He also mentioned Chinram as 'Independent Kiayn Country' (Snodgrass 1827: 320, on map) in his *The Burmese War,* in which he detailed the First Anglo–Burmese War in 1824–26. Sir Arthur Phayer still spelt Chindwin as 'Khyendweng' in his *History of Burma,* first published in 1883 (Phayer 1883: 7). It was in 1891 that the term 'Chin', to be written as CHIN, was first used by Major W.G. Hughes in his military report, and then by A.G.E. Newland in his book *The Images of War* (1894); the conventional spelling for the name became legalized as the official term by *The Chin Hills Regulation* in 1896.

The Myth of Common Descent

Traditional accounts of the origin of the Chin people have been obscured by myths and mythologies that together with symbols, values and other collective memories, are important elements of what Clifford Geertz called 'primordial identities', which so often define and differentiate the Chin as a distinctive people and nationality throughout history (Geertz 1973). As noted already, one such myth handed down through generations describes how the Chin 'came out of the bowels of the earth or a cave called Chin-lung or Cin-lung' (Gangte 1993: 14). According to some, it was located somewhere in China (cf. Zawla 1976: 2), others claimed it to be in Tibet (cf. Ginzathang 1973: 7); yet others suggested that it must be somewhere in the Chindwin Valley since the literal meaning of Chindwin is 'the cave or the hole of the Chin' (Gangte 1993: 14). I shall come back to the debate on the location of 'Chinlung' later.

Almost all of the Chin tribes and clans have promulgated similar but slightly different versions of the myth, which brings the ancestors of the Chin out from the hole or the bowels of earth. The Ralte clan/group of the Mizo tribe, also known as the Lushai, who now live in Mizoram State in India, have a tradition now generally known as 'Chinlung tradition' that brings their progenitors from the bowels of the earth. The story was translated into English and recorded by J. Shakespear in 1912 as follows:

> [Once upon a time when the great darkness called *Thimzing* fell upon the world,] many awful things happened. Everything except the skulls of animals killed in the chase became alive, dry wood revived, even stones become alive and produced leaves, so men had nothing to burn. The successful hunters who had accumulated large stocks of trophies of their skill were able to live using them as fuel.
>
> After this terrible catastrophe, *Thimzing,* the world was again re-peopled by men and women issuing from the hole of the earth called 'Chhinlung' (Shakespear 1912: 93–94).

He described another similar story:

> The place whence all people sprang is called 'Chinglung'. All the clans came out of that place. The two Ralte came out together, and began at once chattering, and this made Pathian (*The Supreme God*) think there were too many men, and so he shut down the stone (ibid.: 94).

Yet another story of the origin of the Chin, also connected with the 'Chinlung tradition' as handed down among the Mara group of the Laimi tribe (also known as the Lakher) was recorded by N. E. Parry in 1932:

> Long ago, before the great darkness called *Khazanghra* fell upon the world, men all came out of the hole below the earth. As the founder of each Mara group came out of the earth he call his name. Tlongsai called out, 'I am Tlongsai'; Zeuhnang called out, 'I am Zeuhnang'; Hawthai called out, 'I am Hawthai'; Sabeu called out, 'I am Sabeu'; Heima called out, 'I am Heima.' Accordingly God thought that a very large number of Mara had come out and stopped the way. When the Lushai came out of the hole, however, only the first one to come out called out, 'I am Lushai', and all the rest came out silently. God, only hearing one man announce his arrival, thought that only one Lushai had come out, and gave them a much longer time, during which Lushais were pouring out of the hole silently in great numbers. It is for this reason that Lushais to this day are more numerous than Maras. After all men had come out of the hole in the earth God made their languages different, and they remain so to this day (Parry 1932: 4).

All sources of Chin traditions maintain that their ancestors originated from 'Chinlung' or 'Cin-lung'. Sometimes the name for 'Chinlung' or 'Cin-lung' differs, depending on the specific Chin dialect – such as Khul, Khur and Lung-kua, – but it always means 'cave' or 'hole' no matter what the dialect. The reason that Chin-lung was abandoned, however, varies from one source to another. Depending on the dialect and local traditions, it is said that Chin-lung was abandoned as a result of an adventure, or because of the great darkness called *Khazanghra, Thimzing* or *Chunmui.* In contrast to the stories above, some traditions maintain that their original settlement was destroyed by a flood. The Laimi tribe from the Haka and Thlantlang areas had a very well-known myth called *Ngun Nu Tuanbia,* which related the destruction of human life on Earth by the flood. The Zophei also had their own version of the story about the flood, called *Tuirang-aa-pia* (literally meaning: 'white water/river is pouring out or gushing'), which destroyed their original settlement. The story goes as follows:

> Once upon a time, all the humankind in this world lived together in one village. In the middle of the village there was a huge stone, and underneath the stone was a cave that in turn was connected with the endless sea of water called *Tipi-thuan-thum.* In this cave dwelt a very large snake called *Pari-bui* or *Limpi,* which seized one of the village

> children every night and ate them. The villagers were in despair at the depredations committed by the snake, so they made a strong hook, tied it on the rope, impaled a dog on the hook and threw it to the snake, which swallowed the dog and with it the fish hook. The villagers then tried to pull out the snake, but with all their efforts they could not do so, and only succeeded in pulling out enough of the snake to go five times round the rock at the mouth of the hole, and then, as they could not pull out any more of the snake, they cut off the part that they pulled out, and the snake's tail and the rest of the body fell back into the deep cave with a fearful noise. From that night water came pouring out of the snake's hole and covered the whole village and destroyed the original settlement of mankind. Since then people were scattered to every corner of the world and began to speak different languages. And, it was this flood, which drove the ancestors of the Chin proper to take refuge in the Chin Hills (Ceu Mang 1981: 12–19).

Many Chin tribes called the Chindwin River the 'White River', *Tui-rang, Tuikhang, Tirang, Tuipui-ia*, etc.; all have the same meaning but differ only in dialect term. Thus modern historians, not least Hutton, Sing Kho Khai and Gangte, believe that the traditional account of the flood story, which destroyed the Chin's original settlement, might be the flood of the Chindwin River. They therefore claim that the Chin's original settlement was in the Chindwin Valley and nowhere else.

THE CHIN CONCEPT OF *RAM*

For the Chin, *Miphun* cannot exist without *Ram*. They therefore define themselves as a *Miphun* with a strong reference to *Ram* – the original homeland, a particular locus and territory, which they all collectively claim to be their own. At the same time, they identify members of a community as 'being from the same original homeland' (A. Smith 1986: 29). The inner link between the concepts of *Miphun* and *Ram* was strengthened in Chin society through the worship of *Khua-hrum* at the *Tual* ground. As Anthony Smith convincingly argues, 'Each homeland possesses a center or centers that are deemed to be "sacred" in a religio-ethnic sense' (ibid.). In Chin society, the *Tual* grounds, the site of where they worshipped the guardian god *Khuahrum*, were the sacred centres, which stood as protectors of both men and land.

For the Chin, the concept of *Ram*, or what Anthony Smith calls the 'ethnic homeland', refers not only to the territory in which they are residing, i.e. present Chinram, but also the 'original homeland' where their ancestors once lived as a people and a community. What matters most in terms of their association with the original homeland is that 'it has a symbolic geographical center, a sacred habitat, a "homeland", to which the people may symbolically return, even when its members are scattered ... and have lost their [physical] homeland centuries ago' (ibid.). Ethnicity does not cease to exist simply because the Chin were expelled from their original homeland, or because they are artificially divided between different countries, 'for ethnicity is a matter of myths, memories, values and symbols, and not material possessions or political powers, both of which require a habitat for their realization' (ibid.). Thus the Chin concept of *Ram* as 'territory' and 'original homeland' are relevant to *Miphun*. The relevance of the 'original homeland' is this:

> Not only because it is actually possessed, but also because of an alleged and felt symbiosis between a certain piece of earth and 'its' community. Again, poetic and symbolic qualities possess greater potency than everyday attributes; a land of dream is far more significant than any actual terrain (ibid.: 28).

I shall therefore trace the history of the Chin's settlements, not only in present Chinram but also in their original 'homeland' in the Chindwin Valley, in the following sections.

Migration Patterns

Chin tradition maintains that the ancestors of the Chin people originated from a cave called 'Chinlung', but in the absence of written documents, it is difficult to locate the exact site of Chinlung. Scholars and researchers therefore give various opinions as to its location.

K. Zawla, a Mizo historian from West Chinram, suggests that the location of Chinlung might be somewhere in modern China, and the 'Ralte group [of the Mizo tribe] were probably one of the first groups to depart from Chhinlung' (Zawla 1976: 2). Here, Zawla quoted Shakespeare and accepted the Chin legend as historical fact. He also claimed that the Chin came out of Chinlung in about 225 BC, during construction of the Great Wall and during the reign of Emperor Ch'in Shih Huang, whose cruelty was then at its height. Zawla relates the story of the Ch'in ruling dynasty in Chinese history in a fascinating manner. He uses local legends known as *Tuanbia* (literally 'stories or events from the old-days') and many stories which are recorded by early travellers and British administrators in Chinram, as well as modern historical research on ancient China. Naturally, this kind of compound story-telling has little or no value in a historical sense, but is nevertheless important in terms of socially reconstructing collective memories as identity creating resources.

Other theories have been advanced in this connection, more noticeably by Sing Kho Khai (1984) and Chawn Kio (1993). Both believe that the Chin ancestors are either the Ch'ing or Ch'iang in Chinese history, which are 'old generic designations for the non-Chinese tribes of the Kansu–Tibetan frontier, and indicate the Ch'iang as a shepherd people, the Ch'ing as a jungle people' (Sing Kho Khai 1984: 53). Thus, according to Chinese history, both the Ch'iang and Ch'ing were regarded as 'barbarian tribes' (ibid.: 21). Gin Za Tuang – in a slightly different manner than Zawla, Sing Kho Khai and Chawn Kio – claims that the location of 'Chinlung' was believed to be in Tibet (cf. Ginzathang 1973: 5; Sing Kho Khai 1984: 10; Gangte 1993: 14). Gin Za Tuang, nevertheless, maintains that the Chin ancestors were Ch'iang, but he mentions nothing about the Ch'ing.

Gin Za Thang simply follows Than Tun's and G. H. Luce's theory of the origin of Tibeto–Burmans and other groups of humans, believed to be the ancestors of the Southeast Asian peoples. According to Professors Than Tun and Gordon Luce,[5] the Ch'iang were not just the ancestors of the Chin but of the entire Tibeto–Burman group, and they 'enjoyed a civilization as advanced as the Chinese, who disturbed them so much that they moved south' (Than Tun 1988: 3). Regarding this, Gordon Luce says:

> With the expansion of China, the Ch'iang had either the choice to be absorbed or to become nomads in the wilds. It was a hard choice, between liberty and civilization. Your ancestors chose liberty; and they must have gallantly maintained it. But the cost was heavy. It cost them 2000 years of progress. If the Ch'iang of 3000 BC were equals of the Chinese civilization, the Burmans [and the Chin] of 700 AD were not nearly as advanced as the Chinese in 1300 BC (cited in Than Tun 1988: 4).

Before they moved to the wilderness along the edges of western China and eastern Tibet, the ancient homelands of Ch'iang and all other Tibeto-Burman groups, according to Enriquez, lay somewhere in the northwest, possibly in Kansu, between the Gobi and northwestern Tibet (Eriquez 1932: 7 8). It is now generally believed that the Tibeto-Burman group and other Mongoloid stock who now occupy Southeast Asia and Northeast India, migrated in three waves in the following chronological order:

1. The Mon-Khmer (Talaing, Palaung, En Raing, Pa-o, Khasi, Annamite)
2. The Tibeto-Burman (Pyu, Kanzan, Thet, Burman, Chin, Kachin, Naga, Lolo)
3. The Tai-Chinese (Shan, Siamese, Karen)

The Tibeto-Burman group initially moved toward the west and thereafter subdivided themselves into several groups. They followed different routes, one group reaching northern Tibet, where some stayed behind, while others moved on until they reached Burma in three waves. These people were:

1. The Chin-Kachin-Naga group
2. The Burman and Old-Burman (Pyu, Kanzan, Thet) group
3. The Lolo group (Enriquez 1932: 8)

This migration pattern theory, as mentioned above, has mainly been adopted by historians like Than Tun and Gordon Luce. However, anthropologists like Edmund Leach believe that 'the hypothesis that the Southeast Asian peoples as known today immigrated from the region of China is a pure myth' (quoted in Lehman 1963: 22). The main difference between the historical and the anthropological approaches is that while historians begin their historical reconstruction with the origins and immigration of the ancestors, anthropologists start with 'the development within the general region of Burma of symbiotic socio-cultural systems: civilizations and hill societies' (ibid.: 22). However, both historians and anthropologists agree – as historical linguistics, archaeology and racial relationships definitely indicate – that the ancestors of these various peoples did indeed come from the north. But anthropologists maintain their argument by saying that 'they did not come as the social and cultural units we know today and cannot be identified with any particular groups of today' (ibid.: 23). Their main thesis is that the hill people and plain's people are now defined by their mutual relationships in present sites, because, for anthropologists, ethnicity was constructed within the realm of social inter-action between neighbouring reference groups.

The anthropological approach could be very helpful, especially when we investigate the pre-historical context of the Chin people, with no written documents existing. Thus, based on ethnic and linguistic differentiation, not on written documents, Lehman demonstrated that 'the ancestors of the Chin and the Burman must have been distinct from each other even before they first appeared in Burma'. He continues:

> Undoubtedly, these various ancestral groups were descended in part from groups immigrating into present Burma, starting about the beginning of the Christian era. But it is also probable that some of these groups were in Burma in the remote past, long before a date indicated by any present historical evidence. We are not justified, however, in attaching more than linguistic significance to the terms 'Chin' and 'Burman' at such dates (ibid. 22).

And he concludes, by saying: 'Chin history begins after A.D. 750, with the development of Burman civilization and Chin interaction with it' (ibid.).

Chin anthropologists like T. S. Gangte seem eager to agree with Leach and Lehman. He rejects hypothetical theories proposed by Zawla and Gin Za Tuang, who locate 'Chinlung' somewhere in China and Tibet, respectively, as myths. 'In the absence of any written corroboration or the existence of historical evidence to support them', he said, 'such hypothetical theories are considered highly subjective and conjectural. They are, therefore, taken with a pinch of salt. They remain only as legends' (Gangte 1993: 17). He nevertheless accepted the 'Chinlung' tradition as the origin of the Chin and even claims that the Chindwin Valley is where Chin history begins.

Similar to Gangte, the 'Khuangsai source of Chin tradition mentions that the location of *Chin-lung* was somewhere in the Chindwin area' (Sing Kho Khai 1984: 10).

The Chin's Homeland of Chindwin

Than Tun claims that Tibeto-Burman groups of the Burman came down into present Burma via the Salween and Nmai'kha Valleys, and reached the northern Shan State before AD 713. But before they were able to settle themselves in the delta area of the Irrawaddy Valley, 'the rise of Nanchao checked their movements soon after 713' (Than Tun 1988: 3). The Nanchao made continuous war with neighbouring powers such as the Pyu who had founded the Halin Kingdom in central Burma. In 835 the Nanchao plundered the delta areas of Burma, and in 863 they went further east to Hanoi. However, by the end of the ninth century the Nanchao power collapsed because, according to Than Tun, 'they had exhausted themselves'. Only after the collapse of the Nanchao were the Burman able to move further South into the plains of Burma.

The Chin, according to Gordon Luce, descended from 'western China and eastern Tibet into the South via the Hukong Valley' (1959b: 75–109), a completely different route than the Burman had taken. Thus Lehman's theory is quite convincing that the ancestors of the Chin and the Burman were distinct from each other even when they first appeared in Burma. There is ample evidence that the Chin were the first to settle in the Chindwin Valley. The Pagan inscriptions dating from the eleventh century onward refer to the Chin of the Chindwin Valley. There is also persistent reference in the legends of almost all the Chin tribes to a former home in the Chindwin Valley. Chin myths uniformly refer to the ruling lineage when speaking of the original homeland in the valley (cf. Lal Thang Lian 1976: 9). Archaeological evidence supports this interpretation.[6]

Sing Kho Khai therefore claims that 'the literal meaning of the name "Chindwin" definitely suggests that the Chindwin area was primarily inhabited by a tribe called the Chin' (1984: 36). Vumson goes even further by saying: 'When the Burman descended to the plains of central Burma, during the ninth century, they [the *Chin* people] were already in the Chindwin Valley' (1986: 35).

Concerning historical evidence of the Chin settlement in the Chindwin Valley, reliable sources come from the Burman inscriptions erected by King Kyanzzittha and other kings during the peak of the Pagan dynasty. According to Luce, an expert on Pagan inscription, 'Chins and Chindwin ("Hole of the Chins") are mentioned in Pagan inscriptions from the thirteenth century' (Luce 1959a: 19–31). The earliest Pagan inscriptions put the Burman in upper Burma in roughly the middle of the ninth century. Professor Luce suggested that the Chin settlement in the Chindwin Valley began in the middle of the eighth century, while allowing for the possibility of a date as far back as the fourth century. Lal Thang Lian, a Mizo historian, also gives the eighth century as the possible date for Chin settlement in the Chindwin Valley (1976: 71).

Before the Chin settled in the Chindwin Valley, kingdoms of the Mon and the Pye existed in the major river valley of Burma, Sak or Thet and Kandu in Upper Burma, and also the Shan in the eastern country, but no one occupied the Chindwin Valley until the Chin made their home there. The Burman fought against the other occupants of the area, such as Thet, Mon and Pyu, but they did not fight the Chin. G. H. Luce writes;

> The Pagan Burman had wars with the Thets (Sak), the Kandu (Kantú), the Mons, the Shans and the Wa-Palaungs, but he called the Chins 'friends'. Moreover, while he pushed far up the Yaw, the Mu and the Irawaddy, he apparently did not go up the Chindwin. I cannot identify any old place of the Chindwin much further north than Monywa. From all this I infer that in the Pagan period the home of the Chin was mainly in the Chindwin Valley above Monyaw (Luce 1959 a: 21).

In his major work, 'Old Kyakse and the Coming of the Burmans', Luce also mentions the Chin settlement in Chindwin and their relation with the Burman as follows:

> If the Chins had joined the Thet peoples in opposing the Burmans, the latter's conquest of the central plains might have been precarious. But the Thets probably hated the Chins, whose descent from the Hukong Valley had cut off their western tribes in Manipur, and overwhelmed their tenure of Chindwin. Burman strategy here was to conciliate the Chins. They advanced up the Lower Chindwin only as far as Monywa and Alone, called the Chins *Khyan*, 'friends', and seem to have agreed to leave them free to occupy the whole Upper Chindwin Valley. There is no mention of any fighting between the Chins and the Burmans; and whereas the Pagan Burmans soon occupied the *M'u* Valley at least as far as *Mliytú* (Myedu) and the *Khaksan*, *Yaw* and *Krow* Valleys as far as the *Púnton* (Póndaung) Range and perhaps *Thilin*, I know of no place up the Chindwin much beyond *Munrwa* (Monywa) and the *Panklí 10 tuik* (ten 'taik' of Bagyi), mentioned in Old Burmese (Luce 1959b: 89).

Based on the Burman inscriptions of the Pagan Kingdom, which refer to the Chin as comrades and allies in the Chindwin Valley, Luce even suggested that the word 'Chin' might come from the Burmese word *Thu-nge-chin* 'friend'. But this is very unlikely, because the word 'Chin' had already been well recognized by the Burman and other peoples, such as Kachin and Shan, even before the Chin made their settlement in the Chindwin Valley. The Kachin, for instance, who never came down to the

Chindwin Valley but remained in the upper Hukong Valley and present Kachin Hills, called the Chin *Khiang* or *Chiang*. So did the Shan. Thus, it is obvious that the term 'Chin' had been used to denote the Chin people long before the Chindwin Valley became their homeland. And the term Chindwin comes from 'Chin', as in 'the hole of the Chin' or 'the river of the Chin', but not the other way around.

Collective Memories of Chindwin

Over the course of time, the Chin people moved up from the eastern bank of the Chindwin River to the Upper Chindwin of the Kale Valley. Although we do not know exactly when and why, the date can be set approximately to the final years of the thirteenth century or beginning of the fourteenth century. Until the fall of the Pagan dynasty in 1295, the Pagan inscriptions continuously mentioned that the Chin were in between the eastern bank of the Upper Chindwin and west of the Irrawaddy River. Thus, it can be assumed that the Chin settlement in the Kale Valley began just before the end of the thirteenth century. The reason is equally unknown. Perhaps a flood destroyed their settlement as oral traditions remembered it; or as Luce has suggested, 'the Chin were left to themselves in Upper Chindwin' (Luce 1959b: 89). As far as linguistic evidence is concerned, traditional accounts of the flood story seem more reasonable than Professor Luce's suggestion. The traditional Chin account from the Zophei group of the Laimi tribe has recounted that the flood from the low valley had driven their ancestors to the mountains on other side of the river, in Chin: *Khatlei, Khalei or Khale.* It is believed that the root word of Kale is *Khalei,* and the meaning is 'other side of the river'.[7]

After their original settlement in the Chindwin Valley was destroyed by the flood, according to the traditional account, the Chin moved to the Upper Chindwin, and some groups such as the Asho went as far as the Pandaung Hills and other hills near the western part of the Chindwin River. Since then the Chin have been broken into different tribes speaking different dialects. Many myths and legends exist to explain why they broke into distinct tribes and speak different dialects. One such story is recorded by Carey and Tuck:

> They (the Chin) became very powerful and finding no more enemies on earth, they proposed to pass their time capturing the Sun. They therefore set about a sort of Jacob's ladder with poles, and gradually mounted them higher and higher from the earth and nearer to their goal, the Sun. However, the work became tedious; they quarreled among themselves, and one day, when half of the people were climbing high up on the pole, all eager to seize the Sun, the other half below cut it down. It fell down northwards, dashing the people beyond the Run River on the Kale border and the present site

> of Torrzam. These people were not damaged by the fall, but suddenly struck with confusion of tongues, they were unable to communicate with each other and did not know the way home again. Thus, they broke into distinct tribes and spoke different languages (Carey and Tuck 1976: 146).

Another story from the Zophei area, known as the 'Leather Book', relates not only the story of the Chins being broken up into distinct tribes but also how their written language came into being:

> In the beginning, when the stones were soft, all mankind spoke the same language, and there was no war on earth. But just before the darkness called *Chun-mui* came to the earth, God gave different languages to different peoples and instructed them to write on something else. While the Chin ancestors carefully inscribed their language on leather, the Burman ancestors, who were very lazy, wrote their language on stone, which was soft. However, soon after they had made the inscription of their languages, the 'darkness' came and the Sun disappeared from the earth. During the 'darkness' the stone became hard but the leather got wet. Before the Sun came back to the earth, and while the wet leather was still very smelly, a hungry dog ate up the leather, and in this way, the Chin ancestors lost their written language.
>
> When the Sun came back to the earth, the Chin ancestors realized that while they had lost their written language, the Burman language which was written on the stone had turned into 'the magic of letters'. Moreover, while the sons of Burman spoke the same language, the sons of Chin spoke different dialects because their common language was eaten up together with the leather by the hungry dog. Thus, the ancestor of the Chin prepared to make war against the Burman in order to capture 'the magic of letters'. Although the Burmans were weaker and lazier, the Chin did not win the war because 'the magic of letters' united all the sons of the Burman. Since the sons of Chin spoke different dialects, their fathers could not even give them the war order to fight the Burman. It was for this reason that the Chin broke into distinct tribes and speak different dialects (Pu Sakhong 1969: 11–12).

Another story connected with the 'magic of letters' comes from the tradition of the Mizo tribe, which was recorded by Shakespear in 1912. According to Mizo tradition, God gave mankind not only different languages but different talents as well: '[T]o the ancestor of the Poi (*Laimi*) tribe he gave a fighting sword, while the ancestor of the Lushai tribe only received a cloth, which is the reason that the Poi tribes are braver than the Lushais' (Shakespear 1912: 95). In contrast to the Zophei tradition, the Mizo story tells that 'the magic of letters' was given to the white man, not to the Burman. Shakespear therefore concludes by saying that 'I was told he (the white man) had received the knowledge of reading and writing – a curious instance of the pen being considered mightier than the sword' (ibid.: 95).

From the Chindwin Valley to Present Chinram

Historical evidence indicates that the Chin lived peacefully in Upper Chindwin of the Kale-Kabaw Valley for at least a hundred years, from the fall of Pagan in 1295 to the founding of the Shan's Fortress City of Kale-myo in 1395. There is no historical evidence that between those years, the Chin's life in the Kale Valley was disturbed either by natural disaster or by political events. During the period, the Chin founded their capital at Khampat in the Kabaw Valley. Lal Thang Lian, a Mizo historian, and M. Kipgen, a Zomi historian, both claim that the Khampat era was 'the most glorious period' in Chin history. 'Most of the major clans, who now inhabit the Chin State of Burma, Mizoram, Manipur, Cachar and Tripura, are believed to have lived together there under a great chief having the same culture and speaking the same language' (Kipgen 1996: 39).

But in 1395 when 'the Shan built the great city of Kalemyo with double walls' at the foot of what is now called the Chin Hills, twenty miles west of the Chindwin River, a century of peaceful life in the Kale Valley came to an end (Luce 1959a: 26–27). The Shan had become the rising power in the region of what is now called 'Upper Chindwin' and 'Central Burma' by the middle of the thirteenth century. Before they conquered the Chin country of the Kale Valley, the Shan had already dominated the regions by conquering the then most powerful kingdom of Pagan in 1295. They continued to fight among themselves and with the Burman kingdom of Ava, which was founded after the fall of Pagan by King Thadominphya in 1364. The Shan finally conquered Ava in 1529. Although Ava was recaptured by the Burman King Bayinnaung in 1555, the Kale Valley remained under the rule of Shan until the British period. In the century after they had conquered the Chin country of the Kale Valley, the Shan also annexed Assam and established the Ahong dynasty, which lasted for more than two centuries.

According to Sing Kho Khai and Lal Thang Lian, the Chin did not leave the Kale Valley after the Shan conquest. The Chin traditions of the Zomi and Mizo tribes, which were accepted as historical facts by Sing Kho Khai (1984) and Lal Thang Lian (1976), mentioned that the Chin lived in the Kale Valley side by side with the Shan for a certain period. Zomi tradition, as noted by Sing Kho Khai, goes on to relate that 'while they were living in the Kale Valley, a prince came up from below and governed the town of Kale-myo. During the reign of that prince the people were forced to work very hard in the construction of the fortress and double walls of the town' (Sing Kho Khai 1984: 43). The hardship of the forced labor was said to be so great, according to Naylor, that 'the fingers of workers, which were accidentally cut-off, filled a big basket' (Naylor 1937: 3). The tradition continues to relate that the Chins, unable to bear the hardship of manual labour, moved up to the hills region to establish such a new settlement as 'Chin New', which was located in the present township of Tiddim of the Chin State in Burma (Carey and Tuck 1976: 127). Historian D. G. E. Hall confirms that the Shans 'drove the Chin out of the Chindwin Valley into the western hills' of present Chinram (Hall 1968: 158).

According to a legend, which Lal Thang Lian accepted as historical fact, the Chin planted a banyan sapling at the site of an altar where they used to worship their *Khua-hrum*,[8] just before they were forced to abandon Khampat. They took a pledge at the sacrificial ceremony to their *Khua-hrum* that 'they would return to Khampat, their permanent home, when the sapling had grown into a tree and when its spreading branches touched the earth' (Kipgen 1996: 40–41).[9]

We do not know exactly when the Chin left Khampat and the Kale-Kabaw Valley to settle in the hilly region of Chinram. But we *can* trace the periods, approximately, from the Shan and the Burma chronicles from the east and the Manipur chronicles from the west. The Manipur chronicles first mentioned the Chin people, known to them as Kuki, in 1554 (cf. Shakespear 1955: 94–111; Lehman 1963: 25). It is therefore certain that the Chin settlement in present Chinram began only after the founding of Kale-myo in 1395, and reached the furthest northern region of their settlement in present Manipur State of India in about 1554.

According to Sing Kho Khai, the first settlement made in present Chinram was called 'Chin Nwe', or 'Cinnuai' as he spelt it. Carey and Tuck, however, spelled it 'Chin Nwe' (1976: 127). The Chin lived together in 'Chin Nwe' for a certain period. But they split into tribal groups because of 'their struggle against each other for political supremacy' (Sing Kho Khai 1984: 41). Economics may have been the compelling reason, because 'Chin Nwe', a rather small, hilly region, could not provide enough land for the self-sufficient agriculturally oriented economic system of peasant society. Thus, one group made their new settlement in 'Lai-lung', located in the present township of Falam, and eventually became the 'Laimi tribe' (Z. Sakhong 1983: 5). Another group who first settled in 'Locom' eventually became the Mizo tribe who now populate part of Mizoram State in India. From 'Chin Nwe' some groups moved up to the north, and they are now known as 'Zomi', meaning northern people, or highlanders. Prior to these settlements, there is no historical evidence that differentiates the Chin into the Liami, Mizo and Zomi tribes, etc.. Only the national name of 'Chin' is represented in the records. Until that time, there were no such tribal names as Asho, Chó, Khuami, Laimi, Mizo and Zomi. B. S. Carey, who knew very well the Biblical story of the fall of mankind,[10] described 'Chin Nwe' as 'the Chin Garden of Eden', which indicated 'before the fall came upon the Chin people', to use the symbolic term (Carey and Tuck 1976: 127).

Some Chin tribes, however, did not move to the hills but remained in the Chindwin Valley, especially in remote areas like the Gankaw Valley and the Kale-Kabaw Valley of Upper Chindwin. They are still today called by their original name but with suffixes like Chin-pun, Chin-me, etc., because of their old-fashioned tattooed faces. Asho groups, as mentioned earlier, split away from the main groups even before they moved to Upper Chindwin. They first lived in the Pandaung Hills and then scattered around the Irrawaddy Delta, Pegu Yoma, Arakan Yoma; some of the Asho tribe even reached the Chittagong Hill Tracks in what is now Bangladesh (Lian Uk 1968: 7). In Arakan and Chittagong they are still known by their old name, 'Khyeng'.

TRIBAL GROUPS AND *TUAL* COMMUNITIES

Historical evidence shows that the Chin were known by no other name than CHIN until they made their settlement in 'Chin Nwe'. However, after they were expelled from their original homeland, the Kale Valley in Upper Chindwin, by the flood as oral traditions recounts – or conquered by the Shan as modern scholars have suggested – the Chin split into different tribal groups with different tribal names and dialects.

Undoubtedly, a vast majority of the Chin people moved over to the hill regions of present Chin State in Burma, Mizoram and Manipur States in India, and the Chittagong Hill Tracts in Bangladesh. But some groups, as mentioned, remained in their original homeland of the Chindwin Valley and later scattered into such areas as the Sagaing, Maqwi, Pakukko and Irrawaddy divisions of present Burma.

Linguistically, according to the 1904 *Linguistic Survey of India*, the Chin dialects are divided into four major groups: Northern, Central, Old Kuki and Southern.

1. The Northern group: Thado, Kamhau, Sokte (Sukte), Siyin (Sizang), Ralte, Paite

2. The Central group: Tashon (Tlaisun), Lai, Lakher (Mara), Lushai (Mizo), Bangjogi (Bawmzo), Pankhu

3. The Old-Kuki group: Rangkhol, Kolren, Kom, Purum, Hmar, Cha (Chakma)

4. The Southern group: Chin-me, Chin-bok, Chin-pun, Khyang (Asho), M'ro (Khuami), Shendus (Yindu), and Welaung (Grierson 1904: 67).

Scholars generally agree that there are six major tribal groups of the Chin, namely the 1) Asho, 2) Chó or Sho, 3) Khuami or M'ro, 4) Laimi, 5) Mizo (Lushai) and 6) Zomi (Vuson 1986: 40).

For the Chin, the term 'tribal group' is a social group comprising numerous families, clans or generations together with slaves, dependants or adopted strangers. In other words, it is a group of the same people whose ancestors made their settlement in a certain place together, after their common original homeland in the Chindwin Valley was destroyed. The Laimi tribe, for instance, is made up of the descendants of the group who made their settlement at Lai-lung, after being forced to leave the Chindwin Valley. Thus, the term 'tribe' as a Chin concept does not refer to common ancestors or common family ties but to a social group of the same ethnic nationality, who settled in a certain place. As the names imply, the tribal groups among the Chin rather denote geographical areas and the ownership of the land; for example, Asho means 'plain dwellers', Cho means 'southerners', Khuami may be translated as 'the native people', Laimi means 'descendant of the Lai-lung' or the 'central people', as Stevenson (1943) defines it, Zomi or Mizo stands for the 'northern people', and so on. The tribal group therefore is not a divisive term, it only denotes how the Chin are split into various groups, having lost their original homeland of Chindwin.:

In the course of time, different tribal groups gradually developed their own tribal dialects and identities, which in turn were integrated through the ritual systems of *Khua-hrum* worship. Because of difficulties in communication between the different groups, different local dialects and customs gradually developed. This level of group can be called a sub-tribal group, or *Tual* community in Chin. The *Tual* community was usually begun by the same family or clan, settling in the same village. However, as the community became larger and the number of newcomers increased, they would also establish satellite settlements and villages, although they all shared the principle *Tual* village when they worshipped their guardian god, called *Khua-hrum.* I shall explore the nature of the *Tual* community in the next chapter. This kind of sub-tribal group, or *Tual* community, was usually ruled by a single chief or the patriarch of the clan and his descendants. The Lautu group of the Laimi tribe, for instance, was ruled by the Lian Chin clan, who worshiped the Bawinu River as their guardian *Khua-hrum.* The entire community of Lautu – some fifteen villages – shared the *Tual* of their principle village Hnaring. Likewise, in the Zophei group of the Laimi tribe, more than twenty villages shared the *Tual* worship of their principal village Leitak, and so on.

The significance of different *Tual* communities is that although they developed their own local spoken dialect, they all used the same 'mother tongue' tribal dialect when composing a song or epic. To give an example, among the Laimi tribe there were several sub-tribal groups, such as the Zophei, Senthang, Lautu, etc. All these groups had their own local spoken dialects; some were quite different from the main Lai dialect. But when they composed traditional songs and epics, called *Hla-do, Hla-pi* and others, they all used their mother dialect, the Lai dialect, and sang in it. However, because of communication difficulties, feelings of close kinship between

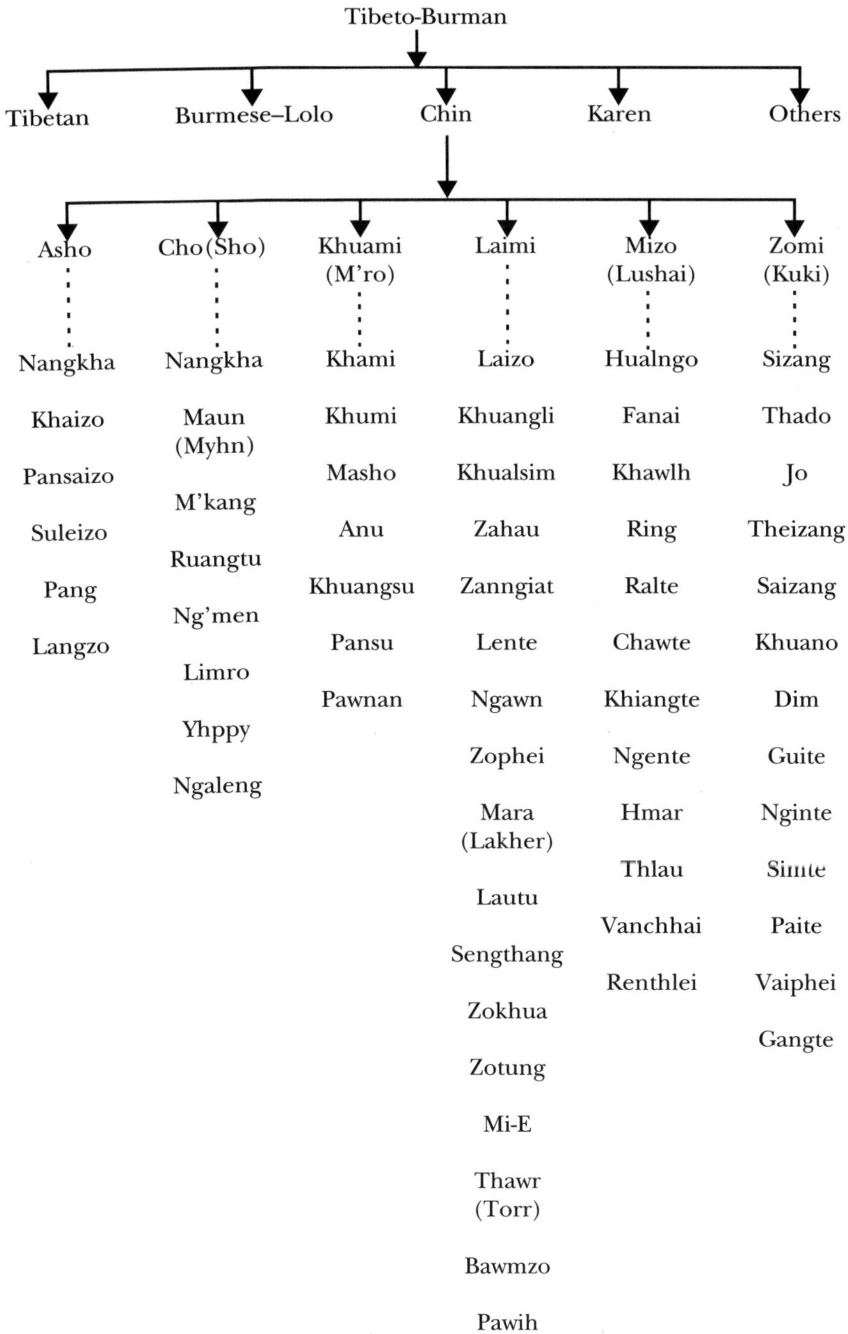

Diagram *of Chin tribal groups which can generally be divided into smaller sub-tribes or groups as shown above.*

tribal groups were no longer strong, sometimes replaced by *Tual* community-oriented sub-tribal group or clan identities. Because of this, the British administrators, as we shall see in Chapter 3, adopted the *Tual* community of sub-tribal groups as the basic structure for what they called the 'Circle Administration'.

NOTES

1 The term 'Chinlung' has been widely used by most of the Chin scholars from the beginning. See J. Shakespear 1912, pp. 93–94; B. S. Carey and H. N. Tuck 1976 [1896], p. 142; N. E. Parry 1976 [1932], p. 4

2 *Re-Unification of the Chin People: Memorandum Submitted by the Paite National Council to Prime Minister of India for Re-unification of Chin People of India and Burma under One Country* (Imphal, Manipur: Azad Printing, 1960). The memorandum was signed by T. Goukhenpau, President, and S. Vungkhom, Chief Secretary, Paite National Council.

3 In Burmese, the combination of k and h is pronounced as ch.

4 Father Vincenzo Sangermano, *The Burmese Empire* (Westminster: Archibald and Co., 1833; reprinted, Bangkok: White Orchid Press, 1995), p. 43. [Explanation within brackets is given by John Jardine, who wrote an introduction and notes when the book was first published in 1893, some 100 years after Father Sangermano passed away. As John Jardine had explained quite clearly, what Sangermano described as 'Jo' is not the 'Jo' group of the Zomi tribe of Chin, but the 'Yaw' people who occupied the Gankaw Valley of Upper Chindwin. This particular point was misinterpreted by many scholars, especially Zomi scholars, including Vum Kho Hau and Sing Kho Khai, sometimes quite knowingly. Vum Kho Hau, for instance, writes in his book *Profile of Burma Frontier Man*, 'From time immemorial we call ourselves Zo (Jo, Yaw). This fact had been admirably recorded by Father V. Sangermano since the year 1783 when he made his headquarter at Ava.' (1963: 238)]

5 Than Tun and Gordon Luce are regarded as the best-known scholars in the study of ancient Burmese history.

6 In his Zo History Vumson mentioned that the 'remains of Chin settlements are still found today in the Chindwin Valley. Two miles from Sibani village, not far from Monywa, is a Chin ritual ground. The memorial stone was, in earlier days, about thirteen feet (4.3 m) high, but now decayed from exposure. The Burmese called it Chin paya or Chin god.' (p. 34).

7 The term 'Kale' is a Burmanized version of Khalei. The literal meaning of Kale or Khale in Burmese is 'Children', which makes no sense for a geographical name. Linguistic study confirmed the Chin traditional account of a flood story, and also the root word of the name Kale Valley.

8 According to M. Kipgen, the banyan tree was 'at the palace site' (1996: 40).

9 This is said to have actually happened in 1916 when Saingunvaua (Sai Ngun Vau) and his party left for Khampat in the Kabaw Valley, where they made a new settlement in order to fulfil the old prophecy. Khampat once again become the centre of the Chin community in the Kale-Kabaw Valley, and more than half of the Kale-Kabaw Valley population are Chins. See Lal Thang Lian 1976: 87–89.

10 B. S. Carey was a grandson of Dr William Carey, a prominent English Baptist missionary to Serampore, India in 1794.

TWO

Traditional Chin Ways of Life: Phunglam

The Chin word for 'ways of life'– *phunglam* – comes from the root word *phung,* which translates literally as 'culture'. The meaning of the term *phung* is deeply religious, almost equivalent to religion itself; as the Chin people express their religious adherence in the words *Lai phung in kan rak um!* 'We practiced Chin traditional religion!' or 'We lived according to Chin traditional ways of life!' or *Krifa phung in kan um!* 'We are Christians!' In this sense, the Chin expression of 'religion' is very close to Clifford Geertz's definition of 'religion as a cultural system' (cf. Geertz 1973: 87–125).

When a group or a community of people practises the same system of symbols collectively, they not only share the same faith but 'ways of life' as well. This is what the Chin call *phunglam,* 'ways of life', the 'powerful, pervasive, and long-lasting moods and motivations in men' which become both the tradition and the pattern of life and also the provider of the foundation of social values. This is an inclusive 'functional' definition, and at the same time clearly structured. This inclusiveness has its own advantage when we deal with tribal society, in which religious, social, cultural and political elements are inseparably intertwined.

Despite these terminological similarities and the same metaphor of expression, the inner meaning of the Chin concept of religion is not just 'a cultural system', nor 'a system of symbols'. Geertz's definition of 'religion as a cultural system' does not refer to any 'superhuman being', 'the sacred' or even a 'transcendental element'. In the Chin concept of religion, on the other hand, it was this transcendent element – including the belief in a Supreme Being and other spiritual beings, souls and life after death – that distinguished religion from politics, and social and economic elements in life, though they were closely linked with each other.

Moreover, religion for the Chin is not just a matter of belief, of internal commitment to the existence of some external power. It also involves ideas that this power is related to human interests. It is accepted that a believer can take the initiative in communicating with this power. This communication, as in many other parts of the world, normally 'takes two forms, verbal and nonverbal' (Firth 1996: 158). In traditional Chin religious practices there was no verbal form of communion without an accompanying nonverbal act, such as offering and sacrifice. In other words, there was no prayer without ritual. This kind of inseparable combination between prayer and religious ceremony was a means to bind the community or congregation together. Communion in Chin traditional religious practice was not just a communion between God and men, but also between men: the chief, who offered the prayer, and the community of people, who followed and shared the meal of the sacrificial ceremony with him.

In this chapter, I shall investigate the conceptual structures of Chin traditional religion and its theology, the interrelationship between religious concept, political structure and the economic system in the *Tual* community. Instead of following conventional methods, which usually separate belief systems and ritual practices on the one hand, and political systems and social structures, on the other, I shall categorize all these elements into a single category of *phunglam*. Instead of separating them, belief systems, social structures and political systems will be treated together in this chapter. Similarly, the sacrificial ceremonies and worship services in Chin traditional religion, and the concept of life on earth as preparation for a period of life after death known as *Mithi-khua,* will be discussed together in the next chapter.

THE PRINCIPLES OF TRADITIONAL CHIN THEOLOGY

In order to understand the basic principles of traditional Chin religion and theology, I begin with the concept of *Khua*.

Khua in Chin thought is a philosophical concept or category, which includes:

- All the spiritual beings: such as *Khua-zing*, Supreme God; *Khua-hrum,* guardian gods; *Khua-chia,* evil spirits; and *Khua-va*ng, shaman or diviner.

- Time and space: such as *kum-khua,* the endless cycle of seasons, unending time or eternity; *minung-khua,* human inhabitants on earth; *Mithi-Khua,* the departed abode in heaven; *vancung-khua,* heaven or space; *khua-ram,* geographical land and space on earth.

- Cosmic forces of rain, wind, heat and fire: such as *khua-hnaw,* the source of rain and the beginning of water or the rain god; *khua-ri,* the source of wind and sound; *khua-tek,* the source of light and fire; *khua-lum,* the source of heat and warmth; *khua-sik,* the source of coldness; *ni-khua,* the Sun, the Moon, the Universe and Nature – including weather.

Khua therefore is all being, all essence within and beyond time and space. In traditional Chin philosophy and religion, nothing can be conceived outside the concept of *Khua.*[1]

Given *Khua* as the basic term of reference, the conceptual structures and belief systems of traditional Chin religion can be categorized into the concepts of

1. a Supreme God: *Khua-zing*
2. guardian gods: *Khua-hrum*
3. evil spirits: *Khua-chia*
4. shaman: *Khua-vang*
5. life after death: *Mithi-khua*

The Concept of a Supreme God: Khua-zing

A fundamental, essential phenomenon in traditional Chin Religion is the concept of the Supreme God, known as *Khua-zing.*[2] The name *Khua-zing* is 'a combination of two separate words: *Khua* and *zing*' (Laisum 1994: 126). I consider the concept of *Khua* and *zing* separately first, and then define the concept of *Khua-zing* as a whole.

The meaning of *Khua* is 'all beings and all essence within and beyond time and space', and the literal meaning of *zing* is 'invisible force.'[3] In other words, *zing* is the vital force or source of life, which includes the vital force of human life and also the forces of all things: the force of spirits, the forces of sun and moon, light and darkness, rain and river, even the forces of time and space. The Chin believed that the source of *zing* itself comes from *Khua-zing.*

Khua-zing therefore is the Supreme God, creator of the entire Universe, including spiritual beings and human beings, who gives *zing* the vital force of life for every being in this Universe. N. E. Parry has defined the term *Khua-zing* as 'the Father of all', 'the meaning of everything' and 'the source of life' (cf. Parry 1932: 349). Since *Khua-zing* is the giver and source of life, Chin belief maintains that He also is the controller of everything in the Universe: darkness and light on earth, sunshine and rain on the hill and the river, the eclipse of the sun and the moon, the fate of man and even spirits.

The Chin expressed their knowledge of *Khua-zing* in myths, stories, prayers, religious ceremonies, songs and proverbs. They expressed the eternity of *Khua-zing* in a simple way; *Khua-zing* lives beyond *Minung-khua* and *Mithi-khua,* has neither been born nor died, but knows who has been born and died. No men can reach him because he is living beyond *Minung-khua* and *Mithi-khua,* but he is very close to human beings because he is *Khua-zing* and they are a part of his being. *Zing* is present in every human being as the vital force of life. Therefore, he knows everything about human beings, even their fate. He can see everything, even in the darkness, because he is *zing* who controls *zing le zan*, day and night, light and darkness. He is absolute because the absolute *zing* is the only *Khua-zing*; human beings have only a small *zing*, which come from the absolute *zing* of *Khua-zing*. The *Khua-zing* lives forever because *zing* never dies, even the *zing* of the human being shall return to *Khua-zing* after the body dies. Without *zing* there is no life and *Khua-zing* is life itself (cf. Sing Kho Khai 1984: 130–140; Parry 1932: 349–353).

The nature of *Khua-zing* can be described as the ultimate source of reality, including humanity and human institutions, but it rarely intervened in natural processes by voluntary acts of care and salvation. *Khua-zing* may be the foundation of moral values, as we will see below, but there was no general expression equivalent to 'God is Love'.

In addition to the nature of God, the Chin had clear concepts about God's holiness and the moral attributes. The Chin expressed the holiness of God as *Pathian*, and even after they became Christians, they called upon God as *Kan Khua-zing le kan Pathian*, Our Supreme and Holy God. Literally, *Pathian* is the one who is holy. In the prayer and concept of Chin traditional religion, the word *Pathian* is not the name of a deity but the attribute of the holiness of *Khua-zing*, which we can clearly see in the *zing-dangh* prayer, which I describe under the heading 'Sacrificial ceremonies and worships' in the next chapter.

The holiness of *Khua-zing* was also indicated by the strict rules when performing the ritual directed to *Khua-zing*. Sacrificial animals, for instance, had to be of one secret colour and the priest or elders who performed the sacrifice had to refrain from sexual intercourse and/or certain foods or activities before and after the ritual. These ritual formalities clearly demonstrate that the people regard *Khua-zing* as sacred.

The Chin expressed the moral attributes and goodness of *Khua-zing* as '*ding-thlu le lai-rel*.' Literally, '*ding-thlu*' is the one who paved the way of righteousness, and '*lai-rel*' is the one who judges correctly and fairly.[4] The goodness of *Khua-zing* was manifest in his averting calamities, supplying rain and providing fertility to people and their cattle (cf. Carey and Tuck 1976: 195–196; Parry 1932: 349–350). Prayer in traditional Chin religious ceremonies clearly expressed the goodness of *Khua-zing*.

The Concept of Guardian Gods: Khua-hrum

As mentioned above, the Chin believed that every human being has the vital force of life, called *zing*, which protects life itself as a guardian spirit. In addition to an individual guardian spirit or god, they also believed in a household god, which protected the family and the clan. The household god was called *chung-um*, the guardian god who resided inside the house. The *chung-um* was usually worshipped at the annual *inn-chiat* sacrificial ceremony.

Functionally, the essential guardian god in traditional Chin religion was neither a personal 'guardian angel' nor the household god, but the guardian god of the village or the community, called *Khua-hrum.* The worship of *Khua-hrum* in Chin society was so important – in terms of religious function and to promote social cohesion – that traditional Chin religion was even known as *Khua-hrum biaknak,* or *Khua-hrum* religion. The difference between *zing,* the guardian god of human beings, and *Khua-hrum,* the guardian god of the village and community, is that while the vital force of *zing* proceeded directly from *Khua-zing,* the power of *Khua-hrum* was believed to consist purely of cosmic forces. In other words, *Khua-hrum* was the guardian god of such earthly phenomena as mighty mountain ranges, big rivers, deep valleys, etc., for which the power of the spirits come from cosmic forces such as the sun, the moon and the rain.

The concept of *Khua-hrum* as the spirit of cosmic forces further implied that as every human being has a guardian spirit called *zing,* every place or location has a god or spirit which guards it. Any place or location believed to be guarded by a benevolent spirit or god was regarded as a good or sacred place. Naturally, such a place was the best choice for human settlement as well, because traditionally the Chin believed that if they made their settlement within the sovereignty of the *Khua-hrum,* they would receive protection and blessing from the *Khua-hrum,* which guarded the location. When they wanted to settle in such a favourable location, the first thing they did was to make offerings to the guardian spirit of that place. They offered a prayer to the guardian spirit: 'Protect us, help us, and bless us' (cf. Pu Sakhong 1971: 27). At the place where they first made the offering and prayer to the *Khua-hrum,* they erected a sacrificial stone, and this was usually followed by the planting of *bungkung* (banyan trees). The reason, as Strait observed, is:

> Banyan trees are the common choice for the abode of the spirits, for they are different from other trees and often take a very peculiar shape in the process of growth. As the growing process outruns the recollection of man, the trees are so to speak without beginning. They are somewhat rare and have about them an element of mystery (no doubt due to the decision of the spirits in choosing the Banyan trees as their desired place of residence), which makes them the natural choice for the dwelling of the village spirit (Strait 1933: 52).

The place where the villagers first made their offering and prayer is called *Tual.*[5] According to tradition, this place should never be abandoned, the sacrificial stone never removed and the *bungkung* never cut – not even its smallest branch; together they all eventually become the symbols of community and sacrifice. When the founder of the village died, they placed the tombstone near the tree. From the religious point of view, the *Tual* was the sacred ground where all kinds of communal sacrificial ceremonies and rites were performed. From a socio-political organizational point of view, it was a cultural centre where individual members of the entire community were collectively linked to each other. It also symbolized the historical link between the founder and the community, the ancestors and the living generation. It was also a place where the living and the dead (ancestors) could meet though sacrificial ceremonies. Sing Kho Khai asserted that the *Tual* was the centre of Chin social structures, and he defined the Chin society as a *Tual*-oriented community, or '*Tual* community' (1984: 205).

The Concept of Evil Spirits: Khua-chia

As mentioned above, the general category (and prefix) of spiritual beings is *Khua*. The evil spirit is called *Khua-chia*. The teaching of traditional Chin religion maintained that there were endless types of *Khua-chia,* and no *Khua-chia* was capable of bestowing a blessing. On the contrary, all were believed to be agents of evil, which cause harm, accidents, sickness and death. The category of *Khua-chia* also includes 'myriad spirits of places, particularly large rocks, springs, and other outstanding topographical features. They, too, cause sickness, but they are clearly identified because they have location' (Lehman 1963: 175).

'One soul having been seized or detained by *Khua-chia*' summarizes the basic view of sickness in traditional Chin religion (cf. Parry 1932: 325). The concept of death can also be formulated as 'either one dies as a war casualty or is killed by *Khua-chia*' (Vaipum 1980: iv). The fear of *Khua-chia* therefore is one of the most important aspects of traditional religious life. Indeed, traditional Chin life was permeated by the fear of irresponsible actions on the part of unpredictable *Khua-chia,* who are all around, moving like the air in any direction. The most dangerous kind of *Khua-chia,* believed to be 'the main cause of sickness, is called *rai*' (Lehman 1963: 175). *Rai* was a moving spirit and could not be identified with a certain place and location. Thus, all kinds of sacrificial ceremonies connected with sickness were generally categorized and known as *rai-thawi,* meaning appeasement to the moving spirit.

The only way of treating sicknesses, according to traditional Chin religion, was to pacify *Khua-chia,* which had imprisoned the soul (cf. Parry 1932: 133; Sing Kho Khai 1984: 184). As soon as a sickness occurred, a sacrifice had to be performed to the *Khua-chia.* Different *Khua-chia* demanded different sacrifices. It was useless to try to propitiate a mithan-demanding *Khua-chia* with a pig or a pig-demanding *Khua-chia* with a cock. When a man fell sick, he sacrificed a young fowl, the smallest sacrificial animal. If he got well, it meant that the *Khua-chia* was satisfied and had withdrawn the sickness, but if he did not recover, then he had made a mistake; perhaps it was a pig, not a fowl that should have been sacrificed. If still sick after sacrificing the pig, then he might try sacrificing a mithan, the largest of sacrificial animals. As long as the sickness continued, he would continue to sacrifice all that he owned (cf. Carey and Tuck 1976: 197).

Since it was very important not only to offer the right sacrificial animal, but also to address oneself to the particular *Khua-chia* responsible for the sickness, the role of *Khua-vang* became vital in the Chin's traditional religious life. The main function of *Khua-vang* in traditional Chin religion was not to cure the sickness but to identify the particular *Khua-chia*, which was believed to be responsible for the sickness.

The Concept of Shaman: Khua-vang

(a) The Nature of *Khua-vang*

The belief in *Khua-vang* is one of the most fundamental and striking features of traditional Chin religion. Literally, the meaning of *Khua-vang* is the 'mysterious power of light', or the 'visible power in the darkness'. According to the traditional Chin religious concept, darkness includes all categories of spiritual beings: *Khua* because spiritual beings are invisible forces, like darkness. The definition of *Khua-vang* therefore is:

- the one who can meet the celestial gods – both *Khua-hrum* and *Khua-chia* – face to face and bring them offerings from human beings;
- the one who can seek the soul of a sick man, which has supposedly been away from his body or has been captured by the *Khua-chia*;
- the one who can show the soul of a dead man to its new abode of *Mithi-khua*, which is called *Khua-vang Mithi-tlawn*. Indeed, *Khua-vang Mithi-tlawn* was one of the most outstanding functions of *Khua-vang* and also an essential phenomenon in traditional Chin religion;
- the one who can foretell the future.

In Chin society, *Khua-vang* were almost exclusively women, but there were a few male *Khua-vang* as well. The helping spirit of a male *Khua-vang* is always a male spirit, but the helping spirit of a female *Khua-vang* may be either male or female (cf. Pu Sakhong 1971: 43; Vaipum 1980: 78). To become a *Khua-vang*, the initiative must always comes from the spirit, known as *Lulpi*. *Lulpi* are celestial gods who made their abode in heaven but not in the same place where *Khua-zing* lives. They are neither *Khua-zing* nor *Khua-hrum* but closely associated with both of them. They are the helpers of human beings, so 'when surprised, a man will often cry out: *maw Khua-zing, maw Lulpi*' (Lehman 1963: 177).

The origin of *Khua-vang* in Chin society can be traced in the story of the best known *Khua-vang*, named *Ni Tial Khua-vang*. The story goes as follows:

> Once upon a time, Mang Lian, the patriarch of the Za Thang clan, was taken away by the whirlwind while he was tending a herd of cattle in the Nawi valley in Haka. The whole community of Haka was in search of him for seven days, but they could not find him. After the

> seven nights, Mang Lian told them in dream 'Do not look for me anymore'. The *Lulpi* took me away by whirlwind. So, I have become a *Lulpi* myself and you will not be able to find me. You will find, however, two cows hanging at Nawi where I was tending the cattle; and from them two *Thing-thli* trees will grow one day. Let my offspring offer sacrifice to me at these *Thing-thli* trees whenever they make a feast such as *Khuang-cawi* and *Bawite-bawi*'. Two *Thing-thli* trees indeed grew from the very spot where two cow hung at Nawi.[6] So, all Mang Lian's descendents of the Za Thang clan offered sacrificial animals at the bottom of *thing-thli* trees whenever they made a feast. They built one large stone altar there and named it '*Mang Lian Lung*': the Mang Lian Sacrificial Stone. The stone became not just a sacrificial place, but the *Khua-hrum* of the Za Thang clan (Pu Sakhong 1971: 45–46).

After *Lulpi* took Mang Lian, he lived in the *Lulpi* abode. He married Dar Za Men, daughter of the *Lulpi* chief. Mang Lian and Dar Za Men were blessed with two sons, Sang Kua and Sang Ling. Sang Ling was the one who appeared in *Ni Tial Khua-vang* as a helping spirit. Ni Tial herself was a sister of the Leitak chief Mang Hnin the first (cf. Pu Sakhong 1971: 47; Siang Uk 1984: 12–22).

Ni Tial Khua-vang was the best known *Khua-vang* in Chin history. Her prophecies about the future proved to correct. No *Khua-vang* could foretell as she did. Some of her prophecies were:

- White monkeys with iron sticks will arrive at Mount Rung Tlang of Haka;
- Silver trees will grow on the Mount Rung Tlang and the strong men will gather the fruits;
- Boar will make a nest at the house site of the Vuang-tu chief.

The prophecies were uttered in the beginning of eighteenth century.[7] The first two were realized when the British invaded the Chinram just before the end of nineteenth century. The British soldiers brought guns with them, and there also came silver coins. Moreover, the British administration changed the economic life of the Chin and established their headquarters in Haka. Since the arrival of the British, Haka became the capital of East Chinram. The third prophecy came true after World War II, when Vuang-tu village was moved to another place and the old site was deserted. At the time of the prophecy, Chief Vuang-tu was so powerful and rich that nobody believed such a thing could happen. The tragedy also was that when the Vuang-tu village was deserted, no direct descendent of the chief's family survived.

(b) The Categories of *Khua-vang*

The Chin believed that the *Khua-vang*'s prestige was determined by the status of the god – *Lulpi* – who appeared in the *Khua-vang* as a spirit helper, known as '*A Khua*'. The chief family of *Lulpi* never appeared as the spirit helper of the commoners, but only of a chief or noble family like Sang Ling in Ni Tial. They were regarded as the most highly respected *Khua-vang*. The main task of such *Khua-vang* was to foretell the future and perform sacrificial ceremonies, particularly for the chief. They were believed to be the chief's helper. The concept of the chief being blessed and helped by *Khua-vang* is known as '*Khua nih a hnuai hna:* they are molded and blessed by the *Khua*'. This means that the power and the wealth that they possessed were the blessings which came from '*Khua*'. No chief in Chin society was fully powerful without the help of *Khua,* especially from *Lulpi* through *Khua-vang.*

The second category of *Khua-vang* included those able to cure the sick and to perform *Mithi-tlawn,* the guided journey to the soul of dead person in *Mithi-khua.* The least respected *Khua-vang* were the one who could cure the sick but could not contact the soul of the dead. For both categories, the helping spirits came from ordinary *Lulpi.* I shall describe the function of *Khua-vang* in Chin society, especially curing through sacrificial ceremonies connected with sickness, under the heading of 'sacrificial ceremonies and worship' in the following chapter.

Life after Death in Mithi-khua

(a) The Concept of Life and Death

In order to understand the concept of life after death in *Mithi-khua,* I shall begin with the concepts of life and death themselves.

As mentioned earlier, the Chin traditionally held that 'every person is believed to have a sort of tutelary deity or guardian angel' called '*zing*' (cf. Parry 1932: 350). As Parry observed, the Chin did not know exactly where the *zing* lived, 'but they say that it is always in close proximity to the being of whom it is in charge, and follows him about wherever he goes' (ibid.). The reason for this was that in the Chin concept, *zing* was like *hu* (breath), which is always 'in and out', but remains very close to man. *Zing* and *hu* were virtually identical but not exactly the same. *Zing,* like *hu,* kept the body and soul together, but the source of *zing* was believed to come from *Khua-zing* and the source of *hu* proceeded from the human being itself. If a *zing* was well-pleased with a person, then his *hu* became strong and healthy, a state called '*a hu-ham a thawng*' or '*hu-ham a ngei*'. '*Hu-ham ngeih*' was the most admirable personality for being a human in the Chin religious concept. Therefore, they perform the *zing-dangh* sacrifice to propitiate their own *zing,* not only for a newborn baby but also for a grown person. I shall describe the *zing-dangh* sacrifices as performed at *Bawite Bawi* and *Khua-cawi* in Chapter 3. If the *zing* was well-pleased, it could make a person 'happy, healthy, and prosperous, grant him children and protect him from accident' (Lehman 1963: 177; Sing Kho Khai 1984: 234).

In addition to *zing*, the Chin also believed in the human soul, *thla-rau*. According to Parry, 'a man's soul resembles his body in appearance and size, but is invisible' (1932: 352). The source of *thla-rau* in the Chin traditional religious concept came from the human being, like *hu*, not from *Khua-zing*. The difference between *thla-rau* and *hu* was that *hu* was a non-being, but *thla-rau* was an existent being, albeit invisible. *Thla-rau* and body are kept together by both *zing* and *hu*. The Chin also believed that the *thla-rau* could leave its body and *zing*, but the link between the *Thla-rau* and body still remains through *Hu*. When *Thla-rau* leaves its body and *zing*, two things could happen. One was sickness and the other was a dream. Sickness occurred when the *thla-rau* wandered away from the body without any protection of *zing*. If this happened, they would perform the *thla-auh* sacrifice, a sacrifice to call back the wandering soul. I shall describe the *Thla-auh* sacrificial ceremony in the following chapter.

The concept of the inner links between the body and *hu* also encompassed, as Parry observed, the reasons for dreaming. He describes the dream concept as follow:

> At night, during sleep, the soul sometimes leaves its body and wanders about: a link called *hu* in the shape of an invisible cord remains, however, between the soul and the body, and on the sleeper awakening the soul returns. It is because souls roam about in this way that dreams arise, and as souls in their wanderings are able to foresee future events, dreams often come true (Parry 1932: 352).

The traditional Chin concept of death was simply the separation of *zing*, *thla-rau* and the body. When death occurred, the *zing* went back to its ultimate source, *khua-zing*, and the *thla-rau* to *Mithi-khua*. The body decayed, and the function of *zing* ended because it had a duty only to protect the body and *thla-rau* together. There is no need to protect the *thla-rau* after death, because they live in *Mithi-khua*. In *Mithi-khua*, there were no *Khua-chia* who could harm the soul.

(b) The Concept of *Mithi-khua*

The Chin believed in the existence of the soul and life after death. Life after death, in Chin traditional religious concept, began the moment that an individual's soul departed to live in another world, called *Mithi-khua*, meaning the 'village of the dead'. *Mithi-khua* was believed to be situated not on earth but in heaven. The Chin concept of heaven has three levels. *Khua-zing* resided on the highest level, *Lulpi* on the second level and *Mithi-khua* formed the third level.[8]

The Chin believed that the departed soul reached *Mithi-khua* through a gate, situated between earth and *Mithi-khua.* A mythical being, an old woman named Saw-nu, guarded this gate. She had a bow made of bamboo and pellets large as eggs, which she shot at passing souls; a victim suffered or became sore for three years or more. The *Saw-nu* constantly shot at souls that passed, but she dared not shoot chiefs, noblemen and warrior heroes, who killed many animals and enemies, celebrated many feasts of merit and festivals, possessed many slaves and domestic animals and had many children.[9]

According to traditional Chin religious understanding of eschatology, two places for the human soul exist in the life after death, or two levels of *Mithi-khua,* namely *pulthi-khua* and *sarthi-khua. Pulthi-khua* was believed to be a place for those who died a normal death at home. In *pulthi-khua,* life would be as normal as here on earth; one could enjoy everything achieved during a lifetime on earth. The chief, for instance, would remain a chief. His worldly house would be the image of his home in *Mithi-khua.* A great hunter and those who celebrated many feasts and festivals would possess many domestic animals in *Mithi-khua* for all the animal souls killed, including domestic animals sacrificed during feasts, festivals and religious ceremonies. And 'if a man obtained many human heads [i.e. from enemies] in this world, he would have many slaves in his future existence in *Mithi-khua*' (Carey and Tuck 1976: 196). Thus, a great warrior would have many slaves in *Mithi-khua* for all the victims' souls he killed during his lifetime. All these worldly things obtained during his lifetime here on earth were ritualized in the funeral rite. Thus, 'the Chin idea of eschatology is connected with the desire to pursue worldly things on earth' (Sing Kho Khai 1984: 234). In short, life here on earth was only the preparation period for the future life in *Mithi-khua.* As Lehman observes:

> If one's status is to be permanent, it must be established in the world of the dead. When the performer dies, he hopes to have those *worldly achievements and sacrificial animals he killed during his lifetime* with him in *Mithi-khua.* But the matter is not so simple. He has assured himself by possessing signs of status and property in the land of the dead hereafter, and he has also pleased the inhabitants of the afterworld by giving or sending these gifts there. Presentations must be constantly sent to maintain one's connection of status with them. The sacrifices are at the same time a validation of status in the eyes of the living public, because these are fed with valuable meat of the slain mythons (Lehman 1963: 179).

A single achievement in this world was more important for the future life than good deeds and bad deeds; achievement remained forever but good deeds and bad deeds did not affect the future life. One went to *pulthi-khua* if one died of natural causes, and one departed to *sarthi-khua* if one died an accidental death (cf. Carey and Tuck 1976: 196). *Sarthi-khua* therefore was a place for the souls of people who had died in accidents or died outside the village or family compound. The misery of *sarthi-khua* was that the soul became detached from everything achieved during a lifetime on earth. The deceased could no longer communicate with his relatives, not even through *Khua-vang*. He would be unable to meet the souls of his loved ones even if they died after him. His ancestors' souls would not identify him as their kin anymore. He would be completely alone. In other words, his soul would be detached from both *Mithi-khua* and *Minung-khua,* life after death and life here on earth.

There was only one way of redeeming the soul from *sarthi-khua.* If a person died as a victim of war or by the hand of somebody else, and if his relatives could avenge his death, then his soul would be redeemed from *sarthi-khua* and could be transferred to *pulthi-khua.* This tenet of faith further implies that the one who died by the hand of the enemy remained in *sarrhi-khua,* disconsolate and restless, until his death was avenged by blood. After his death was avenged by blood, he would enter *pulthi-khua* and be as happy as those who died a natural death on earth. If his death could not be avenged by blood, he remained in *sarthi-khua* until his slayer also died. When the slayer died, the slain soul became a slave of the slayer. But the blood revenge could be carried out even after the slayer died. That was why the Chin took revenge for events that had happened two or three generations previously. The impetus behind tribal war was usually vengeance for past events.

The eschatological concept of life after death in *Mithi-khua* became a basic foundation for the Chin way of life, as they regarded life here on earth as merely a preparation period for *Mithi-khua.* I shall return to the traditional Chin concept of life on earth as the preparation period for life after death, but now we need to explore first the *Tual* community.

THE *TUAL* COMMUNITY

Chin social structure was centered upon the *Tual,* in which the religious concept of god, *Khua-hrum,* the political authority of the chiefs and the economic system of land ownership composed the decisive sources of social power. In this section I shall investigate how religious concepts, political power and the economic system were symbolically and functionally interrelated with each other; and how the interrelationship among them symbolized the functional pattern of 'god–chief–land', as the symbol of imagination, consciousness and meaning and also as the functional structure which defined power and status in the society.

In order to understand the functional pattern of 'god–chief–land' – or the interrelationship between religious power, political structure and land resources, I shall begin with the concept of power and its legitimacy, and the structure of the Chin social system itself.

The Concept of Power and Its Legitimacy

In the traditional Chin concept, power – especially political power – is not a human invention but the mandate of *Khua-hrum*. As mentioned already, every place or location has a god or a spirit, believed to be the guardian of that place. When people made their settlement in a certain place, which was believed to be guarded by a benevolent spirit, the guardian spirit of the place became the *Khua-hrum* as well.

However, establishing a settlement, village or town in a place guarded by a benevolent spirit was not so simple. The Chin believed that a person able to establish a settlement in such a benevolent place had a mandate from *Khua-hrum* itself. This kind of mandate is called *Khua nih a hnuai hna*. They also believed that the *Khua-hrum* granted its mandate only to a certain person, usually the patriarch chief, who belonged to the noble clan called *Bawi-phun*, which was regarded as ritually clean. 'Ritually clean' means not possessing the evil eye (*hnam ngei*) or being a witch (*eih thiam*), a slave or even a member of an ordinary clan (*chia hrin*). Thus, in Chin society, the *Bawi-phun* clan regarded themselves as the chosen clan of *Khua-hrum*. In order to know if any particular clan was the chosen clan, they always consulted with *Khua-vang*, who, as noted already, was believed to be able to communicate with *Khua-hrum*. Only after being certain of their chosen status did they dare to approach the powerful *Khua-hrum* and establish settlements within the sovereignty of *Khua-hrum*, who eventually became the protector of themselves and their subjects.

In traditional Chin society, all of the aristocrat clans of *Bawi-phun* thus claimed that their power was a mandate from the guardian god *Khua-hrum*. Almost all of the aristocrats were usually the direct descendents of the founder of a particular clan or particular settlement, and they therefore regarded their political office and institution – in reality their own patriarchal lineage and clan – as a sacred office. This interrelationship between the *Khua-hrum* and the *Bawi-phun* was maintained continuously by means of various rituals. In fact, they were the ones permitted to carry out sacrificial ceremonies to the *Khua-hrum*. This kind of religious authority and the right to offer to *Khua-hrum* was known in Chin as *Do-dang-tu* or *Khua a do tu le a dang tu*. To illustrate the links between the *Khua-hrum* and *Bawi-phun*, and how one became the *do-dang-tu* of the community by inherited right, I shall present as an example the Za Thang clan of the Haka aristocratic family.

The patriarch of Za Thang was a founding father of the Laimi tribe, who first settled at Lailung (lit. the cave of Lai or Laimi) after their original homeland in the Kale Valley was conquered by the Shan in the thirteenth century. In Lailung, the leading families were Hlawn Ceu, Za Hau, Za Thang, Thian Hlun and Mang Sui (founding father of the Senthang), Lian Bawi (founding father of the Zophei), Lian Chin (patriarch of Lautu) and others. After living together peacefully for a period of time in Lailung, some groups moved to Zotlang. From Zotlang, some groups led by two brothers of the Za Thang clan, named of Sui Hlei (Seo Hle) and Hlawn Za (Hlwa Sha), moved further to Phaileng. In Phaileng, misfortune struck the group because of the sin of their leaders: the elder brother Sui Hlei committed adultery with the wife of Hlawn Za and, when anger arose uncontrollably, Hlawn Za killed Sui Hlei. According to Carey and Tuck, the reason for the murder was not adultery but a power struggle between the two brothers.[10]

According to the Chin ritual system, the elder brother Sui Hlei was the only one with the right to perform sacrificial ceremonies to the *Khua-hrum* of Phaileng, as a *do-dang-tu.* The *Khua-hrum* did not recognize Hlawn Za, so when Hlawn Za made his sacrificial offering to the *Khua-hrum* of Phaileng as leader of the community after Sui Hlei's death, he received not a blessing but a curse of his brother's blood. As a result, his children and all of his domestic animals began to die, one after another. Hlawn Za had no choice but to move out of Phaileng with all his followers and establish a new settlement called Haka, at the foothill of Mount Rungtlang. He was convinced that the mighty Mount Rungtlang, as a guardian god, could protect him and bless him abundantly. He thus composed the following song of prayer:

> *Rung cin e, van deng hauka in,*
> *Van Ni thla za awi rih lang,*
> *Ka semmawng nih hrinzung to lai hna maw*
>
> (At the top of the mighty mountain of the Rungtlang, /Which almost reaches the gate of heaven,/Let me offer my prayer to the Heavenly Spirit,/So that my offspring shall not perish any more,/And let them enjoy the tree of life under the protection of Rung le Met)[11]

In addition to this song of prayer, there is another verse of prayer which the Haka people used to recite whenever they made the offering to their *Khua-hrum* called *Rung le Met.* It always started with the same verse:

> *Aha! Maw Rung aw, Maw Met aw,*
> *Rung cin e, Van deng ka hauka in e,*
> *Kikawg dothla in kan biak e,*
> *Ka co law, ka hlang law,*
> *Tluang law, zaa law…*[12]

According to tradition, the Mount Rungtlang *Khua-hrum,* called '*Rung le Met*', had blessed the Za Thang family with abundant life. They increased in great numbers, performed all kinds of feasts and festivals and ruled all the villages and communities located within the range of Mount Rungtlang, which covers almost all the central part of the present Chin State of Burma.

In theory, anyone who wanted to settle within the realm of Mount Rungtlang must worship '*Rung le Met*' as their *Khua-hrum.* The Za Thang clan was the first family to settle at Haka and worship the *Khua-hrum* of *Rung le Met.* They were therefore the *do-dang-tu* of *Rung le Met.* As long as they could maintain their lineage, and as long as they were able to offer the proper sacrificial animals to *Rung le Met,* the religious power of *do-dang-tu* which they had inherited from their ancestor could always be retained by the Za Thang clan. As *do-dang-tu,* they also had the right to collect many kinds of religious and other taxes from each household of every village, who cultivated, hunted, fished or otherwise made their livelihood within the range of Mount Rungtlang. Haka, where the ruling chief lived, became the principal village of the *Tual,* because the *Tual* ground of Haka was the only sacred place from which the *Khua-hrum* of *Rung le Met* was to be worshipped. The surrounding villages and all their satellite communities became the *Tual* community of Haka.

Like the Za Thang clan, the ruling house of the Haka *Tual* community, there emerged several 'lords of the soil' and *Tual* communities in Chin society. The Lian Bawi clan of Thang Nget Chung, who made Mt. Bawipa-tlang their *Khua-hrum,* ruled almost all the Zophei area under the range of Mt. Bawipa-tlang. The Lian Chin clan, who was the *do-dang-tu* of the Bawinu River, ruled the whole Lautu area in the Bawinu Valley, and so on. In this way, the hereditary nature of chieftainship and the hierarchical power structure of the *Tual* community were established in Chin society.

SOCIAL STRUCTURE AND DISTRIBUTION OF LAND AND POWER

As noted already, it is not easy to distinguish between religious, social, cultural and political elements in tribal society. It is even harder to distinguish between the political realm, social structure and economic system. They are all inseparably intertwined, and deeply rooted in religious concepts. I shall therefore discuss the interrelationship between the social structure and economic system in this section, instead of attempting to isolate these two related aspects of social life in Chin society.

The Chin social system can be categorized into three social classes, namely:

(1) *Bawi-phun* (noble class)

- (a) Chief (*Ram-uk*)
- (b) Priest (*Tlang-bawi*)
- (c) Nobility (*Phunsang*)

(2) *Mi-chia* or *Zaran* (ordinary class)

(a) *Mi-chia phunthiang* (ordinary, ritually clean)
(b) *Mi-chia phunthiang lo* (ordinary class, believed to be possessors of unclean spirits, the evil eye or witchcraft.)

(3) *Sal* (slaves)

(a) *Innchungkhar sal* (household slaves)
(b) *Innlak sal* (slaves who lived in their own houses but worked for their master)

(1) Noble Class

From a religious point of view, the noble class (*Bawi-phun*) was composed of the families, lineage and the clans of '*khua nih a hnuai mi hna*', who received the mandate of power and privilege from the guardian god: *Khua-hrum.* Since they were '*do-dang tu*' they were the chosen clan and, at the same time, ritually clean. From a political point of view, the *Bawi-phun* was made up of the families, lineage and clans of the chief and headman, who received tributes and services from the people. It also included those members who were by birth noble, but who lived independently and did not pay tax or receive tributes. There would be several such families of nobility in a community, but usually only the ruling chief had the privilege of receiving tributes and services. (Even today, although the special right of the nobility has been revoked, a sense of clan consciousness still remains, and many are still regarded as natural noblemen and leaders.)

The noble class can be categorized into three groups, as follows:

(a) Chief

The chief in traditional Chin society received the mandate of religious power and political authority from the *Khua-hrum.* Because of this mandate, he and his descendents had the privilege and the right to control the entire land within the sovereignty of *Khua-hrum,* which was believed to be the guardian of the land. From the religious point of view, the guardian god *Khua-hrum* was the real owner of the land and the chief acted on behalf of *Khua-hrum,* called in Chin: '*Khua nih a hnuai mi hna an si.* Because the chieftain clan received their mandate from the guardian god, or were exercising their powers on behalf of *Khua-hrum,* these powers should not be diverted to any other clan or family, but must be handed down generation after generation. The chiefs in Chin society therefore were usually direct descendents (*Nuhrin fa* or *Nuhrin chung*) of the founder of a particular clan or settlement, and they maintained and preserved their lineage almost as a holy institution. Powers and privileges, 'status rank and class rank', all that they had received as a mandate from the guardian god, 'were firmly anchored in the lineage system' (Lehman 1963: 144).

On the practical level, however, the real powers of the chief in Chin society stemmed from the control of the main economic resources of the land. H. N. C. Stevenson defines the term *Ram-uk* as 'the lord of the soil' (Stevenson 1943: 82). Literally, *Ram-uk* means the ruler of the land, or the country, and 'they are the tribal chiefs' or 'country chiefs', as Lehman puts it. As 'the lord of the land', the chief had the right to distribute all the lands to the community, which usually included:

- the right of distribution of unoccupied plots (*lo-lawng*);

- the right to spilt up large occupied plots into smaller units if and when the increase of the population or the shortage of land required such action;

- the right of partitioning rotational fields (*lobung*) into plots (*lopil*);

- the right to delineate and alternate the boundaries of rotational fields; and

- the right of granting permission to cultivate specially rescued areas such as *tuklaw* [*Khua hmawng*] lands (ibid.: 82).

Theoretically, the chief as a *do-dang-tu* owned all the lands within the sovereignty of *Khua-hrum.* He distributed the *lopil* and parceled out the *lobung* to the families of the entire community, and every family had the right to pass the *lobung* down from the father to the son, and the chief collected a certain percent of the crops as a tax, called *lo-nga.* In Chin society, the basic unit in the village was the household, so the *lo-nga* was levied on each household, though for service involving manual labor all able-bodied individuals generally turned out and took part. In addition to *lo-nga,* the chief also imposed a religious tax, known as *Tual man,* which was supposed to be used for the community's sacrificial worship to *Khua-hrum.* I shall come back to religious taxation in the next chapter.

Since the chief was the 'lord of the soil', he not only owned lands but also the forests, the woods, the rivers and all living beings residing in his territory. He therefore received a flesh levy called *sa-haw,* when any villager shot a wild animal, consisting of either the saddle or a hind leg or in some cases both of those animal parts, and in the case of an elephant, he took one of the tusks. Moreover, 'whenever a domestic animal is sacrificed for any private purpose, he gets one hind leg and one fore leg' (Lehman 1963: 148). The chief also received one hind leg of every animal sacrificed by any of his men for religious purposes, as a religious tax, called *sa-khua.* And if he ruled more than a single village, he collected taxes called *Khua-man,* from all the domain villages. The *Khua-man* was levied collectively upon the village, not the household.

As 'the lord of the soil', the chief was not only the owner of the lands but 'the head of the village. He was the leader in war, the protector of his people' (Parry 1932: 250). Being the *do-dang-tu* of the *Khua-hrum,* he also was a defender of the faith. And the chief had full powers of control over his villagers, 'he can punish them by fines, and in the last resort, if a villager refuses to obey the chief's order, the chief can refuse to allow the offender to cultivate his land any longer, and can turn him out of the village' (ibid.: 250). However, as Parry noted:

> Though in theory possibly the chief is a despot, and though the chief can and on occasions doubtless does commit tyrannical acts, the basic relationship between the chief and his people is one of mutual benefit and mutual help. The chief must protect his people, let them use his land to cultivate, and help them in time of famine or other distress, and in return his people must pay certain dues, render him certain services, and come to his aid when called upon by him for assistance (ibid.: 249).

A most valuable and honourable services rendered was the chief's house – usually regarded as the centre of the political institution for which he stood. It was built at by the community of his subjects who received no payment for this, and when the chief and a member of his family traveled, some villagers always accompanied him to carry his loads for free. When the chief or a member of his family died, the villagers were expected to dig his grave, erect his gravestone, called *lungdawnh,* and sacrifice a mithan as *ruak hngah* to accompany the dead chief to *Mithi-Khua* (ibid.: 252).

(b) Priest: *Tlangbawi*

Community Priest. The most important religious leader in Chin society was the chief himself, who was the *do-dang-tu* of the *Khua-hrum,* the political leader and also the head religious official. In many cases, especially when a single patriarch ruled an entire community, the chief combined both the offices of priesthood and chiefdom. When the community increased in number, however, a special religious office was created, and a priest, called *Tlang-bawi,* was appointed – usually the younger brother or youngest son of the patriarch of the tribe or clan. His descendants automatically inherited his religious office. In Chin society, the chief's eldest son became chief after his death, and the youngest son usually became the priest. The priest received feudal dues as did the chief.

The *Tlang-bawi* had to take a *nutak,* a woman of the priest or chief class, as his major wife. Before a major wife was taken, or even afterwards, the *Tlang-bawi,* like the chief himself, might take a minor wife, a *nuchun,* from any class he chose, as long as he did not perform a *zing-dangh* sacrifice called *Nutak-he-zing-dangh,* which made the union a legal marriage (cf. Head 1910: 15; Strait 1933: 67–68). If he failed to marry a woman from his own class or the noble class, or if he performed a *zing-dangh* sacrifice with a minor wife, he definitely lost his office of priesthood. Consequently, his descendents no longer had the right to hold the priesthood office. The election of a new priest in such a case would then be held among the noble class.

The functionary role of *Tlang-bawi* in the *Khua-hrum* worship at the *Tual* communal ceremonies, which was regarded as a confirmation of the political legitimacy of the chief, was something of an assisting role. He stood side by side with the chief in the communal sacrificial ceremony to guide him during prayer. He also prepared the holy water for the sacrifice. But sometimes he played a leading role. For example, the *Tlang-bawi* would not only perform sacrificial ceremonies for the chief and the noble class, such as *Bawite-bawi and Khuang-cawi,* he would also lead communal sacrificial ceremonies connected with cultivation and sickness, such as *lopil-nam* (for cultivation and harvests) and *Tlang-rai-thol* (for sickness, epidemics and other ill health).

Household Priest. The traditional Chin society also had the so-called 'household priest'. This priest did not hold religious office, but was very important even for the *michia* class. He was the head of his own household, or the elder of the clan, whose duty was to perform sacrificial ceremonies for himself, his household and his clan, especially sacrifices connected with sickness and *zing-dangh.*

The first kind of sacrifice that the household priest had to perform was the marriage *zing-dangh,* also known as *nutak-he-zing-dangh,* meaning 'the sacrifice with the major wife', to legalize the marriage. The fundamental concept of marriage *zing-dangh* is the one-ness of husband and wife, and their *zing.* They therefore cooked the sacrificial meat on a three-legged stove in three pots stacked one on the top of the other. Then they collected the distillate from the still. In fact, the distillate, which came from sacrificial meat, was the main course of the sacrificial meal eaten by the husband and wife alone. No one else was allowed to partake or even enter the house. To indicate that they were performing the holy sacrifice, they hung the green leaf called *kaw-dai-hnah,* meaning 'green-peace-leaf', on the gate of the compound and the entrance of the house. The Chin believed that partaking this sacrificial meal together, the *zing* of husband and wife became one in this ceremony.

When the couple was blessed with a baby, the husband, who was also the household priest, performed a second kind of ceremony, the *nua-zing-dangh,* a sacrifice for a newborn. I shall discuss *nua-zing-dangh* in further detail in the next chapter. Thirdly, he performed every sacrificial ceremony for his household, in time of both fortune and misfortune, generally called *do-le-dangh.* A variety of different objects of worship call for *do-le-dangh* sacrifices: *Khua-hrum* or *Khua-chia,* depending very much on the nature of the sacrifice. Fourthly, if he was of the chief or noble class, he performed household sacrifices for his whole family and clan in *Bawite-bawi* and *Khuang-cawi,* which they called *cing-la-tawn-ni.* I shall study *Bawite-bawi* and *Khuang-cawi* closer in the next chapter. Finally, if he was the first-born son or the descendant of the first born son, called *nuhrim-chung*, no matter whether he was of the noble class or not, he performed the sacrifice for the whole clan, called *Phun*, in the *Khua-hnaw-kai* festival, which will be discussed in the next chapter.

(c) Nobility: *Phunsang*

The *Bawi-phun* included those members who by birth belonged to noble families, but who lived independently and did not pay or receive tribute. Several families of this category lived in a community, but only the ruling chief had the privilege of receiving tributes and services. The *phunsang* in a community, who did not belong to the same clan as the ruling chief, would even pay religious tax, such as *Khua-man* and *Tualman*. But the *phunsang* clans were usually exempt from manual service, which all members of the community had otherwise to render to the chief. In time of war, members of the *Phunsang* played a leading role, for their blood-ties with the chief made them a loyal tribal army – though not a standing army – for the chief. Moreover, the chief might select and appoint the members of his ruling community council from the people who belonged to *phunsang*. In a nutshell, the *phunsang* played an important role because lineage and kinship played a major role in traditional Chin society.

(2) Ordinary Class: Mi-chia le Zaran

The ordinary class can be divided into two groups, *mi-chia phunthiang* and *mi-chia phunthiang lo*. *Phunthiang* means that the clan is clean, and *phunthiang lo* indicates an unclean clan. Unclean clans were those who possessed the evil eye and witchcraft. Clean clans were free from such signs of evil spirits. *Phun thiang lo* were regarded as evil people and traditionally treated as outcastes. The Chin believed that these people could cause illness and death to others. When a person was ill and the illness was believed to have been caused by a person possessed by the evil spirit or *hnam*, then the family of the sick person usually made an offering to the *hnam*. The patient would also offer something to the person who he believed had caused his sickness, to please his *hnam*, evil spirit. However, the offering should not be obvious or made public. If it was made public, it amounted to a direct accusation, in which case the other party could take legal action for libel. The fine for falsely accusing anyone of being *hnam* has always been very high and is called *phun thawh*. (*Phun thawh* means the price of the clan, which is usually paid as a 'marriage price'.) Anyone who was *hnam* belonged to an unclean clan, *phungthiang lo*, and if a woman was believed to be *hnam*, nobody from *Bawi-phun* or *mi-chia phunthiang* would marry her. Indeed, *hnam ngei* or *phunthiang lo* social statuses were even lower than those of the slave. They were the genuine outcastes of society.

The *mi-chia phunthiang* clan, classified as those people who were neither noble nor slave, paid tributes to and provided services for the chief. They were peasants. They held second-class status in society. Nevertheless, intelligent and wealthy people of this class often stood next to the chief in the leadership of the community. An able man could rise in social position by holding important feasts, like *vok-thau thah* and *Bawi-te bawi.* However, none of these feasts provided license to demand equality with members of noble clans, even if they earned the respect of the people. While feasts could not change an individual's clan status, if a rich and ambitious man was serious about improving his social status, he could always marry one of the chief's or noblemen's daughters.

Traditional Chin society was patriarchal, but the place of women was very important. Through marriage a woman couls enhance a man's clan and provide leverage for improving her husband's social status. If an ordinary class (*chia hrin*) man married a woman of the noble class (*Bawi-phun*) and both his son and grandson also married a woman of *Bawi-phun,* then his great-grandson became a nobleman, '*chan thum hnu bawi*' meaning 'he who became noble after three generations'. Thus, by successive, strategic intermarriage over three generations, the great-grandson of a man of an ordinary clan could ritually become a nobleman, equal in every aspect with other noblemen except for the fact that he could not become a chief, because both clan and class rank are firmly anchored in male lineage. In a clan (*phun*) and class (*Bawi-Chia*) conscious society, to become a nobleman from a humble beginning is no small achievement, and almost impossible for most. His ancestors had to be strong, intelligent, ambitious and rich to be able to scale the stone wall of class, or the *Bawi-Chia* system of Chin society. The marriage price of a woman from a noble clan was five times higher than that for an ordinary woman. This barrier in most cases prevented the leveling of social class.

(3) Slaves: Sal

The slaves in Chin society arranged themselves in three groups:

(a) war captives or those captured in raids;
(b) debtor slaves; and
(c) slaves who had become so voluntarily or who had been made slaves by their relatives as payment for gambling debts (Brown 1960: 19).

The first kind of slave was the most common in Chin history. Like other peasant societies, agriculture was the main profession and manual labour was the basic means of production. Moreover, the Chin practised demanding sacrificial worship ceremonies and religious feasts. All required a tremendous amount of manual labour. The chief and the noble class, therefore, used to raid slaves, mainly from Burma and Bengal, in order to maintain their power, social status and privileges. Chin folklore was full of stories like 'Thli Chawn and Hlawn Ang', the story of a Chin hero who carried out many slave raids and captured hundreds of Burman slaves (cf. Thla Peng 1981: 15–20). Carey and Tuck claimed that the British had invaded Chinram in order to stop the slave raids committed by the Chin within the newly occupied Chindwin Valley and Bengal Valley. I shall describe further details about the connection of slave raids and the invasion of Chinram by the British colonial power in Chapter 4.

There were two levels of slaves. One was the domestic slave who lived with his master in the same house. Such slaves, known as *inn chung khar sal* were the lowest slaves in Chin society. Hardly considered human beings, they were treated almost as material possessions of their master. They could be sold and one male slave's value equaled two or three mithan or a piece of real estate. They could be given away as part of a marriage price, or they could be given to the bridegroom by the bride's father as a dowry. *Bawi-phun* of the highest noble class used to require a 'slave price' (*sal man*) as the marriage price for their daughters. It was also called *thut hmun rolh.* When the nobleman of Lungo in the Zotung area married a cousin of the Aibur Chief Sang Bill, they gave a grown woman slave by the name of La Kung Nu for *thut hmun rolh* to Chief Sang Bill. That was only two generations ago.[13] A despicable slave could be buried alive with his master when the latter died. Although this practice rarely occurred, it is recorded that when the son of a Vuangtu chief died, one of the slaves was buried alive with the dead child.

Ordinary slaves called *inn lak sal* lived independently of their lord in their own houses and formed the second level of slaves. They worked for their master just as the domestic slaves did, but they could not be given away as a part of a marriage price as a dowry. They could not be sold to others, yet their status was still lower than the ordinary class. They were anonymous persons without a clan name (*phun*); as the Chin would say, *phung ngei lo, hram ngei lo,* 'A man without a clan name is like a tree without roots.'

TUAL ORIENTED POLITICAL SYSTEMS

Traditionally the Chin developed three different political systems: the *Khua-bawi* system, *Ram-uk* system and 'Tlaisun Democratic Council' (Stevenson 1943: 213).

The Khua-bawi and Ram-uk Systems

In the *Khua-bawi* system, an independent village was ruled by the hierarchical local chief, who had the power and the duty to control and manage every aspect of the political, religious, social and economic affairs within his domain. *Khua-bawi* means 'the ruler of the village'. The village ruler had the right to appoint the village council, the right to distribute land and the right to collect taxes from his subjects. He had the right to make *Tual Khua-hrum* sacrificial offering ceremonies on behalf of the entire community of his subjects. In short, he was the 'lord of the soil' in his own village. The nature of the chief's power and duty in the *Khua-bawi* system is generally the same as the chief in the *Ram-uk* system. Historians such as Frederick Downs call this kind of political system an 'independent village-state' (Downs 1983: 11). In reality, a completely independent village-state was very rare. One way or another, they made war alliances with other villages and communities, or with stronger tribal chiefs. In some cases they even paid tribute to the war ally chief. However, as long as they performed the *Khua-hrum* worship at the *Tual* ground independently, we categorize them into the *Khua-bawi* system of independent village-states.

The *Ram-uk* system is a political system where the hierarchical tribal chief ruled either the whole tribe, part of the tribe or at least more than two villages or the entire *Tual* community. This kind of political system can be called tribal feudalism. The *Ram-uk* system was the most important and common political system in traditional Chin society. In principle, the *Ram-uk Bawi-pa* must rule at least two villages, but he usually ruled the entire tribe – as the Sailo chief ruled the entire Mizo tribe of present Mizoram State in India, or as many as 300–500 villages – as Chief Hua Cin Khup ruled the entire Tiddim and Tawnzang townships of present Chin State in Burma.

The power and the duty of *Ram-uk* have already been described above. I shall discuss in the next chapter how the chief's powers in the *Khua-bawi* and *Ram-uk* systems were legitimized and confirmed at the worship of *Tual Khua-hrum.*

Tlaisun Democratic Council

In addition to the *Khua-bawi* and *Ram-uk* systems, the Chin traditionally developed a democratic political system called the 'Tlaisun Democratic Council', which had 'a democratic constitution' and 'ruled by the council of the elders, who were presumably all aristocrats' (Lehman 1963: 141).

Actually, the Chin developed such a democratic council almost by chance. The Tlaisun, like the Sunthla and Haka, were the descendents of Lai-lung, the original home of the Laimi tribe. During the sixteenth and seventeenth centuries, the Haka became very powerful, so the power struggle between the Haka on the one hand and the Sunthla and Tlaisun on the other became severe. War broke out several times. First, the Sunthla and Tlaisun appeared to be the winners, but during the reign of the Haka chief Lian Nawn '[he] attacked and destroyed both villages [Sunthla and Tlaisun] and scattered the inhabitants, who for years lived in settlements as fugitives. However, in the course of time the ShunKlas [Sunthla] made their peace with the Hakas, but Tlaisun found a new settlement, the present Falam' (Carey and Tuck 1976: 141). Old Tlaisun was not deserted but Falam became their capital.

After they lost that war, the Sunthla never again played an important political role in Chin history. The Tlaisun, however, re-emerged in a very different form of political system. They formed a democratic council instead of a traditional *Ram-uk* system. Different historians have given different reasons for the foundation of such a democratic council. According to Carey and Tuck, they were unable to choose a paramount chief, *Ram-uk Bawi-pa,* because their community was composed of aliens, which meant that the aristocratic headmen were not only from the Hlawn Ceu clan, as in their previous capital, but some villages were ruled by the Za Hau clan and even the Mizo tribe of Hualngo. Another reason, according to Za Peng Sakhong, was that they 'simply refused to comply to the terms of defeat by forming a new settlement and a democratic council' (1983: 9). If they continued to live in the old capital and worship the same *Khua-hrum,* they would have to pay the *Khua-man* and other tributes to the Haka who had defeated them. In tribal war, if one party lost the war, it meant not only that the people were defeated, but the *Khua-hrum* as well. Thus, after defeat, their own guardian god, *Khua-hrum,* would no longer protect them or bless them because their *Khua-hrum* was already subordinated to the winner's *Khua-hrum.*

Instead of complying in defeat and degrading their *Khua-hrum* into subordination, the Tlaisun revolted against the entire system, thus reforming it. Not only did they abandon their capital and their *Khua-hrum*; they also abandoned the idea of a paramount chief *Ram-uk* system altogether. If they had maintained the old system, they would have remained permanently under the rule of the Haka chief. They therefore cleverly formed a Democratic Council, which appeared successful and strong. Within the course of a single century or so, they conquered almost all of their neighbours' areas, except Za Hau and Haka. They even conquered Sizang and Sokte of the Zomi tribes, and Hualngo and other Mizo tribes (cf. Carey and Tuck 1976: 143). When the British conquered the Chinram in 1896, Carey and Tuck reported that Tlaisun were 'the most powerful' in the whole of Chinram. Chin historians, such as Za Peng Sakhong and Vumson, agree that if the British had not appeared on the scene, the Tlaisun might one day have united all the Chin people under their domain (Z. Sakhong, 1983; Vumson 1986: 103). Before I move onto these dramatic developments, we have to look more closely into the traditional patterns of worship and rituals, another significant dimension of *phunglam.*

NOTES

1 Among Western scholars, Lehman is the only one who paid enough attention to the meaning of *'khua',* in his now classical work *The Structure of Chin Society.* He also defines the term *'khua'* as 'a general category of spiritual beings'. See Lehman (1963), pp. 172–174.

2 The spelling of *Khua-zing* can be different based on various Chin dialects. While Parry, for instance, spells '*Khazangpa*' in Lakher dialect, Carry and Tuck spelled '*Khozin*' in Lai dialect.

3 David Laisum translated *Zing* as 'dark or darkness'. However, 'it is hardly used in a literal sense' (1994: 126). His statement, in my view, is both right and wrong. Depending on various Chin dialects, *Zing* can be translated as both darkness and light. In Zophei dialect, *Zing* is darkness, but in Lai, especially in the Haka area, *Zing* is morning, which is light. In both senses, *Zing* is an invisible force, which gives the source of both light and darkness. Darkness, therefore, is not the literal meaning of *Zing* but the symbol of *Zing*.

4 Chester U. Strait reported *Ding-thlu le Lai-rel* as a different deity from God, perhaps from his misunderstanding of the Chin words (Strait 1933: 57).

5 Here I follow Sing Kho Khai, and use Zomi dialect. In Mizo dialect it is called *Mual,* but in Lai dialect it is called both *Tual* and *Mual.* In Zophei and Lakher dialects *Tual* is called *Tlenlai,* and the sacrificial stone therefore is called *Tlenlai Lung.*

6 The two *thingtli* trees still stand in Nawi Valley.

7 Interview with Mang Lung of Aibur, 29 March 1986.

8 I have interviewed several people about this concept, and I usually got a comparative answer. They compared three levels of heaven with mountains – the highest with *Khua-zing*'s abode, the second highest with Lulpi's abode, and the third one with *Mithi-khua*; while the earth with the valley, and *Khua-chia's* abode was underneath of the world. See also Sing Kho Khai (1984), Hre Lian Kio (1972), Pu Sakhong (1971) and Lian H. Sakhong (1988b).

9 See the myth of the journey to *Mithi-khua* and *Khua-vang Mithi Tlawn* in Lian H. Sakhong (1988b), especially Chapter 6; and also Hre Lian Kio 1975, especially Chapter 5.

10 Carey and Tuck 1976: 152–153. Here I follow local oral tradition, according to Capt. Sui Mang's *Haka Thawhkehnak.*

11 The meaning of *Rung* is 'supreme, mighty, awesome', and *Met* is power, especially divine power, mysterious power; *Met* can also be translated as the central part of life, the core of life, the cream and the best part of life. *Hrinzung* or *Hrinzung Kung* is the 'tree of life' that comes from a deep theological concept in Chin traditional religion. Once there was a 'tree of life' called *Hrinzung Kung* planted here on earth by *Khua-zing.* But one day, while a dog which guarded this tree was asleep, the Spirit Being from heaven stole the tree and planted it on the moon. Although the tree of life was already on the moon, the Chin tradition maintained that they could still received 'source of life', called *Zing* from *Khua-zing* as a blessing, *Van-ni,* through the sacrificial ceremony called *Zing-dangh,* and the worship of *Khua-hrum.*

12 It is believed that the original term 'Haka', capital of the Chin State in Burma, comes from the Chin traditional religious concept of 'Hauka' and its theological meaning is the 'Gate of Heaven'.

13 Interview with Pi Par Cin, daughter of Chief Sang Bill of Aibur, on 25 March 1986.

THREE

Rituals as Confirmation of Power and Social Status

In this chapter, I shall to explore two additional aspects of the traditional Chin way of life encompassed by the concept *phunglam* – ritual categories and human rites of passage. Rather than categorize ritual ceremonies and rites of passage conventionally, I shall analyse how sacrificial ceremonies and worship services in traditional Chin religion can be viewed as the confirmation of power and status, and how rites of human passage and the meaning of life on earth itself were viewed in traditional Chin society as preparation for life after death in *Mithi-khua*.

SACRIFICIAL CEREMONIES AND WORSHIP

Did They Worship Khua-zing?

As mentioned earlier, the Chin believed in the existence of a Supreme God called *Khua-zing*, the creator of the universe and life giver to every living being. They also believed that, '*Khua-zing* is good' (Strait 1933: 23), and he 'gives health, richness, children, and other human wishes' (Vumson 1986: 16). However, despite their belief in the good nature of *Khua-zing*, who controls 'good fortune and bad fortune ... human life ... [and] eternal destiny' (Laisum 1994: 127), the 'Chin do not worship *Khua-zing*[1]' (Carey and Tuck 1976: 195). The reason, according to Vumson, is that, 'God is good, never cruel and never harms people. Therefore, they never sacrifice or make offerings to appease God' (1986: 16).

Although the Chin did not worship *Khua-zing*, 'the name of *Khua-zing* is usually mentioned in the process of prayer' (Strait 1933: 16–107). Strait points out that the Chins indeed mentioned the name *Khua-zing* in *zing-dangh* sacrificial ceremonies – especially in *nau zing-dangh* (a sacrificial ceremony for the newborn), *nutak he zing-dangh* or *Arnakhman* (a sacrificial ceremony which legalized marriage) and at other important sacrificial ceremonies including *Bawite Bawi* and *Khuang-cawi*. The prayer they offered at the sacrificial ceremony for a newborn reads as follows:

Aha! Kha-zing aa, Pathen aa,
Tui then e, saw then se; A maa a zing se, ka ca dang e!
Si sen lai ce, vaw sen lai ce;
A ni aa e, a khua aa e; A chin aa e, a zen aa e;
Ung pui la, can pui la!
A ni hli ta ma sii sen lai e; A hu ta ma caw sen lai e!
Ngen dang kha e, hnaw tlan kha e;
Ni nawn ta la, thla va ta la; Lin la, saw la![2]

(Oh! *Khua-zing*, you who are Holy [*Pathian, Pathen* in Zophei dialect], I offer you this unblemished sacrificial animal and holy water. I offer you this for [name of the child] and his/her *Zing*. Be pleased with him/her and keep him/her well. Do not let his/her soul escape from him/her, do not let his/her shadow escape from him/her. Do not let his/her breath [*hu* in Chin] stop. Be warmed and pleased with him/her, so that he/she may be healthy and strong.)

The theological concept behind this prayer holds that every person was believed to have a source of life called *zing*, which comes directly from *Khua-zing*. However, the Chin traditional belief maintains that the *zing* of the child was not acquainted with the body of the child at birth. The *zing-dangh* sacrifice was therefore necessary in order for the soul and body of the child to dwell together. Parry therefore mentioned that all Chin, based on his study of the Lakher, 'sacrifice to *Kha-zang* at least once in their lives' (Parry 1932: 350).[3]

Though they mentioned the name *Khua-zing* in the process of prayer at *zing-dangh*, it seems that the ceremony itself did not involve worship directed to *Khua-zing;* rather it encompassed the worship of *zing*, or the 'tutelary deity or guardian angel', as Parry defined the word. Thus, the *zing-dangh* sacrificial ceremony cannot be categorized as the worship of God, but a sacrifice to the guardian god. The link between the two concepts, however, is that since the *zing*, or source of life, comes directly from *Khua-zing*, the prayer for *zing* must go to *Khua-zing* himself. Since the context was the worship of *zing*, not *Khua-zing*, the *zing-dangh* sacrificial ceremony was sometimes known as *Arnak-hman*, or *Zanchunh*, literally 'black hen sacrifice', or 'the sacrifice at night' respectively, because the sacrificial ceremony had to be performed after the sunset, and the sacrificial animal they used for the ceremony had to be completely black (cf. Strait 1933: 153–154). The symbols of night and black in traditional Chin religious concept stood for the invisible force of *zing*, to whom the suppliants made their offering.

Traditional Chin religion was not the religion of *Khua-zing* but the religion of *Khua-hrum* and *Khua-chia,* as almost all sacrificial ceremonies and worship services, as we will see below, revolved around the two concepts of *Khua-hrum* and *Khua-chia.*

THE WORSHIP OF *KHUA-HRUM*

(a) *Sacrificial Ceremonies Connected with Cultivation and Rain*

The Chin believed that it was in the power of the *Khua-hrum* to give them good or bad crops. It is not surprising to find that each agricultural phase was therefore marked by a specifically associated sacrifice intended to placate the spirits of the hills and the fields. The first of these sacrifices, called *lopil-nam,* was performed by all the villagers together when the fields (*lo*) were beginning to be cut. The object of the sacrifice was to prevent the cutting edges of the knives and axes from getting chipped and blunted, and to prevent people from cutting themselves by accident when harvesting the fields.

After this sacrifice, the fields were burnt, and then followed another sacrifice called *lopil-luh. Lopil-luh* was an annual sacrifice offered jointly by groups of people who had their fields on the same slopes, in order to induce the spirit of the slope to give them good crops and good health, to prevent wild animals from eating the crops, and to provide them good luck in hunting. The *Tlang-bawi* performed the sacrifice, and usually a pig was used for this sacrifice.

The next sacrifice was called *lohring-khen,* performed between the sowing of the seeds and the second seeding called *belh-tuk.* It was also intended to secure good health, to ensure good crops and to prevent them from being eaten by wild animals. Offered to the *Khua-hrum* of the field, it was performed in front of the field hut. On the same day that *lohring-khen* was performed, another sacrifice, called *lo-ar-thah,* was also held. This sacrifice was directed to the spirit who dwells on the borders of the field and the forest and who can stop wild animals from entering the field to eat the crop, if successfully appeased.

Moreover, in relation to *Khua-hrum,* another sacrifice called *lo-hma-thawh,* was performed in the villages at the beginning of the harvest, after all the rice plants had been pulled up and before the grain had been gathered in. Near the field hut a threshing mat was constructed, at one corner of which a wooden post was erected. A small basket containing paddy (irice in the husk), millet, maize, and in fact every kind of plant cultivated, plus some flour was placed at the foot of the wooden post. A red hen was sacrificed, and the seeds were anointed with the hen's blood. The priest chanted the following prayer.

Loi li sazai va tlun la, Loi lu sa zai va tlun la!
Nga-ti kai aa va kai la, Nga-len kai aa va kai la!
A sawn vaa tlin la, a zaw va tlin la!

(Oh! paddy, from the bottom of the field come; Oh! paddy, from the top of the field come. Swarm together like the *nga-ti* and *nga-leng*. Fill a hundred baskets full, fill a thousands baskets full. [*Nga-ti* and *nga-leng* are small fish that swim about in large schools.])

At the end of the prayer the priest blew a bamboo whistle to call the spirit of the paddy, and he killed the fowl.

(b) *Khua-hnaw Kai or Khua-do*

The most important sacrificial ceremony connected with cultivation in traditional Chin society was called *Khua-hnaw Kai* in Zophei, *Khua-do* in Zomi, *Thlai-thar* in the Lai dialect. Literally, *Khua-hnaw Kai* means 'the offering to the rain god'. The *Khua-hnaw Kai* was usually held between the last week of October and the middle of November, after the wet crops had been harvested. This really was a sacrifice to the *Khua-hrum* and the rain god, *Khua-hnaw*, with hope and thanks that the *Khua-hrum* had granted and would continue to grant the village general prosperity – especially good harvests, good hunting, many domestic animals, health and fertility. The sacrifice was also intended to please the spirits of rice and maize.

On the given day, the village's gate was erected if the village was a new settlement or renovated properly if it was not.[4] In the evening the chief, the priest and all the elders proceed to the *Tual* ground. The priest sacrificed a red hen[5] at the foot of the *Tual* tree, where they had already placed a handful of seeds of every kind of food crop. The seeds were anointed with the fowl's blood. The priest set aside the fowl's tongue, liver and the seeds anointed with blood as *phangva,* meaning 'food for the spirit', on the *Tual* stone where the *Tual* sacrifice was performed.

Before killing the hen, the chief, guided by the priest in saying the prayer, invoked the spirit of paddy in the following prayer:

> *Oh! paddy, I sacrifice a fowl to you, increase and endure, if you increase and endure I will give a great feast Bawi-te bawi and Khuang-cawi, so that my Khua-hrum shall please you. Remain with me from year to year, winter to rainy [season]. Endure and increase so that our Khua-hrum may eat you. Oh! paddy (Parry 1932: 423).*

Meanwhile, a small gong called *baleu,* a big gong called *dar-khuang,* and a drum called *paren* were hung up on the *Tual* tree, and a ceremonial dance called *I-zaw-n* was performed. The first song of the dance always proceeded:

Zai si bawn sawn a sun pa can mawi sa le,
Sa paw chen ta za dai hen bu lai aa e,
I-zaw, I- zaw, I-zaw.

(We wish the blessing of our harvest in abundance. We wish all the graces to not stop untill many generations to come).

After this, the villagers filed around the chief and priest three times. When the villagers finished marching, the chief and priest put themselves at the head of the line, followed by the villagers playing gongs and drums, and all proceeded to the fence erected outside the village. There they sacrificed another hen and offered up other prayers, calling upon the wild animals to come to the village from north, south, east and west. While the chief and priest prayed, the villagers again filed around them three times. Having finished his prayer, the priest killed the sacrifice. Small pieces of the intestine, liver and comb were placed at the foot of the fence as *phangva.*

The next morning the chief, priest and their assistants, each carrying a red cock, waited at the fence outside the village, where they were joined by the villagers with all the guns they could collect. The armed men set off to shoot – and those without firearms to trap – any bird or animal they could find. The chief, priest and assistants followed the villagers into the jungle and, having found a suitable spot, erected a shelter where they called upon all neighbouring *Khua-hrum* by name to send them stags with fine heads, boars with heavy tusks, bears, tigers and all kinds off wild beasts, and then they sacrificed two red cocks. The hunters then spread out through the jungle, trying their best to shoot game, if necessary camping in the jungle for two nights, but never more.

As soon as anyone shot an animal, a message was sent to the village and those who remained behind spread mats on the *Tual* ground, passing time by drinking *Zu* and playing on gongs and drums until the successful hunters arrived. When they reached the entrance to the village, the hunters sang hunting songs called *vawhla* or *hla-do* and fired their guns. The man who shot the game, the priest and his assistant, who sacrificed the red cocks, and the people carrying the game meat proceeded to the *Taul* ground. The women of the successful hunter (or hunters if they shot more than one wild animal), all dressed in their best, came to meet him. Every house in the village got a share of the meat. All night the successful hunter(s) performed a *sasawm-tuk* sacrifice. Over it they sang a song and danced the entire night. In the Tidim area of the Zomi tribe, instead of performing a hunting expedition they collected honeycomb for the celebration of *Khua-do puai.*

The next day saw the beginning of the main *Khua-hnaw kai* ceremony. The men (women were excluded) divided themselves into groups by clan. A clan group of six brought one pig and one pot of *zu,* a clan group of ten brought two pigs and two pots of *zu,* a clan group of fifteen, three pigs and three pots of *zu,* and a chief or noble clan might even bring a mithan. Each clan group separately offered their sacrificial animals to the *Khua-hrum* at the *Tual* ground. The head of the clan offered the sacrificial animal to the chief, and the chief, assisted by the priest, sacrificed the animal to the *Khua-hrum.* This was not only a time for clan solidarity and kinship reunion, but a time to annually renew a kind of contract, the right to cultivate land from the *Khua-hrum* and the chief. In this way, they legitimized not only the political and religious powers of the chief, but also confirmed his ownership of the land by offering their sacrificial animals to *Khua-hrum* through the chief.

As indicated already, only the men partook of the sacrificial meal. It was *aanaw,* or taboo, for women to attend the ceremony. According to patriarchal Chin society, only the men had the right to use their clan name, called *phun.* Thus the women – who had no right to bear the clan name – were excluded from the ceremony. In fact, the sacrificial ceremony was a kind of family or clan communion, which included their chief and guardian gods, *Khua-hrum.* During the ceremony, it was taboo for a stranger to enter the village. The village entrances were all closed and a large bunch of green leaves placed at the entrance to show that the village was *zarh,* observing a holy day.

RITUALS AS CONFIRMATION OF POWER AND STATUS

In traditional Chin thought, rituals were both a religious encounter with a sacred, transcendental reality, and also a 'symbolic statement about social order' (Leach 1954: 14). According to Edmund Leach, 'Ritual serves to express the individual's status as a social person in the structural system in which he finds himself for the time being' (ibid.: 11–12). 'Ritual makes explicit the social structure. The structure, which is symbolized in ritual, is the system of socially approved "proper" relations between individuals and groups' (ibid.: 15).

Chin social structure centred around the *Khua-hrum* worship at the *Tual* ritual ceremonies, in which the chief would discover his place in the cycle of generations as he offered sacrifices to both *Khua-hrum* and the spirit of the ancestors on behalf of his clan and the entire community. The ordinary men would discover themselves as part of a well-knit family, as they performed sacrificial communion with their god and their chief. Each individual would discover the reality of life within his household, clan and tribe as he transcended the individual self and joined together with others in the enduring and true forms of community through ritual. In this way, ritual made Chin social structure, called the *Tual* community, 'explicit', as ritual, political power and social status were created and confirmed.

In this section, I shall illustrate how religious ritual functioned in Chin society as what Frederic Barth calls 'the media of social interaction' between individual and group, household and clan, chief and community, and political power and economic forces.

The Worship of Khua-hrum: Tual Dangh

Edmund Leach explains that 'ritual makes explicit the social structure' (1954: 15). In traditional Chin religion, the most important ritual, the one that created a distinctive social structure, was the worship of *Khua-hrum* at the *Tual* ground, called *Tual dangh.* At the *Tual* ritual ceremony, Chin social structure was symbolized in accordance with the functional pattern of 'god–chief–land', in which the power of god, the political authority of the chief and the economic system of land ownership were all confirmed and legitimized together. In so doing, they not only ritualized the confirmation ceremony but also the myth of common descent, a shared memory of a collective past which always is associated with the original homeland; at the same time they emphasized the solidarity of the clan, tribe and community, for they all believed that they were the offspring of the same ancestor. As they expressed the myth of common descent, the *Tual dangh* always began with the worship of the ancestors, called *lamnai-cheih,* in which the living and the dead were made reunited through ritualized sacrificial ceremony.

(a) *Lam-nai Cheih*

Literally, *Lam-nai cheih* in the Zophei dialect means 'cleansing the road to the graveyard', which is described by N. E. Parry in the Mara dialect as *Laliachhia* (Parry 1932: 445–447). The essence of the *Lam-nai cheih* ceremony can be summarized thus: It was a farewell festival for those who died during the entire year, and the worship of ancestor spirits, especially the founders of the clan, the village and the community. Parry observed that they offered the sacrificial animals to 'the spirits of their ancestors to induce them to help to make crops good, the domestic animals healthy and fertile, and to give good hunting, and blessings to their children' (1932: 445). The ceremony was held in December, when the moon was in its final waning phase.

Very early in the morning the road to the graveyard was cleaned, and all the villagers marched up and down it, beating gongs and drums to send off all the souls of the dead. They then visited the graves of all people who had died within the last three years, and placed handfuls of every kind of food and flour, called *chang-vut,* on the graves for the spirits of the dead to eat. Then they tidied up the graves and surroundings.

After this, they marched to the houses of the families who had lost their loved ones within the last three years, to offer condolences. Then they continued to the house of the priest. He provided *zu* for all who came to his house, and after the sun set and all had gone home, the priest sacrificed a red hen at the foot of the main post at the back of the house, where they had already placed handfuls of seeds of every kind of crop. The seeds were anointed with the bird's blood, and a prayer was offered to the ancestors and the *Khua-hrum.* Traditionally, this ceremony marked the end of the year.

(b) *Tual Khua-hrum Dangh* in the *Khua-bawi* System

In traditional Chin religion, the most important sacrificial ceremony for the whole community was the *Tual Khua-hrum dangh* or *Tlen-lai dangh* or *Khua-hrum rai-thawi,* meaning the 'sacrifice to the guardian god of the community'. In the Zophei area, *Tlen-lai dangh* or *Tual Khua-hrum dangh* was held the day after *Lam-nai cheih,* the first day of the waxing moon. This ceremony actually was a new year festival for the Chin people.

The sacrifice was always performed at the same place in the village, on the *Tual* ground, under the *Tual* tree, usually the *Bung-kung,* which was planted in every village the first time the sacrifice was performed on that village site. Under the tree, a flat stone called *Tlenlai-lung* was laid on the ground and an upright stone was erected at its head. The flat stone was used for laying out the *phangva* – sacrificial meal offered to the guardian god. The sacrificial animal was usually a mithan, the most important sacrificial animal in Chin society. Before the sacrifice, all the fires in the village were extinguished. The old fire was regarded as defiled, having been used for cooking funeral meats and having also been present through any illness that may have taken place in the house, so it had to be put out, including all its ashes, before the sacrifice.

The sacrificial ceremony began soon after sunset. The chief and the priest stood in front of the altar, called *Tual-tung,* followed by the noble class, then the ordinary class and the slaves who would stand in the back row. Every male in the community, especially the head of each clan and household, had to attend this ceremony. There was no excuse for absentees; any household absent during this ceremony would be forbidden to cultivate the land the following year, or they might even be expelled from the village and community. Thus it was the most important religious ceremony of the year, one which united the entire community and consolidated the powers of the chief.

Before the chief and priest offered up prayer to the *Khua-hrum,* they invited the spirits of the ancestors by calling their names chronologically from the name of founding patriarch to the father of the present chief. In some cases, however, they only mentioned the names of ancestors who had made significant contributions to their history. They asked for the blessings of their ancestors and for permission to perform the sacrifice. Only after they had permission from their ancestors did they offer the prayer to *Khua-hrum:*

Oh! our Khua-hrum, we sacrifice this mithan to you,
Accept it without anger and be pleased with us,
Grant us sons and daughters, and let them be clever and comely;
Bless our pigs and cattle, and cause them to multiply,
Watch over us in illness and save us from death.

Bless us in all our works and deeds.
Watch over us and keep us from harm
We cannot pray to you as well as our patriarch did
But if we have made omissions, forgive us our mistake (Parry 1932: 363).

As soon as the sacrificial animal had been slaughtered, the chief lit the new fire on the *Tual* ground and the villagers came with torches, which they lit at the new fire. As they lit their torches, they pronounced their own names and their clan name, and presented themselves to the chief as his loyal subjects. The slave who did not have a clan name would mention only his own name and his master's name. And then they went back to their houses to kindle the new fire on the hearth. The sacrificial meat was distributed among all villagers, and they cooked on the new fire. The social meaning and theological essence of this sacrificial meal was communion between the *Khua-hrum* and the community, as well as the chief and his subjects. In this way, they started the New Year with a new fire and a new sacrificial meal of communion, which had already been offered to *Khua-hrum.* The next day was *zarh,* celebrated as a holy day by the entire village. No stranger was allowed to enter the village; it was *aanaw,* or taboo. Here too they placed the green leaves at all the gates of the village.

(c) *Tlang Khua-hrum Dangh in Ram-uk System*

The *Tual Khua-hrum* worship was thus the most important religious activity in Chin society. In a community made up of a single village, the chief received confirmation from the people attending the ceremony of the legitimacy of the political power that stemmed from *Khua-hrum.*

However, in communities where the ruling chief governed an entire tribe, or the entire *Tual* community, or at least more than two or three villages, the single-village-oriented *Taul dangh* sacrificial ceremony was insufficient to legitimize and confirm the chief's power. Thus, where a single chief ruled two or three villages or more, another level of *Khua-hrum* worship was needed, in which all his subjects from different villages could participate. This level of *Khua-hrum* worship is called *Tlang Khua-hrum dangh.*

The *Tlang Khua-hrum* worship was held tri-annually, usually combined with the *Tual Dangh* at the principal village where the ruling chief lived. For that particular purpose, the chief had the right to collect a religious tax called *Khua-man* (meaning 'price of the village') from very village within his *Tual* community domain. Since it was always combined with the *Tual Khua-hrum* worship of the principal village, the ceremony would be attended by the residents of the main village, the headmen, who represented their respective villages, and the heads of every clan within the entire *Tual* community. Thus, as it was being grander than the *Tual Khua-hrum,* the

Tlang Khua-hrum worship was officially the most important public event in the religious life of the Chin. However, since it was always combined with the *Tual dangh* of the principal village, the expression '*Tlang Khua-hrum*' was rarely used, and the term *Tual* remained the primary identifying criterion of the *Tual* community.

Rituals, Powers and Exclusive Isolationism

As indicated above, the Chin traditional religious practice of *Khua-hrum* worship did not allow for the presence of any foreigner or outsider during the ceremony. If any stranger were present at the *Khua-hrum* worship (both *Tlang Khua-hrum* and *Tual Khua-hrum*), the Chin believed that the presence of the outsider would pollute the sacrifice and the *Khua-hrum* would withhold blessings. Strangers were not welcome, but should they intrude, they had to be captured and enslaved. If the villagers could enslave the stranger, the sacrifice would no longer be polluted because the stranger was, after all, only a slave, and a slave did not need the recognition of the guardian god *Khua-hrum*. In this context, the slave was the property of the chief, and thus anonymous, without a clan name (*phun*). If he could compensate for his transgression and pay all costs of the ceremony, he would be released and the village would hold another ceremony. But to compensate for such a costly ceremony was almost impossible for a single person or family. If he escaped, then he would be pursued to be recaptured or even killed. This fate applied to his extended family as well. Historically, this caused many inter-village and inter-tribal wars among the Chin.

During the *Khua-hrum* sacrificial ceremony, the villagers themselves were not allowed to leave the village. If any villager, especially the head of a household, did not participate in the ceremony of the *Khua-hrum* sacrifice and went out through the village gate, he would not be allowed to cultivate his field for the next three years. For the Chin, the *Khua-hrum* sacrificial rite was not just a ceremony but an act of communion between the villagers, headed by the chief and their guardian god *Khua-hrum*, and also between the chief and his subjects. Thus, everyone had to be present at the ceremony in order to receive recognition from the chief and *Khua-hrum*. Without such recognition no blessing could be expected from their guardian god, resulting in famine and epidemics according to traditional Chin concepts. Without protection from the chief and the *Khua-hrum*, the lives of ordinary people were believed to be unprotected and went open to attack by evil spirits to the soul and enemy to the body. The chief was bound to take revenge for his loyal subjects in time of war and death. If somebody wanted to be absent from at such a ceremony, it demonstrated a lack of loyalty and rejection of the authority of the chief and the *Khua-hrum*. Therefore, attendance by the whole community at the *Tual Khua-hrum* worship in the *Khua-bawi* system and *Tlang Khua-hrum* worship in the *Ram-uk* system was vital.

Such religious belief systems and ritual practices in Chin society required extreme exclusiveness or 'exclusivism', bordering on isolationism. Since they did not allow strangers during these religious ceremonies, they closed the village gate if the worship was for the whole community, or the gate of their compound if the worship was only for the household. Through ritual practices, they excluded others and isolated themselves. Paradoxically, this exclusivism could only be maintained through traditional instruments of integration, such as wife-giving and wife-taking, clan composition, tribal solidarity, etc., supported by fundamental isolation from other societies. Thus, for the traditional Chin, defending exclusivism and isolationism meant defending the fundamental doctrines of their religion and the foundation of their society. They therefore defended exclusivism and isolationism as part of their own way of life and resisted any kind of social structure change. In this way, the traditional Chin isolated themselves from other peoples for many centuries.

With such strong convictions, the Chin defended their land with their lives and resisted all alien intrusion, including the British invasion. Even after the British officially declared the occupation of Chinram and promulgated the Chin Hills Regulation in 1896, the Chin continued to revolt against the British whenever the opportunity presented itself.

THE WORSHIP OF *KHUA-CHIA* AND THE ROLE OF *KHUA-VANG*

Sacrificial Ceremonies Connected with Sickness

As mentioned earlier, the Chin believed that sickness was caused by *Khua-chia*, and practically 'the only means of averting or curing sickness is performing the appropriate sacrifices and ceremonies' (Parry 1932: 455). Laura Carson, the wife of Arthur Carson, the first missionary to Chinram, wrote that sacrificial ceremonies connected with sickness were 'their only system of medicine' (Carson 1927: 161).

As soon as sickness occurred, the patient consulted the *Khua-vang* called '*Litin-thiam-nu*', who could identify the particular spirit responsible for the sickness. *Litin-thiam-nu* was a *Khua-vang* but one whose functional category was only that of a medicine woman, and who could identify the sickness. She could not, however, foretell the future or perform a visiting journey to *Mithi-khua*. Nevertheless, she was as much a spirit possessor as any other *Khua-vang*. Neither *Khua-vang* nor *litin-thiam-nu* cured the sickness, but they did 'enter into a self-induced trance with the help of a local beer called *zu*, in which state they find out which spirit has caused the sickness or whether some other agent such as *hman*, the evil-eye and witchcraft, had done it and what kind of sacrifice the patient must [use to] placate the *Khua-chia*' (Hre Lian Kio 1972: 57).

The sacrifice of animals ranging from chicken to mithan would be offered to the *Khua-chia*, hoping that the *Khua-chia* might be pacified and restore the patient's health. These sacrifices might be repeated in a series until the patient got well. The *Litin-thiam-nu* might even recommend an extravagant feast of celebration as a last resort. Strait reported that the nature of sacrifices to *Khua-chia* was 'more a bargain transaction' (1933: 108). Hre Lian Kio also argued that, 'For some kind of disease the sacrifice may be given in the form of ransom for the patient's soul who is believed to be kept in captivity by the *Khua-chia.* In such a case, in addition to the propitiatory animal sacrifice, offerings of a mithan, gongs, spears, strings of beads, copper or brass vessels (all of which are highly valued by the people) may be given' (Hre Lian Kio 1972: 58).

Superstitious fears of the *Khua-chia* remained a part of traditional Chin religious life as long as it flourished, and sacrifices to the *Khua-chia* in order to gain back a person's health were one of the most important aspects of this tradition. The main issue in this religious practice was not sin or forgiveness, but personal health. The healthy man was strong; the strong man could work hard and get rich; and wealth was good not only for the well-being of the person here on earth, but also for the soul in life after death in *Mithi-khua.* This simple logic was the basic functional concept of Chin religious life. Sing Kho Khai also argues that '... the traditional function of Chin religious ritual is essentially aimed at maintaining the physical body of man to be in a state of healthy condition, and thus it can be linked to the function of medical service in the modern age' (1984: 159).

(a) *Sacrifice for the Individual: Thla-auh*

The *Thla-auh* sacrifice was among the most familiar sacrifices for the individual person. Literally, *thla-auh* means 'to call back the soul'. If a hunter or a traveller fell ill, it was believed that his soul had been caught in the jungle by some *Khua-chia,* and a sacrifice had to be performed to call it back. The household priest usually performed the sacrifice after consulting with the *Litin-thiam-nu.*

> [The] sacrifice is chicken, either a hen or cock, and is made about dusk when people come into the villages for the night and all travel has ceased until the next day. The one performing the sacrifice made out of mud a miniature woman, a mithan, a tom-tom, and some mud beads. Having sacrificed the animal, he took these mud images, and a spear, a Chin blanket called Cong-nak-puan, and a small bamboo whistle, and goes to the village gate through which the sick person made his last exist (Strait 1933: 156).

The priest lit a fire on the spot, so that 'the sick man's spirit may see the smoke and come back to the place of sacrifice' (Parry 1932: 467). He then sacrificed the cock, and prepared the cock's entrails, liver, comb, etc., on the flat stone for the *Khua-chia;* this is called *phangva,* a 'meal prepared for spirits'. He then blew a whistle three or four times and called the *Khua-chia,* saying he had prepared the most delicious food for them, and he also had 'a woman, a mithan, a tom-tom, some beads, a spear, and a blanket, and summon[ed] them to release the captive's soul' (Strait 1933: 165). Two pebbles were taken from the place where the cock was sacrificed to represent the sick man's spirit, and the sacrificer went to the sick man's home, blowing the whistles as he walked, calling out:

> Come back, Chan Mang's spirit (mention the sick man's name), I call you back with a fowl; I have already paid a ransom of *mython,* tomtom, beads, spear, and even a woman for your release. Let them free you, and come home (Parry 1932: 468).

When he reached the family home, he stopped on the step and asked if the sick man's soul had returned. One of the patient's family replied, 'Yes, it came back a while ago'. Then the priest entered the house and placed the fowl and two pebbles at the foot of the main post at the back of the house and sat down there. After again calling to the sick man's soul and blowing his whistle, he took stones in his hand and a little water and sprinkled the sick man with water, after which he replaced the pebbles at the foot of the post. If, when the sick man was sprinkled with water, he jerked a little as the cold water struck him, it was believed that he would recover quickly. The next day was observed as a holy day, *zarh.* No stranger was allowed to enter the house or even the compound. As a sign of the sacrifice, they put *kaw-dai-hnah* at the gate of the house.[6]

(b) *Sacrifice for the Community during an Epidemic: Tlang-rai Thawl*

The *Tlang-rai-thawl* ceremony was solely intended to stop an epidemic from entering the village, and was performed whenever the neighbouring villages were afflicted. Traditionally, disease and *Khua-chia* were the same element in different forms. The term disease, *tlang-rai,* and *Khua-chia* were used interchangeably. *Tlang-rai-thawl* was performed in various ways. Here I shall present the procedure as commonly carried out in the Zophei, Lau-tu, Senthang, and Mara (Lakher) areas.

As soon as it was heard that an epidemic was raging in a neighbouring village, the chief and priest would set a day to hold the *tlang-rai-thawl.* Members of each house made small bamboo baskets and filled them with samples of every kind of food. At the adge of the village, a bamboo fence was erected with a bamboo archway which spanned the road. The baskets of food were placed outside this fence. A dog and a fowl were killed by having their throats cut, and were left for the *Khua-chia* to eat. The dog's intestines were taken out and stretc.hed between the fence posts, arching over the roadway, which was called *ui-ril-zam* or *ui-ri-ten.* Then the villagers, led by the priest, appealed to the spirit of the disease, saying:

Maw! Tlang-rai aw!
Kan khua lut hlah law, Kan ram lut hlah law;
Mi zun a thu e, Mi ek a nam e;
Kawl deng ah zam law, Vai deng ah kal law;
Ti ral ah kal law, Thing par ah zuang law;
Thli sin ah kal law, Mei sin ah zam law;
Na khua tlung law, Na inn hlam law.

(You! the spirit of *Tlang-rai.* Do not enter our village, do not enter our country. Go to *Kawl* [Burman] country, go to *Vai* [Indian] country. Go beyond the river [of Chindwin], go over the trees. Go to the storm, and go to the fire [... to die!]. Go to your village, go to your home. Here is a smell of human excrement and wind.)

They then all spit into the basket. The sacrifice was both minatory and propitiatory. The spitting into the basket was designed to frighten away the *Khua-chia.* The food in the baskets and the offering of a fowl and a dog were intended to please the *Khua-chia,* and induce it to spare the village (cf. Parry 1932: 455–456).

LIFE ON EARTH AS A PREPARATION PERIOD FOR *MITHI-KHUA*

As mentioned earlier, life here on earth was only a period of preparation for future life in *Mithi-khua.* The traditional religious life of the Chin was not confined to rituals related to the cycles of nature – such as ceremonies to renew the earth's fertility or related to the harvest, sickness or the annual *Khua-hrum* worship at the *Tual* ground. The traditional religious life of the Chin included rites of passage, that is, the passage of a person through the life cycle, from one stage to another over time, from one role or social position to another, integrating the human and cultural experiences with biological destiny, birth, reproduction and death.

Human rites of passage share aspects of 'normal, natural and social life' almost everywhere in the world. What makes them different from one society to another is the varying religious and social significance, symbolized through rituals and accepted as social norms. When rituals and rites of passage are symbolized and accepted as social norms, they become 'ways of life', *phunglam,* in the Chin context.

In this section, I shall explore how rites of passage in traditional Chin society served as thresholds between social classes and between child and adult, male and female, living and dead.

Nau-Zing-Dangh: Sacrificial Ceremony for the Newborn

The rites of passage in traditional Chin society always began with the sacrificial ceremony for the newborn, called *Zing-dangh,* which brought the soul and body of the child together permanently. The sacrifice was usually performed by either the child's grandfather (if still alive) or the father who served as the household priest.

The *zing-dangh* sacrifice was always followed by the name giving ceremony, called *nau chuah ven le minsak.* Feasts to celebrate the birth of children were mostly held by the noble classes. After a child was born to a couple, the family or families on the mother's side, who had received the marriage price for the mother, sent word to the couple that they planned to see the child. On an appointed day, they brought food and curry (some families killed pigs and some noble families might even kill mithan), enough to make a feast for the relatives of the couple and neighbours. When the birth of a son was thus celebrated, the uncle and aunt gave a blessing to the child: 'May he live, may he have good health, may he bear the honour of the family, may he give feasts, may he be famous, and may he live long' (Pu Sakhong 1971: 11).

Usually the sacrificial feast was held the evening of the day that the name was given to the child. The family invited all neighbours and relatives (both paternal and maternal). They prepared meats of pig or mithan and rice for the feast. Pots of *zu* were offered. They sang traditional songs and epics, often passed on for generations, which told the story of the family's fame and honour. They related stories of the family's achievements: the feasts they held, the wars they won, the animals they killed and so on. The purpose of relating the family's greatness was a prayerful hope that the child might grow up to be as strong, famous and rich as its forefathers. To clarify the point by example, here is the song of Chief Mang Hnin of Leitak.

> *Kha hlan e, Kapu le cu,*
> *Min than e, Laa bang an peh tawn e,*
> *Ai rawl e, kho lai ding ah,*
> *Mang Chum nang tal chuncha bang dir law.*
>
> (The good name and honour of our forefathers continued in strands like cotton yarn. To replace and claim this honour, stand up Mang Chum, like a rainbow across the country.)

The first son of a noble family was traditionally given the name of the grandfather or a forefather on the paternal side. In the same way, the eldest daughters were entitled to be given the names of their aunts. But the names of other sons and daughters were picked up mostly from the songs that bore witness to the glory of the family. For instance Chief Sakhong of Aibur had five sons. The eldest son was named Rum Thang according to the custom. But Za Peng, the second son, was named from the following song:

> *Lawng pi kut tial raa lia luai, Tual kil in vun tiang ai uh!*

Mirang bawi cawn thiam awk lo Za Dun val cawi har,
Vun tawn ter law Za Peng mi kian lem lo chung

(Erect high walls of *Lawng* in front of my house; even emperors cannot imitate our splendour, other men of ordinary birth cannot do the same as we can, the Lord Za Peng who is higher than the rest.)

Ordinary people, with no deeds of honour in their heritage to merit songs of praise, usually asked the noblemen to give names to their children. If the child was a girl, she was obliged when married to give a token of her marriage price to the person who named her.

Sasawm-tuk

Becoming a famous hunter was clearly regarded as a great accomplishment for men and to kill many animals was one of the main goals of a man's life. Parents wanted their sons to become skilled hunters and prepared them to be so from childhood. When a small boy killed his first bird with a slingshot, the parents would be so happy that they magnified the achievement by killing a chicken – or even a pig if they were wealthy – to celebrate his skill.

When the boys grew old enough to handle a gun, the family tried to acquire a rifle for him. Before the young man handled the gun or went into the jungle for hunting, the parents prayed for him and traditionally killed one red rooster, making the following supplication:

Hi ka mei thal hi e,
A li khaun ter law, a tek au ter law,
Rilisa ka peknak law, Thing haung sa ka peknak law.

(Let this gun work properly and be successful, let it kill the animals on the land and the animals under the water.)

When he came back from the hunt with his first game, the family would celebrate his kill with a pig. But they again used a red rooster for ceremonial purposes. While killing the rooster the priest said:

Aha! Sakhi paw aa,
Kaa ca lawn se, ka ca lui se;
Na u a va chui la, na nawi a va chui la,
Va rih hlui paw a va chui la, Sakhi ce paw a va chui la,
Kii chaw a va chui la, nua chaw a va chui la;
Ka tlun na ti ta la, ka kai an ti ta la;
A chi lai ce, a pe lai ce:
Ui hmai min lai ce, Aa hmai min lai ce,
A kung ma va rawn lai ce, a ni ma va rawn lai ce;

(Oh! Deer or buck [they mention the name of the animal] we rejoice in getting you/Bring your kin/Bring the fowls of the sky with you/Bring the barking deer/Bring the beautiful antelope/Bring animals with good teeth and horn/Consider this is where you usually come to/Do not stumble on the way/ Do not get lost/Do not mind the criticism of dogs and chicken/ Do not hesitate/Do not lets months and days pass by.)

If a hunter killed a deer or smaller game, he did not sing songs of praise for his skill. But if he killed big game and ferocious animals he would surely compose a song or just sings songs of praise called *vaw-hla* or *hla-do*. The songs were sung in unison with those who accompanied him. When they reached the entrance of the village, they fired some shots to inform the villagers of their success. To illustrate this, I include the *hla-do* of Chief Sa Khawng of Aibur, when he killed a tiger:

Na dir e, tlang sang bo in e,
Duh nawl e, zei khua dah na ruah tawn e,
Na tlawn e, lennak khua le hmawng tha e,
Chiah bang na khawlh theu e.

(Standing on the top of the mountain, what do you consider? You collect herds as tax, from the jungle you roam. Now, you, the mighty beast of the jungle, bring all your high esteem and honours to me!)

Here is another illustration of a song by Chief Mang Hnin of Leitak:

Fung tian e, Ki vang cu e,
Val ran tum ngam awk si hlah ngai e,
Arfi e, van puak bang di ling e,
Leng in a chuak e.

(Getting the beautiful horn of a gaur is beyond the dream of ordinary men. The horns are now herein my house, twinkling like the stars in the sky.)

A great hunter was an honourable man in traditional Chin society, respected by all. Anyone who killed a dangerous animal, such as bison, gaur, tiger or bear, was considered head and shoulders above any other man in the community. The triumph lasted beyond his own lifetime. The honour followed him into the life after death, for the traditional Chin believed that the spirits of the animals killed during one's lifetime accompanied the hunter into the next world, *Mithi-khua*. This concept is clearly indicated in Chief Mang Hnin and Chief Sa Khawng's songs. Mang Hnin's song says:

Ka me e, rem cang lai,
Far khan tlin dang ka ih ah,
Mi tial e, min hngalh nak ah,
Tum lian kal law ka hmai ah.

(When I consider the forest of my hunt, I will lie in peace, for I am not an ordinary hunter. Let the large gaur march in front of me into the next world).

In this song, he wanted to say that when he marched into the world of the dead, all the dead (before him) would recognize him because he had killed such rare animals as a bison and a gaur. Pu Sa Khawng, Chief of Aibur, also composed the following song before he converted to Christianity.

Fung tian e, Ki vang aw,
Nang nih tung lam tu law,
Sa bawi aw, hrum ruang mang law,
Lung rawn lai khua kan tlerh lai e.

(Gaur with beautiful horns, make a trail in front of me. Let the tigers roaring announce that I, the lord and mighty, have come. And let them [his subjects who died before him] prepare a place for me.)

Soon after famous hunters killed highly regarded animals, their minds turned to the world of life-after-death in *Mithi-khua.* Chin did not hunt only for lifetime pleasure, they also hunted for triumphant entry into *Mithi-khua.* For many hunters the highest honour was to kill a tiger and a gaur, which are rare, dangerous and difficult for the ordinary hunter. For rich or poor, killing a tiger or a bison would always be followed by a *sasawm-tuk* celebration of the kill. However, the celebration was for the glorification of the hunter's name in the present life. The significance of the celebration therefore was two-fold: to receive honour within the society and to receive the blessing of the guardian god, *Khua-hrum,* for increased skill to kill more game.

The *Sasawm-tuk* celebration involved the hunter going to a sacred place in the jungle and erecting on the ground the same number of stones as the heads of the animals he had killed. In those days, the Chin put the skulls of the animal heads on the porch of their house. This enabled them to count the number killed. The hunter sacrificed a red rooster following the prayer on the spot where he erected the stones.

Lasi bawi nu aw, Arhlisen in kan biak e,
Sasawm ka rel hi e, nang mah pek mi a si e;

Lung bang hmunh ter law, thing bang dir ter law,
Ki tha tlun ter law, ho tha tlun ter law,
Fung tian tlun ter law, Ki vang tlun ter law,
Vom khuai tlun ter law, Ngal hriang tlun ter law,
Sazuk tlun ter law, Sakhi tlun ter law,
A sawm tlin ter law, A za tlin ter law.

(Oh! Lady fairy, I offer you a red rooster and celebrate my kill. It is what you have given to me. Let it last eternally like the rocks I erected, and let them stand like a tree forever. Let the animals in my possession be complete. Let them have big horns, let them have mighty tusks. Let the bison and gaur come, let the bear come. Let the boar enter my gate, let the deer beat a trail to my house.)

After the ritual ceremony in the jungle, he and his company came home with songs of praise called *hla-do*, and several usually fired their guns at the gate of the village. At the sound of the gunfire, the women would meet them at the gate. At the hunter's house, they killed a mithan and held a great feast lasting the entire night and the next day. They sang songs honouring him and danced. This ceremony marked that the man had come of age, ready to lead his people in good times and bad.

Marriage

In traditional Chin society, marriage had always been regarded as the most important event in the life of a man and a woman. The woman that a man married could make him respectable or she could pull him down to a low and degraded position. Therefore, parents looked carefully for the spouses of their eldest sons. Especially the nobles sought women of noble birth or their equals for their sons. If a man could marry a higher class woman, it was a great honour for him and his family. But if he married a maiden of a lower class, then the other young men of his class would look down on him. For example, among three brothers, if the eldest married a woman of low class while his younger brother married a woman his equal, and the youngest brother married a lady from a higher class, then the eldest son lost his social position to his younger brothers. It was therefore essencial that the eldest son, who was in line to receive the family's name and inheritance, married a woman at least of his own class. If he failed, his brothers, upon marrying women of their equal or higher class, could take away his birthright (cf. Head 1910: 15–16).

A nobleman would not accept the participation of another man of his own class in any ceremonial event, if the latter married a woman of lower class. Two significant social symbols in traditional Chin society were *Bawi-te Bawi* and *Khuang Cawi*. The first was a lesser ceremonial feast, which every nobleman presented to show his ability, power and wealth. The latter was the highest feast, which could only be given by men of considerable wealth. In both cases ceremonial beasts, young mithans, were loosely tied with a rope to the ceremonial post. But the sacrificial beast was also held by other ropes by men of the host's own class. After ceremonial prayer, the man who was giving the feast would shoot the ceremonial beast with bows and arrows. If any man of his caste fell to a lower level on account of marriage, he could not hold the rope of the ceremonial beast. He simply lost his former status. It was for that reason that every parent tried his or her best to get a woman of high class for the eldest son in order that the honour of the family should not be reduced by marriage. However, a man was entitled to marry any number of women of any class as lesser wives. The formal wife, called *nu-tak* or *nu-hrim,* was the one who held all the ceremonial activities such as propitiation or sacrificial feasts to the household god. The sacred utensil called *Khothlaban* used in these ceremonies was hung from the ceiling above the couple's bed. The couple would eat ceremonial food together from these sacred utensils. Only the wife who shared all these was entitled to be the formal wife, *nu-tak* (cf. Strait 1933: 79–80).

If the formal wife died and the man remarried another women who was his equal, then she shared all the ceremonial functions with him; she became his second formal wife, called *nu-tlai.* A son by the formal wife was entitled to inherit his father's properties, especially the house and the land. If he had no son with his first wife, the son of the second wife *nu-tlai* then had the right to the inheritance. The sons of any lesser wives, called *nu-chun,* were not entitled to his inheritance.

Another significant point about marriage involves the marriage price. The marriage price was a special Chin custom. When a man married a woman he had to give a marriage price to 1) the bride's father, 2) one of her brothers, 3) one of her aunts and 4) the brother of her mother. The value of the marriage price depended on the caste of the bride. If the bride was from the noble class, a marriage price would be very high. A woman of this class was worth five times more than an ordinary woman.

Fanu-ah le Puan-phah-kai: Feast in Honour of the Bride

For the noble class, the marriage feast was usually followed by *Fanu-ah,* the feast in honour of the bride. Just as it was an honour for a nobleman to marry a woman of noble blood, it was also an honour for the daughter of a nobleman to marry a man of her own class. For one thing, the nobleman could easily afford the marriage price. Maintaining or enhancing social status was important. Among two or three sisters, for example, the one who could marry a nobleman was highly regarded, and the one who failed to wed a man of her class was looked down upon. Even their children were discriminated against in the ceremonial context. For instance, Cheif Mang Hnin of Leitak had four daughters. The second daughter married Chief Sa

Khawng of Aibur. When Sa Khawng held the *Bawi-te bawi* ceremonial feast, only the husband of the youngest daughter, Hre Mang, was invited to hold the rope of the ceremonial animals. The husbands of the other two daughters were barred from this act on account of their lower caste.

It was an honour for a woman to marry a man of high social standing and the parents and her brothers took pride in it. So it was common to honour their daughter or sister by giving a feast, called *Fanu-ah.* But only noblemen and men of great wealth could afford a *Fanu-ah,* which included laying blankets and various cloths on the path from the house of the bride's father to her husband's house, so that she and her husband could walk on the cloth-paved path between the families' homes. If the bride was from another village, then the house where her parents were being put up for the occasion was used. This ceremonial walk on the expensive blankets was called *puan-phah-kai.* This procession was the highest honour the girl's parents could bestow upon their daughter, for she had no right to receive any inheritance upon the death of her father.

The ceremonial attire for the woman included a headdress called *lukhimh-tung,* a lady's dress ensemble, *cong-nak, kawr* and *hni-tial,* a silver bangle, a string of brass beads and a string of silver beads. The procession would be led by the couple. The husband would dress in a ceremonial men's costume. Whereas the wife was shining beautiful and graceful, the husband was dignified and grand. On some occasions animals such as horses, cattle, etc., given in the lady's honour by her parents and brothers were also led in the procession. The procession would be accompanied by the sound of drums, gongs, cymbals and bangles with intermittent cracks of gunfire. Silver coins were put on every piece of cloth before the procession begins. The silver coins and cloths were gifts to the lady. But the lady did not keep all these for herself. She had to distribute them to the aunts and sisters of her husband. This was illustrated by the song of Sui Hluan of Aibur in honour of her daughter Par Cin:

> *Phaw hna zoting hni tha cu na tlem than te lai,*
> *An ser mirang bawi pum cuang ngun tia sawm he,*
> *Na min rit lai sang thluai phak lo mang ran nih.*
>
> (Give freely the cloths to them, you will receive them again, give them also silver coins and bangles, you will receive a good name in the highest, which no ordinary person can be bestowed.)

The parents and brothers who arranged this kind of feast to honor their daughter and sister also killed many animals, as they would for any other feast. When Chief Mang Hnin of Leitak gave a feast in honour of his daughter Men Tang, he let his retinue kill seventeen mithans and buffaloes alone; including pigs, the number of animals killed added up to more than thirty. The meat was usually distributed to every household, besides the portion used for lunch and dinner during the two days of feasting.

The procession ended at the front porch of the couple. When the couple reached their house, the parents of the husband welcomed them from the porch. While they climb up the ladder leading into the house, the parents welcome them with the following words: 'Come with fortune, enter the house with luck. Bring in the daughter, bring in the sons, let our clan progress, and let our lineage multiply. Come in wholesome and with good health.' They were then pulled into the house by the handle of a hoe.

Bawi-te Bawi

The highest feasts in traditional Chin society were *Bawi-te bawi* and *Khuang Cawi. Bawi-te bawi* (a lesser feast than *Khuang Cawi,* which I shall return to later) was the highest and most important feast that most men could give. So it was the dream of every nobleman in his lifetime. It was recognized by all as evidence that the man was of considerable means, a step higher than other noblemen. The honour it incurred did not end with death. That was the reason why every Chin wanted dearly to give this feast. If a man held this feast in a village, the news spread far and wide. So the villagers took pride in it, and therefore gave help in every way to make the feast successful.

The *Bawi-te bawi* feast lasted for five days; however, for those who could afford it, it could be extended to seven days, during which time the whole village enjoyed the feast. The first day of *Bawi-te bawi* was purely an observance of thanksgiving to the household's god or the guardian god for all the blessings they enjoyed, and to pray for ongoing protection. For the occasion they killed a mithan which was free from blemish or any questionable mark. It should also be virgin and, the whole body should be black, as the black was the colour that symbolized the invisible force of *zing.* The mithan had to be killed inside the compound of the house by the very man who held the feast. Before he killed the pigs, the village priest guided him to say the following prayer word by word:

Aha! Zing-nu aa, Zua paw aa,
Sia doi tlaw se, Ka zing kaa ca dang e,
Pa cawi la, pa hlan la!
Ka khawn tun ma ngai sua lai ce,
Ka daa tun ma ngai sua lai ce;
A pa lawn la, a pa lui la,
Vaw aa nawn ta la, vaw a chah ta la;
Ung ti se la ta la, cang ti se la ta la,
Thla chua pa phu la, Ven ni pa phu la,
Ka tu hmui lai te, ka hrui haw lai te,
Ka lawi thlu lai te, ka thing phu lai te,

Ka va kawi lai te, ka aa kawi lai te,
Bon pa pi la, me pa pi la,
Tlawn la, za la!

(Oh guardian god! I offer you the unblemished mithan. Please accept my offering. Please do not take offense for beating my gongs and drums. Please make yourself happy with me. Please enjoy the feast with me. Please say how pleasant it is. Please say how nice it is. Please say let there be more feasts. Let my house be full of fortune. Let my axe be full with blessings. Bless my house. Bless my plough. Bless my harvest. Give blessing to the animals I keep. Let them multiply. Let them be prosperous.)

After this prayer, he spat out liquor at the *mython* (mithan). He blew a bamboo flute a couple of times. He put some rice powder called *vaidang* into the mouth of the *mython.* Then he took an arrow and shot it into the heart of the sacrificial animal.

When he performed this sacrifice, his major wife beside him held the sacrificial articles for him and handed them over when needed. The sacrificial meat was cooked in *kho thla ban* pots. After the food was ready, a little of the liver, tongue, a little of the gravy and a little of some other parts of the meat were placed in the *kheng* (wooden plate) as *phangva,* the sacrificial food. This sacrificial food was partaken only by the man, his major wife (*Nu-tak)* and their children. The *nu-chun* (minor wife or wives) and their children could not share this sacrificial food. According to Chin customary law, only the children of the major wife had the right to partake of this sacrificial food and could claim to inherit their father's house.

The second day was called *cing le la tawn ni.* At this point, only the relatives and members of the clan had the right to participate in the ceremony. They settled their differences or any outstanding debts so they could stand in unity and show their strength. The Chin also believed that in giving such a big feast, the blessing was not only for the man who gave the feast, but could be shared by all his relatives and members of the clan. In order to share this blessing and honour, they gathered together and reaffirmed their oneness. On the same day, the family or families of the man's sister or sisters (sisters-in-law) would send rice and meat to him. This rice and meat, was called *Arsa-thah.* This was a sign of traditional respect in fulfilment of the honour due to him from a sister's house. In some cases, they killed as big an animal as a mithan, but it was usually a pig. Pots of rice, beer or traditional liquor accompanied the rice and meat, and the second day was just a feast for the relatives. After dinner, they sat in a circle and sang the songs of their forefathers. The songs usually related how their ancestors had brought honours to the clan. There was no dancing, however.

The third day, called *Sia chunh ni,* was the most important day. It involved the whole village and guests from other villages. The guests from other villages for this feast were called *thing thiah mi.* It was most important that the ceremonial animals be slaughtered this day. The family set up an elaborately sculptured post, called *Bawi tung,* in front of the house. It was the ceremonial post. The sacrificial animal, a young mithan, was tethered to the post. The people gathered around this centre spot – the post and the mithan. Skilled male dancers performed a ceremonial dance called *Sarlam* round and round the sacrificial animal. The leader of the dancers dressed in a traditional battle costume and held a spear to match the occasion. It was in essence a war dance. The ladies held their hands and stood in background; they danced to the music of the drums, cymbals and horns, with gongs in the background, and the hisses and shouting of the male dancers were the only songs. The ceremony was both grim and graceful. It was the main attraction of the feast as well as a performance of a war dance in which the males were entitled to take part to show the men's supremacy as protectors of the whole society.

The ceremony of slaughtering the sacrificial animal was held in the evening after the sunset. The sacrificial animal must be without blemish. It must be black without spots, and it must be unblemished (virgin), as they testified the symbols of the invisible force and holiness of *zing.* The sacrificial animal was not tethered to the ceremonial post when the ceremony began. It was tethered with ropes held by several men of the host's class. Men below his social status were not qualified for this privilege. The man must be elaborately dressed in a ceremonial costume. He held a bow and arrows. His wife wore a lady's ceremonial dress and stood by him, holding rice and holy water for the ceremony. Beside the man was the village priest who guided him to say the prayers. He repeated the prayer word by word after the priest:

A ha! Zing-nu aa, Zua paw aa,
Kikong doi tlaw se, Ka Zing kaa ca dang e,
Pa cawi la, pa hlan la,
Sa nu pa thla la, pa saw pa thla la,
Mui du pa thla la, san sui pa thla la,
Tha chawn pa thla la, zang chawn pa thla la,
Mi sin pa thla la, mi va pa thla la,
Nui pa pi la, run pa pi la,
Sa zai pa pi la, tin ni pa pi la,
Pa cui la, pa sang la,
Thla chua pa pi la, ven ni pa pi la,
Ka u ti kha e, ka nawi ti kha e,
Ka nu zua kha e, ka pa tung kha e,
Ngen dan pa pi la, hnaw tlan pa pi la,
Tlawn la, za la!

(Oh my guardian *Zing*! I worship you with unblemished mithan, Give me your blessing and long life! Bless me with sons and daughters. Give me sons who are strong and manly. Give me daughters who are beautiful and clever. Give me a blessing to keep herds of animals. Give me wealth, give me riches, give me an abundance of grains. Put me on the top, put me on the high level. Bless me and all my brothers. Bless me and all my relatives. Bless me and my in-laws from both sides. Bless me with good health. Be kind and be nice to me).

After the prayer, his wife handed over a cup of liquor to the host. He took a mouthful of liquor and spat it upon the sacrificial animal. Then he blew a flute twice and offered rice powder called *vaidang*. He shot an arrow straight into the heart of the animal with his bow. When the animal fell to the ground those who performed the war dance came forward and surrounded the animal. They sang the following song in unison and danced around the animal nine times: '*Anu thaw, Apaw thaw, ping – ping*' (The spirit of mother, the spirit of father – ping – ping). The meaning of the song is that the spirit of the forefathers and mothers were also welcome to attend the feast. Let them be happy and contented. And in so being, let them bless the man who gives the feast for the happiness of the living. The sacrificial meat should be cooked, but before the meat could be eaten, they offered the *phangva* to the household god, and to appease the evil spirits.

After dinner was served, husks of the rice were poured on the ground for wrestling. Wrestling, an outstanding traditional sport of the Chin, was not only a sport event, it was a ritual event which always found its place at every feast and festival. A champion and those who won several rounds on an occasion like this were highly regarded although they did not actually win prizes. The guests spent the night dancing and singing.

The fourth day was called *thangzam ni*. By the fourth day, all the *mython*s reserved for the feast had been slaughtered. The meat was distributed to every household. If they killed ten mithans, for instance, the meat was distributed ten times. The grandeur of the feast was measured by the number of mithans killed. The meat here was not used for dinner served during the feast days. The record of a feast remained for years and those who could afford to slaughter many animals were remembered with songs sung to their praise for many years to come. They were regarded higher than others of their own rank. The song of Sui Hluan, who composed it on the occasion of the great feast given by her husband, Chief Sang Bill of Aibur, illustrates this idea.

Sal pa tuan kam chuak uh, La lang samthun he in,
Tlang zam vui hra a hlei kai lai Kawng thai par ah,
Val nih nan zuam ngam lai maw,
Khua Tin Thang Nawl Hmung.

> (Oh! Slaves, commoners and all my subjects: Come and get your share of *Tlangzam*. The lord your chief has prepared a great feast for you, and you will receive *Tlangzam* ten times over. Who can compete with him, whose fame is spread all over the country!)

In the evening, a classy event was held only for men of honour. A group of men who had hosted at least one of the feasts, *Bawi-te bawi*, *Khuang cawi* or *Cakei fim*, gathered on an elevated platform and partook of liquor and meat together. This meat was called *Sareu ei*. These were men of honour and as they enjoyed drinking and eating they also competed to see who could drink the most and eat the most meat. The meat for that purpose was usually pork fat – thus it was a forbidding competition. This competition was called *Sa siar tleih*. They did not go down to the ordinary feast place and mix with the others. They sang songs about their glory and deeds. They did not dance.

In the meantime, the ordinary groups also enjoyed their share of drinking, eating and singing a dance on the floor. This was the place, which separated men of honour and noblemen from the rest of the people. But the ordinary men and ladies enjoyed their day by singing and dancing all day, and they did not care who sat on the elevated platform. This was their day to enjoy themselves. In the evening after dinner, there was another round of wrestling. Though this was exclusively an event for men of strength, it was the greatest attraction of the feast and everybody watched the contests with great interest. After this event, everyone returned to drinking, singing and dancing, group by group.

The fifth day was *Sia lu sum ni*. The head of each mithan was put on the post to which the animal had been tethered before being killed. But the heads did not always remain on the posts. They were removed to the porch of the house where they were properly arranged according to size. On the fifth day of the feast, the heads of the animals were put on the wall of the front porch in rows. This was the last day of the feast and there was no more dancing. The day was observed by sitting in circles and singing songs the entire day. In the evening, the host and hostess again performed another sacrificial ceremony with a pig. The procedure and prayer of this ceremony were the same as the previous ceremony performed to propitiate their household god. But some people observed this ceremony on the sixth day, if the feast was to continue for seven days. This ceremony was essentially intended to turn back to their god, asking again for care and blessings. In short, this was the day of cleansing. In some communities, especially in the Sen Thang area, instead of these formal ceremonies, they went to the river for fishing and bathing. Customs varied from one community to another, but the idea remained the same.

Khuang Cawi

Of all the feasts that could be given by any Chin nobleman, *Khuang cawi* was the highest both in honour and order. It usually lasted for nine days. Only those who had already hosted a *Bawi-te bawi* were qualified to give this extraordinary feast.

Generally the procedure and ceremonial order were much the same as *Bawi-te bawi.* The only significant additions were:

(1) Building a cabin-like structure called *Khuang* in which the hostess would stand in majesty. This cabin was held aloft by men with bamboo poles raised above their heads. The crowds danced with the cabin, in the middle of the arena. The hostess raised her hands in splendour to the crowds amidst their cheers, singing and dancing. This was, in fact, the essence of *Khuang cawi.*

(2) To add glory to the feast, the host usually asked other neighbouring villagers to attend the feast as special guests. Besides according them the privilege to enjoy the feast, the special guests, called *thing thiah mi,* were offered gifts either in cash or in such things as strings of beads. The guests would have the honour to lift the *Khuang* and dance with it.

(3) The manifestation of *Khuang cawi* was the erection of *lawng,* a tall wall erected at the end of *Khuang cawi* in front of the house. It was built with artistically carved planks, to make the house beautiful and grand. If a man held *Khuang cawi* twice, they made an oval-shaped entrance hole in the middle of the *lawng.* If he held one three times, they made two holes. So the *lawng* and its corresponding number of entrance holes signified how many times the man (the owner of the house) had held *Khuang cawi.* The *lawng* could not be erected to beautify the houses of ordinary people without holding this enormously expensive feast. In the Zophei area, Mang Hnin, the Leitak chief, held it three times, and Lian Hram of Khuapi held it five times.

The Chins placed an enormous degree of honour and esteem on this feast. This was reflected in the song of Sui Hluan of Aibur when her husband Chief Sang Bill gave this feast:

> *Tum Lian Kawng Thai le Lawng Tial tonh ni,*
> *Sung khun kan Lairam par ah,*
> *Sang Lian Mang Thai khun hrin bang ka ti rua,*
> *Kan Mang hawi iang lo*
>
> (To kill a mithan and erect a wall of decorative *Lawng* is the highest honour in our land. We are indeed higher than other nobles who can not compete.)

The origin of the *Khuang cawi* feast can surprisingly be traced back to demons. One day the slaves of Chief Mang Sui of Phaizong in the Senthang area, shepherded herds of cattle, mostly mithan, on the bank of the *Bawi-nu* River. In the meantime, they saw a cabin of logs dancing up and dawn, round and round the river. They also heard music. The cabin was dancing to the rhythm of the sound. It was so fascinating to them that they watched the entire day. When they regained their senses and drove the herds on, it was nightfall. Mang Sui was naturally worried about them. When they got home, they at once reported their observation to the chief. So, Mang Sui accompanied them to the river the next day. When he witnessed the same event himself, the idea came to him that he would like to do likewise to honour his beautiful wife, and thus he began to plan a magnificent feast for his people. He built the box-like cabin for his wife to stand on. He killed many animals and held the great feast, which was the greatest feast ever. After that, every nobleman tried to host such a feast to prove that he was capable of such extravagance for the enjoyment of the people.

Holding this feast required a great deal of preparation. The host must have had a large number of animals, especially mithans. Although buffaloes, cows and pigs could all be used in the feast, they were not regarded as sacred animals, and only mithans were acceptable for ceremonial purposes. Also, the host had to collect enough grain to feed hundreds and hundreds of people every day for nine consecutive days. Furthermore, he had to ask the villagers to help prepare for the *lawng* which took many skilled workers and many days to do. In short, only the very rich could give this kind of feast, and those who did provide such feasts were entitled to great honour, proclaimed customarily in specially written songs.

The prayers and procedures of *Khuang cawi* from the first day to the third day are the same as for *Bawi-te bawi.* The fourth day, which was different from *Bawi-te Bawi,* was called *Lawng bunh ni* or *Sia chunh ni.* On this day the curved wood was brought in from the village gate where it had been kept until that day. This was welcomed with songs and music. Mang Hnin, Chief of Lei Tak, praised his achievement in the following song:

> *Lawng pi kut tial raa lia luai in,*
> *Tual kil in vun tiang ai uh,*
> *Mirang bawi cawn thiam awk lo*
> *Za Dun val cawi har,*
> *Za Peng mi kian lem lo chung*
>
> (Erect a high wall of *Lawng* in front of my house, which even emperors cannot imitate in splendour; other men of ordinary birth cannot do the same except us, the Lord Za Peng who is higher than the rest.)

The fourth day was also known as *Sai chunh ni,* for on this day the sacrificial animal was slaughtered. The procedure of the ceremony for slaughtering the animal was the same as in the *Bawi-te bawi.* Meat was distributed at least twice during the day. After the ceremony, a wrestling event was held which usually lasted until midnight.

The fifth day was *Khuang cawi ni.* This was the day that they lifted the *Khuang* – the box-like cabin in which the hostess would stand in all her glory. The cabin was decorated, covered on all sides with traditional ceremonial blankets called *Cawng-nak.* The hostess must be wearing a ceremonial costume. She would be wearing silver bangles on her ankles and wrists. The strings of beads and necklaces must be on her neck. And she also would wear bright earrings and silver belts around her waist.

It took about fifteen people to lift the *Khuang.* All of them were seated in position. After the hostess got into the cabin they held the eight handles or poles and rose up slowly with the *Khuang.* Indeed they lifted the *Khuang* with great dignity. They sang the customary song in unison: '*Khuang cawi cu e, Bawinu hrawng Mang Sui se a zung e zal te e*', which in plain language means: '*Khuang Cawi* originated in the river of *Bawinu* and Mang Sui perpetuated it as the highest feast.'

While one group of men and women lifted the *Khuang,* the rest held hands and encircle it; just as the first group lifted the *Khuang* high up over their heads and danced, the rest also danced around the *Khuang* to the same rhythm and movements. To glorify the ceremony, they would only sing classical songs about *Khuang cawi.* One group was entitled to lift the *Khuang* thirteen times and dance with it in the arena. At the end of the thirteen lifts, the hostess threw various articles into air such as coins, combs and mirrors. The guests would try to grasp these things as souvenirs. The host also put on a ceremonial costume and equipped himself with bows and arrows like a warrior. In grand style he strolled around and watched how his guests lifted and danced the *Khuang* with his wife as the center of attention. Every time they lifted the *Khuang,* he shot one arrow into the sky until he had shot thirteen arrows before the interval. The significance of shooting arrows this way was connected with the belief that there were evil spirits around that may woefully harm the hostess or destroy the happy occasion out of malice. He thus shot the arrows to kill these evil spirits. In response to the prestigious honour of being lifted in such a manner, Sui Hluan of Aibur, wife of chief Sang Bill, sang:

Khuang cin sung thing thiah maw al sung deuh lai,
Ngan riat sakei tial lian lu?
Zei tu al sung deuh si lai, kan cuan tawn ai u!
Tum Zing Sui Tial vor rih lang, bual rei par ah!

(Which is more honourable, *Khuang cawi* or *Ca Kei Fim,* whichever you choose, watch in awe as I ride on the wave of *Khuang,* let me throw coins for you.)

In the evening there would be another round of wrestling. For the Chins, champion wrestlers were also heroes. Wrestling was a traditional contest of men of strength as well as a serious occasion to wrest the title from many possible champions.

The sixth day was *Thing Khuang cawi ni.* On that day the special guests would have the honour to lift the *Khuang,* one group after another. This was also a contest of sorts; the outcome determined by how beautifully they could lift the *Khuang* and dance gracefully with it. The things thrown from the *Khuang* and this occasion were exclusively for the special guests and included expensive things, like garments, brass pots and gongs. Usually the special guests were from nearby villages. But when chief Mang Hnin of Lei Tak held *Khuang cawi,* he had fifty Burmans from Gankaw in the Magwe Division as one group of special guests. After they lifted the *Khuang* and danced with it during the afternoon, they entertained with a Burmese orchestra, a *saing,* in the evening. This was regarded as extraordinary as the Burmese guests had to walk about 200 miles to attend the extravagant feast. The host was the envy of all as the memorial song below hints:

> *Ka thi e, rual mi hna,*
> *Zu din zan ah ka khawng hngin hlah u,*
> *Mang ram tluk lo ka chum tual cung ah,*
> *Vai lian sawm nga ka tlaih hna.*
>
> (Men of any class and age, do not compete with me at any occasion, I am beyond the reach of ordinary noblemen. Behold! I brought fifty Burmans to my dwelling as special guests.)

His wife Khuang Cin also composed a song about the event:

> *Sung thai seh tiah thlang vai rang hniven na thiah hna,*
> *tlang hra dan ah,*
> *Mang fam nak kan ngeih hlei tha-hni Ngunpum tha he,*
> *Tawn ter hna law zarenh kan chum tual par ah!*
>
> (You hire the men of '*Longyis*' to add grandeur to the occasion. They came across many hills. So let them mix and mingle, to enjoy the wealth of ours, in front of our estate.)

On the night of the sixth day, the special guests and the local folks competed in wrestling again. After that, singing and dancing followed throughout the night. On the seventh day, they pulled down the *Khuang* and buried it underground. So the seventh day was called *Khuang phum ni.* As stated earlier, *Khuang cawi* was adopted from the devils' own feast and therefore the Chin did not want to keep the things of *Khuang* for fear that they might provoke an unpleasant reaction from the devils. They hired a man (usually from India) for this purpose. They dressed him with rags and made him carry a basket full of bones, animal skins and broken things. He performed an imaginative dance for the demons. Then he sat down on the *Khuang* (before it was pulled down). Some blank shots were fired at him, signifying that the demon was shot at and killed, along with all evil things. After that the *Khuang* was destroyed and buried in the ground. That day, the group dance was also a war dance, a celebration of the 'death' of the demon.

The eighth day was called *Sialu sum ni.* Just as during *Bawi-te bawi,* this was the day they took down the heads of the animals killed. There was no dancing. They would sing songs seated. The ninth day was *Zingpi dangh ni,* the cleansing day, the same as *Bawi-te Bawi.*

The *Khuang cawi,* the highest and best feast for the Chin, was quite the same throughout the Chin nation with few exceptions. Some communities south of the Zophei area, namely the Lautu circle, celebrated the feast differently because of the following incident, which occurred many years ago.

It is said that Za Kheng II of the Hnaring was holding *Khuang cawi.* While his wife was being honoured and lifted high in the *Khuang,* she shouted to those in charge of the cooking: 'Don't eat all the livers. Keep some for me!' When Za Kheng heard these words from his chief wife, being honoured as she was on the *Khuang,* he was shocked and ashamed of her. He was unable to control his anger and immediately drew his bow and shot her to death with a well aimed arrow. Since then, they avoid *Khuang cawi.* Instead, they erect *lawng* without performing *Khuang cawi.* Among the animals they kill they included one pregnant mithan. They took the foetus from the womb and wrapped it with ceremonial blankets, called *Conlopuan,* and buried it. This was also to earn a blessing or good name in the life after death.

Cakei Fim

Another significant feast in traditional Chin society was called *Cakei Fim,* the feast for the tiger's head. The dance and other performances were said to have been adopted from the dance of the tiger itself. The story tells that once upon a time a tiger killed a woman in the Zokhua circle. The tiger was trailed into a cave where it kept the woman's body. The pursuers saw the corpse placed properly in a certain manner. Then they saw the tiger dancing round and round the corpse. So, when the men killed the tiger, they also danced around the tiger's body, or head. This important feast was born from an imitation of the animal. Only those who had performed *Bawi-te Bawi* and *Khuang cawi* could perform *Cakei fim.* Because tigers are so powerful, only those honourable men who had performed at least a *Bawi-te bawi* were considered able to overwhelm the power of the tiger and were thus permitted to hold this feast.

Tigers are classified into three categories by the Chin, but all the large cats – the tiger, leopard, jaguar, panther, puma and lynx – share the same name. In the three-level hierarchy, the genuine tiger is highest, followed by the leopard and panther in the second category and the jaguar, lynx and puma in the lowest category. No ordinary man could celebrate the killing of a tiger, only those who performed *Khuang cawi* could hold this *Cakei fim.* But any man who had performed *Bawi-te bawi* could celebrate the killing of a leopard, panther or jaguar.

If a man killed a tiger or a leopard, but had not held any *Khuang cawi* or *Bawi-te bawi* feasts, he had to give the honour of *Cakei fim* to someone who had. Chin customs called for men to wear their hair knotted around their forehead, and the women keep their hair up with hairpins. A man who killed a tiger would be expected to untie his hair knot and released his hair, letting it fall like a woman's. The Chins considered tigers to be vengeful beings whose kin would seek revenge for the death of a family member. The Chins also considered the tiger to be a noble animal that would not harm a female, not even of the human race, despite the legend recounted above. To prevent another tiger from killing the hunter, the successful hunter (and the man who may help him celebrate the deed by holding the feast) were expected to wear their hair like a woman's to deceive the vengeful tigers. When a tiger (or any of the large cats) was killed, the hunters did not bring the body into the village. They hung it on a post at the gate to the village and then buried it near a spring.

The celebration usually lasted three days. On the first night, the tiger was brought to the village gate. The man celebrating the occasion killed a pig as a welcoming ceremony. After that preparation, the actual *Cakei fim* began. The first day was called *Hauka sathah ni.* On that day they killed animals such as pigs, buffalo, *mython* or cows at the village gate, where the ceremony was supposed to held. At the ceremonial place, only males were allowed to participate in the drinking, eating and singing. All the women gathered at the home of the host and enjoyed their own share of eating and drinking.

The second day was called *Sawmpi ni.* On the second day nobody was allowed to go outside the gate. Confined to the village, the adults gathered inside the house and sang songs about prestigious feasts while seated. There was no dancing. The reason no one was allowed outside the gate was that the Chin believed that the kin of the tiger were rampant on that day as they searched for their dead relative.

The third day was *Kei lam ni,* the tiger-dance day, held at the village gate. It was the main feast day. The host put the skin and head (or skull) of the tiger on a small, raised platform. He and his spouses danced around that platform, the host dressed in female ceremonial garb. He would be equipped with all the Chin ladies' normal day-to-day attributes, such as spinning spools. His wives also dressed in ceremonial clothes. Ancient Chin custom allowed a man to take as many wives as he could afford, depending on his ability and his conscience. All the spouses took part in this feast. He was in the middle of the troupe. His wives guarded him left and right. They danced a war dance called *Sar lam,* closely adapted from the legendary tiger. Now it was beneficial to have as many spouses as he could afford to show off to others.

Other guests were not supposed to take part in the tiger dance. They encircled the celebrating family; they beat gongs, drums, cymbals and bangles and watched the ceremonial dance. This was a beautiful dance different from other dances. This was revealed in Sui Hlawn's song when her husband, Chief Sang Bill of Aibur, hosted their feast:

> *Lucung sawm-ai ri-phiar sam khih-ing,*
> *Hawi rang hni thawi thum he in,*
> *Limnu chamrang aa mer hnuaimai bang*
> *Mang Chum chuak, Ngun par khi khiang tial he in.*
>
> (Wearing their hair in female fashion, and adorning lady's garments in this quiet dance, with motion like a python's grace, come forth, lords of the land, with bangles and glittering ornaments.)

Chief Hmung Khar of Lei Tak also boasted of his concubines in his song:

> *Va ca kai bang hruai nuai mai in,*
> *Athai sun run nih la bang zel hna,*
> *Tial Er bual rei hrum nuai mai mi cawn thiamlo in,*
> *Za Kim chia-hrin nan nun hnak lo kan cuan u.*
>
> (He draws himself calmly as a peacock, with concubines in attendance. Like cotton yarn in the arena, no other can imitate him, watch him you commoners.)

At the end of each round of their dance, the host spiked the tiger's head with a spool of thread, and they sang the following song:

> *Kai ta pu pu law, kai ta reng di,*
> *La hmui sung su law, then phai sung su,*
> *Ca hnih bing nge, ca thui bing nge,*
> *Thing hmui na pe, rua cawng na pe,*
> *Sang hmin kai pui, dui vaw zung hnen,*
> *Naa tlang kai law, kai tui ka hrawn.*
>
> (You tiger! lord of the jungle, you misstepped into the trap of spools. I don't know you, I don't hear you. You simply get entangled with the stumps of trees and bamboo poles. Take

> away its spirit to the sky and put it in Venus. When you climb the mountain, I will walk down the valley.)

The idea behind the song was that the host had not killed the tiger. The tiger itself had accidentally stepped into a trap and got itself killed. It was the women's trap. The mighty tiger should not die such a lowly death. But nobody could help it. It was just unlucky. So, do not take revenge on men for its death. Just send its spirit to the sky to Venus. So, let us not meet face to face in the future. When I go along the high plain, roam about the valley; when I go into the valley, you stalk the high mountain. Let no such missteps happen in the future. Do not kill men. Do not kill men's plants and cattle. Do not cross each other's paths.

The next day was the last day, called *Sam hriah ni*. It means the day of 'combing hair'. The host killed a pig and only relatives and neighbours would attend the event. On that day he cleaned himself and combed his hair and he wore it above his forehead in a knot in the normal male manner again.

Funeral Rite: Ruak-hngah

The funeral rite called *Ruak-hngah* is the last *rite de passage* of traditional Chin society. As the rite of *Ruak-hngah* occurred at a great moment of anxiety and sadness, it also was a time of great paradox, a moment when life here on earth ended, and a moment when the life of eternity in *Mithi-khua* began. It was a time of paradox, when a deceased person was beginning to enjoy the new life in the company of his or her ancestors, and the living were crying for the departed one. As the old Chin saying puts it: *A thi cu an ni nuam, A nung tu kan ttap!* (While the deceased are happy, the living are crying!)

If the deceased was a highly esteemed nobleman, as soon as death occured, a gun would be fired off, so that 'the dead man's spirit may take the spirit of a gun with it to *Mithi-khua*, and also that the villagers may know the sick man had departed' (Parry 1932: 399). After this, the body of the deceased was prepared for burial through a process of washing (*thianter*, 'purification') with warm water, which was performed by 'some close relation; the hair is greased and properly tied, and the body is dressed as in life' (ibid.). The deceased was dressed in all the best clothes: a man in hin noble *Pawngpi, Cawngnak puan, Pahrih-bui*, a woman in *Lukhimh-tung, Cawangnak puan, Hni-ttial.*

For a nobleman, the funeral service lasted two or three days but for ordinary people it usually lasted only one day. During the funeral service, 'a wake is held, which is attended by all the deceased's relatives and friends' and the family of the deceased would give a funeral feast called *Ruak-hngah*, for which they killed as many animals as possible, such as mithan, water buffalo, cows and pigs – believing that the spirits of all the animals they killed will accompany the deceased to *Mithi-khua*. During the funeral feast, if the deceased was a nobleman and a warrior, a warrior dance called *Sarlam* was performed. For a noble lady, a special dance called *Rua-kha-tlak* was performed. And the songs that tell the glory of his lifetime achievements and those of his ancestors, were sung repeatedly. Thus, it was a celebration of paradox: a time of mourning for the loved one and a big feast converge. For the Chin, celebrating the past and experiencing the present reality of humanity were tied together through ritual.

There were three kinds of graves in traditional Chin society. Chiefs and important persons and clans generally had family vaults, called *thlanpi* in the Lai dialect, *tlapi* or *longang* in the Mara dialect. *Longang* literally means 'stone house'. These were usually situated within the family compound. A graveyard called *thlan-mual* was for ordinary people and usually situated outside the village. But for the person who died an accidental death, and who was not supposed to go to the ordinary *Mithi-khua* but to *Sarthi-khua* after death, there was another graveyard called *sar-thlan.* At the *thlan-mual,* every grave was marked by a memorial stone called *thlan,* but at the *sar-thlan* no memorial stones were erected.

For ordinary people, erecting a memorial stone, *thlan ser,* was the end of the funeral rite. But the chiefs and noble class, who could afford to give another costly feast, erected a special memorial stone called *lung-dawnh,* which they usually erected at the *Tual* ground or at the entrance of the village gate. In this way, ancestor worship and sacrificial ceremonies at the *Tual* ground were inseparably intertwined with each other, especially at the ceremony of *Lamnai-cheih* and *Tual dangh,* which marked the end of the year.

CONCLUDING REMARKS

In this chapter, I have explored the ritualistic dimensions of traditional Chin religion. In so doing, I have given special emphasis to life here on earth as the preparation period for life after death in *Mithi-khua,* and how these belief systems and ritual practices moulded traditional Chin society, which functioned in accordance with a unitary functional pattern of 'chief–land–god'. In other words, religious, economic, and political institutions in Chin society were inseparably intertwined. The distinction cannot easily be made between religious, social, economic and political elements in such a society. Religion, however, was the main component of all aspects of life, and the ruling traditional tribal and local chiefs were the heads of the society, that is, the heads of every social institution in Chin life. Established forms of belief and ritual systems were seen not only as tradition and/or ways of life, *phunglam,* but even as a giving legitimacy to the chief's powers. In such a religiously oriented society of *Tual* community, traditional religious beliefs and ritual systems were very important in maintaining the stability of society which, in turn, could only be maintained through the traditional instruments of integration and fundamental isolation from other societies. In order to maintain their traditional ways of life, *phunglam,* the Chin isolated themselves from other societies and the outside world for many centuries. However, the arrival of British colonizers and American Baptist missionaries challenged traditional Chin society at its basic foundation. As a result, major changes came quickly. These changes and their results will be studied in Part Two.

NOTES

1 My correction; Carey and Tuck originally spelled the word *Khozin.*

2 The prayer here is Zophei dialect. *Kha-zing* in Zophei dialect stands for *Khua-zing,* and *Pathen* for *Pathian.*

3 N. E. Parry did his fieldwork among the Lakher (Mara), and he used the Lakher dialect *Khazang* or *Khazangpa* for *Khua-zing,* and *Zang* for *Zing.*

4 This sacrifice also included the village's guardian god, and the purpose was part of their defence system.

5 Red is the symbol of grain in Chin traditional religion.

6 C. U. Strait has given a slightly different account of *Thla-auh* sacrifice; he spelled *Thla Kawh* instead of *Thla-auh.* See Strait (1933), pp. 155–156.

PART TWO

COLONIAL POWER, CHRISTIAN MISSION AND THE CHIN RESPONSE

FOUR

The British Annexation of Chinram

The Chin people were divided into different tribal groups and spoke different dialects ever since they first settled in what is now Chinram. These tribal and linguistic differences were strengthened by their traditional religion of *Khua-hrum* worship, and although they all followed the same conceptual pattern of belief systems, the ritual practices of *Khua-hrum* were mutually exclusive and could not serve to unite the entire race under a single religious institution. Moreover, they never had a chance to re-unite under the single political leadership of a national kingdom, as they once had done in Khampat, the ancient Chin capital. Tribally and locally, they were independent. None of the surrounding powers, such as the Bengali Indian or Burman, ever conquered the Chinram. As a result, Buddhism, Hinduism and Islam had not reached the Chin when they eventually encountered Christianity.

At the turn of the twentieth century, however, Chin society was quickly transformed by powerful outside forces of change. The British conquered Chinram, and Christian missionaries followed the colonial powers and converted the people. My primary objective here is to explore this development and analyse what was involved in this process of change. In so doing, this chapter will pay special attention to the historical development of the British invasion of Chinram, from the first encounters in the 1870s to 1896 when the British unilaterally promulgated the Chin Hills Regulation. I shall study this development from the local Chin point of view, and highlight how the British annexation became an important mechanism through which the Chin people were awakened to the self-awareness of national identity, or 'Chin-ness', mainly because of the common perception of a threat to their continued existence.

The British annexation of Chinram has been documented by a number of British military and political officers who served during the campaigns. Although I rely on existing documents and publications, I shall approach my study from the perspective of the Chins' resistance. On the basis of contemporary British reports, I shall highlight the British strategy and military actions as well as – and more importantly – how the Chin responded to the intrusion of the British.

THE BRITISH INVASION OF CHINRAM

First Contacts

The first contact between the British and the Chin occurred indirectly, when the British East India Company occupied the Chittagong region, neighbouring Southwest Chinram, in 1760. As this contact was only an indirect consequence of the occupation, the earliest reference on record to the Chin was first made after seventeen years of occupation in 1777, when Ramoo Khan, probably a Chakma chief, rebelled against the authority of the cotton farmers employed by the East India Company. He called 'into his assistance large bodies of Kookie [Kuki] men who lived far in the interior part of the hills' (Carey and Tuck 1976: 12). As we recall, 'Kuki' is the Bengali word for 'Chin' but it referred particularly to the Zomi tribe of Chin in what is now Manipur State of India and some part of the Chittagong Hill Tracts in Bangladesh.

In 1761, the East India Company annexed the plains area of Tipperah (Tripura) but the hills remained unoccupied and became known as Independent Tipperah, governed by a rajah until 1872. The Tripura Hills was a kingdom, which directly bordered the Lushai (Mizoram) country of West Chinram. From 1785 to 1824, the Chin made several raids into the territory of the Tipperah Hills. In about 1810, Lalul (Lallula), the Lushai chief, was reported to have become 'formidable to his neighbours' and, by pressing towards the frontiers, he 'drove the dispossessed chiefs into British territory' (Elly 1893: 1). Even so, the first recorded encounter between the Chin and British subjects occurred in 1824 when enterprising traders from the British occupied the Tipperah Plains, penetrating Chinram along the Dhalleswari River to collect bamboo and timber. They were killed by the Chin for refusing to pay tribute according to the Chin custom of levying taxes upon those who passed through Chin territory. In that same year, a series of raids were made by those who were then called 'Poitoo Kukies', that is, the Paite group/clan of the Zomi tribe 'who were said to be the most formidable and turbulent of the hill tribes' (Carey and Tuck 1976: 14).

Meanwhile, in 1824, the First Anglo–Burmese War broke out. A peace treaty between the British and the Burman, known as the 'Yandabo Treaty', was signed on 24 February 1826. As a consequence of this treaty, the Burman, for the first time in their history were forced to yield portions of their country – namely Arakan, Assam, Manipur and Tennesarim – to a foreign and non-Asian power.[1]

After the occupation of Arakan, Assam and Manipur, the areas under direct British administration were brought closer to the boundary of

Chinram: Arakan from the southeast, Assam from the west, ar from the north, which in the long run presented a threat of h the Chin.

Ironically, the first serious threat was provoked by the Chin not the British. After the First Anglo–Burmese War, the Chin had greater access to firearms. Although they possessed some firearms already, the Chin could now easily buy firearms from defeated kings, rajahs and tribal chiefs who were willing to sell their arms, at virtually any price, in order to circumvent the British policy of complete disarmament in the newly occupied territories. Thus, from 1826 onwards, the number of raids made by the Chin on British territory increased, culminating in what was called the 'Great Kuki Invasion of 1860' in which fifteen villages in Tripura were 'burnt and plundered, 185 British subjects killed, and about 100 captives carried off' (ibid.). Between 1834 and 1854, the Chin made at least nineteen raids into Chittagong in which 107 people were slain, 15 wounded and 186 carried into slavery. The British could not tolerate the challenge to their sovereignty involved in this harassment, killing and capture of their subjects, and 'it was not to be supposed that those under our protection should escape', as R. G. Woodthorpe wrote in his book *The Lushai Expedition*, published in 1873.

The British Indian government therefore was 'forced to take steps for their protection' (Woodthorpe 1873: 1). However, 'insofar as they saw no opportunity for economic gain in administering' Chinram, 'they first sought to control the raiding through the Tripura and Poang Rajas, who were made responsible for border defense' (Kipgen 1996: 130). Moreover, with a view to protecting their 'border from the aggressions of the hill tribes in the east', the British formed a district called the 'Chittagong Hill Tracts' and placed a Superintendent in charge in 1859. With the appointment of a Superintendent of Hill Tribes in the Chittagong Hill Tracts,[2] the area under direct British administration was brought closer again to Chinram from its southwestern border. The reason for the appointment of a Superintendent and the formation of the Chittagong Hill Tracts District, according to T. H. Lewin, was 'the supervision of the independent tribes, and ... the preservation of the peace of the frontier' (1870: 31–32).

The Tripura and Poang rajahs were ineffective in defending British territories against raids from the Chin. Between 1861 and 1871, seven raids were recorded in Tripura. These were more systematically organized and determined in character than previous incursions (Elly 1893: 6–7). In the year 1871, within a period of thirty days, from 23 January to 23 February, the Chin conducted nine raids on the Cachar plain and attacked the British tea planters 'who had intruded into their territorial hunting ground' (Lian Uk 1996: 3). Simultaneously there were raids into Sylhet, Tripura, Manipur and Chittagong. The Chin also attacked the police stations at Moneirkhal and Nugdigram and took many weapons and ammunition. In fact, by this time the Chins were highly motivated in their raids by the politics of nationalism, not just economic and religious gain. I shall come back to this later. In the process, some villages were burnt and at least one tea plantation destroyed. Upward of 100 people, both police and civilians, were killed, many were wounded and 61 people were taken into slavery. The Chin retreated from the Cachar front only after sustaining the loss of 25 of their own men. An accident connected with these raids in Cachar, which had significant influence upon subsequent British policy, was the 'attack upon the tea plantation at Alexandrapore on 23 January in which the planter, Winchester, was killed and his six-year-old daughter, Mary, carried off as a captive' (Woodthrope 1873: 217).

The First Invasion of Chinram (The Lushai Expedition)

Events in 1871, especially the Chin attack on the British tea plantation at Alexandrapore, led to the first invasion into Chinram by the British. The government of British India now decided to send an expedition into the Lushai country of West Chinram during the cold season of 1871–72. Plans for the campaign were made and in December 1871 the expedition set off in two columns, the right column advancing from Chittagong and the left from Cachar. General Brownlow, C.B., commanded the former, with Captain Lewin, Superintendent of the Chittagong Hill Tracts as the Civil Officer. General Bourchier, C.B., assisted by Mr Edgar as the Deputy Commissioner and Mr Cachar as the Civil Officer, was in charge of the left column. In addition, a contingent of Manipuris accompanied by General Nuthall, the Political Agent of Manipur, carried out a demonstration march across the southern border of Manipur in support of General Bourchier's operation.

Although the entire political and military conduct of the expedition was placed in the hands of the military commanders, the British India government did not intend, as Col. Woodthorpe wrote, to 'exterminate these frontier tribes, but convert them into our allies' (1978 [1873]: 6). Thus, the government gave the following special instructions to the military commanders:

> [T]he object of the expedition was not one of pure retaliation, but that the surrender of the British subjects held in captivity should be insisted on, and that every endeavour should be made to convince them that they have nothing to gain and everything to lose by placing themselves in a hostile position towards the British government (Carey and Tuck 1976: 16).

The Cachar column, which consisted of half a battery of artillery, a company of sappers and 500 rifles, started on 15 December. After 'encountering and overcoming considerable resistance and penetrating very difficult country, General Bourchier destroyed the chief villages of the offending tribes and imposed conditions of peace. Hostages were taken and a fine of arms and produce was levied' (ibid.: 16–17). The column reached Cachar on its return on 10 March.

The Chittagong column, comprised of about the same forces as the one starting from Cachar, advanced from Demagiri to deal with the Lyloos and Howlongs. Punishment was inflicted on these tribes and their full submission on suitable terms was secured. The restoration of all captives and an engagement to keep the peace in the future were among the conditions on which 'the submission of the tribes was accepted' (ibid.: 16).

Indeed, the operations of both columns were successful and most of the Lushai (Mizo) chiefs in West Chinram were reduced to submission. Mary Winchester was surrendered, and the fines imposed were paid. The guns which had been taken at the police stations of Moneirkhal and Nugdigram during the Cachar raids in February 1871 were surrendered, hostages accompanied the forces on their return and a guarantee of free passage in the future to agents of the government was paid. In addition, a topographical survey of 6,500 square miles was made. All this was 'accomplished within the short span of three months' (Woodthorpe 1978: 217).

After the expedition, the question of frontier defence was reconsidered. Despite the advocacy of a 'forward policy' or the 'complete military occupation' of the territory by the frontier officers, the existing policy of 'loose control and conciliation' was reaffirmed in the belief that the expedition had convinced the Chins of the inadvisability of further raids on British subjects (cf. Reid 1893: 32ff). In order to make the policy more effective, cordons of frontier posts were built to protect the newly delineated boundary, and bazaars were opened to encourage the Chin to trade. The frontier officers endeavoured to establish good relations with the Chin chiefs, occasionally giving them feasts and presents for this purpose. Perhaps the most significant efforts were those made by T. H. Lewin who, by 'his sheer courage, insight into the nature of the tribal world, and love and sympathy for the tribal life and culture, won the respect of the Chins. He referred to them as "my hill people", and they, in return, called him 'Thang Liana' (their version of Tom Lewin)' (Cited in Kipgen 1996: 134).

To some extent, the policy was successful. Assam and Manipur enjoyed comparative peace until 1888–89. But in Chittagong, the period of peaceful relations was already broken in 1882. A group of Chin warriors numbering 250 or 300, headed by a Mizo chief named Huatsata attacked a village in Chittagong killing 29, wounding seven and taking 92 persons into captivity or slavery. Two more raids were conducted during the following five years. Moreover, a new encounter between the Chin and British occurred on the Burma side of Chinram's eastern border after the British occupied upper Burma in 1885.[3]

New Encounter after the Third Anglo–Burmese War

The British conquered the Kingdom of Burma or 'Myanmar' in the three Anglo-Burman Wars fought in 1824–26, 1852 and 1885. There was little organized military opposition during the Third Anglo–Burmese War, and the British India Army entered the royal palace of Mandalay without serious military resistance. On 3 December 1885, King Thibaw was dethroned and exiled to India, and the Kingdom of Burma or Myanmar was declared a province of British India. After their king was deposed from the throne, some of the Burman armies reorganized themselves into small defiant groups, which roamed the country and fought as guerrilla bands, but without much success.

In 1886, the possessions of the king of Burma – including the Kale, Kabaw and Yaw Valleys of semi-independent countries, which paid tribute to the king – were transferred to the British (cf. Crosthwaite 1968: 288). As we recall, the Kale and Kabaw Valleys were the original homeland of the Chin people, but were conquered by the Shan in the thirteenth century. Thus, until the colonial period, the Shan ruled both the Kale and Kabaw Valleys, and 'like the states on the Shan Plateau, it was governed by *Sawbwa*' (ibid.: 288). At this time, the Kale Valley, or what the Chief Commissioner of Burma, Sir Charles Crosthwaite, called the 'little Shan State of Kale,' was paralyzed by a civil war being waged between the two rivals for the Sawbwa-ship – an uncle by the name of Maung Yit and his nephew Maung Pa Gyi – and also 'exposed to raids from the Hillman, [who] for a long time past had suffered much from the Siyin (Sizang) group, who were the most frequent and barbarous raiders, burning villages, slaughtering the peasants, and carrying off many slaves into the mountains' (Carey and Tuck 1976: 21–22). In 1886, the old *Sawbwa*, Maung Yit, was in power but the British deposed him, and his rival, Maung Pa Gyi, was appointed *Sawbwa* in his stead on 1 January 1887. The British also arranged the administration of those newly occupied territories: the Kabaw Valley was placed in the Upper Chindwin District, the Kale State was placed under the control of the Deputy Commissioners of that district and the Yaw Tract was included first in the Pagan, and later in the Pakokku District. With this administrative arrangement, Chinram was now completely surrounded by the newly occupied territories of the mighty British colonial power in 1887.

Knowing that the Chin had made several raids into the newly occupied territories, the British authority sent Captain Raikes, Deputy Commissioner of Upper Chindwin, to discuss boundary issues and border relations with the Chin. He was instructed and 'warned to be careful, not to frighten or excite the Chins' (Carey and Tuck 1976: 22). When he arrived at Kalemyo, he sent a message to the Zomi tribe of the Chin inviting the Sizang chiefs, who were the most powerful among the Zomi tribe. These refused to go themselves but sent four warriors: Tunsun (Tunsuang), Howsun (Hausuang), Dowson

(Dosang) and Htensan (Teisang). A meeting was held at the Kale *Phogyikyang* (Monastry School) on 17 December 1887.

The Chin representatives were informed that the British had decided to recognize Maung Pa Gyi as 'Governor of Kale', a small Shan State on the edge of Chinram. As the *Sawbwa,* Maung Pa Gyi 'had behaved in a loyal manner and had also paid tribute to the British government'; Captain Raikes was authorized to inform the representative of the Sizang and Kamhau of the Zomi tribe that 'all raids within the Kale territory must cease and that, if in the future any raids by members of the Siyin (Sizang) tribe occur, they would be considered acts of hostility towards the British government, and the chiefs of the tribe would be held responsible for them unless they turned over the raiders in custody to the Deputy Commissioner and caused all prisoners to be given up free of ransom' (ibid.). In short, Captain Raikes warned the Chin that they should in the future make no more raids within the Kale territory and no more slaves should be taken. But they could keep slaves already taken in their previous raids when 'the Kale country was in a state of anarchy and before the *Sawbwa*'s authority had been formally recognized by the British government'. Moreover, Captain Raikes was

> most anxious to encourage trade between the Chins and the Shan of the Kale State, and informed the chiefs that if they ceased to commit raids and border acts of aggression in Kale territory, they might rest assured that the British government would not interfere with them in any way; on the other hand, the Deputy Commissioner, as representative of the British government, wished to meet the principal chiefs from time to time and maintain friendly relations with them (ibid.).

Captain Raikes also told the Chin representatives that the British government had intended to open the trade route to Chittagong from Kale, and also intended to visit the Letha range. Thus, the 'enquiry was made whether the Chins would object to the advance of a column of British troops through their country; also whether, in the event of their having no objection to such an advance, they could supply transport coolies and guides' (Woodman 1962: 380).

Tunsuang, as speaker of the Chin delegates, replied that they were willing to have friendly relations with the Governor of Kale and would not attack him since he was installed. The route to Chittagong would have to pass through the Laimi tribe territory, not the Zomi tribe territory, and this had to be discussed with the Laimi tribe. Tunsuang said that he strongly opposed any exploration by the British in the Sizang area. Should the British visit the Letha range, they should inform the Sizang prior to the visit; otherwise, 'as the people had never seen a white man before, the British visit could have a disastrous effect. Their women and children would panic at the sight of a strange white man' (Vumson 1986: 114). The meeting ended rather peacefully because the Sizang delegates accepted the gifts from Captain Raikes.

Captain Raikes then travelled from Kalemyo to Indin. He invited Cawn Bik (Sonpek), a member of the Tlaisun Democratic Council, and of Laimi tribe, to meet him there.[4]

Cawn Bik came down to Sihaung, a border village, but refused to go to Indin, which was located outside his territory. So Captain Raikes went to Sihaung to meet Cawn Bik. The meeting was held on 3 January 1888 at Sihaung. For four hours they discussed the British recognition of the Governor of Kale, Sizang raids into Kale territory, the opening up of a trade route between Chin country and Shan (Kale) on the east, and Chin country and Chittagong on the west and, as a preliminary step, an exploration through the Tlaisun tract.

Cawn Bik had no illusions about the British desire to occupy his country. He therefore asked for time to discuss the matter with other members of the Tlaisun Democratic Council and other chiefs. When the meeting ended, Major Macgregor, who was present throughout the meeting, allowed the Gurkha Army, who accompanied the party, to fire two volleys and five rounds of independent shots at a target five hundred yards away as a gentle reminder to Cawn Bik and his colleagues of what they would have to face if they did not submit. This action annoyed Cawn Bik so much that it induced Chin resistance far more quickly than Captain Raikes and other British officers believed possible (cf. Dawt Sung 1976: 56). I shall come back to the Chin resistance in the following section.

After this meeting with Cawn Bik, Captain Raikes marched down to the Gankaw Valley, intending in like manner to negotiate with the Haka and Zokhua of the Laimi tribe. But when he heard that Captain Eyre was already in communication with them, he turned around and marched through the Kale State to the Kabaw Valley. Captain Eyre's negotiation with the Haka and Zokhua of the Laimi tribe, however, ended abruptly and disastrously, 'for the three men whom he sent up to the hills to call the chiefs were arrested by Yokwas [Zokhua]; two were killed and the third, Shwe Hlaing, escaped, found his way to the Arakan Hill Tracts, and thence returned to Captain Eyre at Pagan' (Carey and Tuck 1976: 25).

It was unfortunate that Captain Eyre, the Deputy Commissioner of the Pagan District, had sent Shwe Hlaing as a messenger to the Zokhua and Haka. He was a son of the Zokhua chief by a lesser wife, a Burman woman who was captured as a slave during a raid in the Gankaw Valley. Since Shew Hlaing was not just a son of a lesser wife, but also a slave, he had no future in terms of hierarchy among the ruling Zokhua aristocrats. He thus went down to Yaw country to make his own fortune there and joined the newly created Police Force of the British India Province of Burma. To his father's surprise, he returned home not only with many gifts from Captain Eyre but as an agent of the British authority. Without knowing much about his homecoming, his father gave a big feast for his successful son. But during the feast, while he got drunk, Shwe Hlaing declared that the British intended to install him as a chief of the Zokhua. This was enough reason for his father, a powerful chief of the Zokhua, to issue an order to arrest his own son and two other messengers. But Shwe Hlaing was released secretly by his mother at night. The next morning, his two colleagues were executed by the Zokhua chief (cf. Dawt Sung 1976: 61). This episode not only damaged Anglo–Chin relations but caused a large-scale invasion of Chinram by the British, as I shall discuss below.

Meanwhile the Haka, one of the most powerful among the Laimi tribes, made two raids into the Yaw country, killing eight and taking 28 people into slavery. Despite their frontier negotiations with Captain Raikes, the Sizang (Sizin) under the leadership of chief Khai Kam (Kaikam) attacked a party of Shan in the Kale Valley, killing one and taking four into the hills as captives. Tlaisun (Tashons) also conducted two serious raids in the Kale Valley in early 1888. During October, the Tlaisun conducted one more raid, the Sizang five and Kamhau one. Within twelve days, 122 Shan were taken into slavery, 12 killed, and 14 wounded. Moreover, the town of Khampat was entirely destroyed and Kalemyo lost 35 houses by fire (cf. Carey and Tuck 1976: 26).

To make matters worse for the British, one of the Burman resistance leaders called 'Shwe Gyo Byu Mintha', was granted asylum by the Tlaisun chiefs. He was a man who, although he had no pretensions to royal blood, assumed the title of 'Shwe Gyo Byu Mintha' (Prince of the White Dove) and succeeded in raising a rebellion in October 1887 within the confines of the Chindwin, Mingyan and Pagan Districts. In the operations, which followed, Captain Beville, Assistant Commissioner, and Major Kennedy were killed. When Shwe Gyo Byu Mintha was defeated, he fled into the Yaw country and then into Chinram. The British authority thus sent an ultimatum to the Tlaisun chiefs, ordering them to deliver up the Shwe Gyo Byu Mintha and his followers. An ultimatum was also sent to the Sizang to deliver up Khai Kam, together with the captives whom he had carried off.

The Chin flatly refused to comply with the ultimatums. Instead, they prepared for further attack in the Kale, Kabaw and Gankaw Valleys. Ignorant of the fact that they were now surrounded by the mighty British Army, who could easily subjugate them and occupy their land, the Chin prepared for war because they thought they could easily eliminate those few white chiefs whom they had seen. Moreover, since the Burman king was dethroned and a Burman prince had taken asylum under their protection, the Chin thought that now was the right time to recapture their original homeland of the Kale, Kabaw and Gankaw Valleys from the Shan. The Chin knew that when the Shan lost the Burman's protection, they could easily be eliminated. As noted already, the Shan were more powerful at the beginning of the thirteenth century and could expel the Chin from their original homeland, but once 'the Chin gained access to firearms, the balance of power between the two peoples was reversed and, the Shan in the Kale and Kabaw Valleys were forced to seek the protection of the Burman king from the lowland' (Crosthwaite 1912: 23).

The British policy toward Chinram until that time was not annexation but 'merely protecting the valley' of the newly occupied territory. As Carey and Tuck noted, 'sending a costly expedition into' what they called, 'unprofitable, barren hills' was not in the interest of the British imperialists from the very beginning. But 'the hands of the government were forced by the Siyin Chins, who came down to the plains burning, killing, and capturing with the utmost daring; nor were the Siyins [Zomi tribe] the only tribe on the warpath, for the Soktes [another group of the Zomi tribe] also swooped down on the Kabaw Valley and the Thasun [Tlaisun of Laimi tribe] ravaged the Kale Valley' (Carey ad Tuck 1976: 26). At the same time, Haka and Zokhua of the Liami tribe attacked the Gankaw Valley. By this time, the Chin had put aside all their own internal feuds, tribal and clan wars, and were united in blood, as descendants of a single father, when they faced a greater cause and a bigger enemy. This was reinforced when the British launched the Second Invasion of Chinram.

The Second Invasion of Chinram

In November 1888, the British made plans for their second invasion of Chinram. The British invading forces were led by Sir George White, V.C., and General Faunce, and included Captain Raikes and his assistant Mr Hall. They 'busied themselves with obtaining intelligence regarding the Chins, their villages, and the routes to their country' (ibid.: 26). The coolies were obtained from Assam. The 42nd Gurkha Light Infantry, Assam troops, and the Punjabi military police[5] were sent through Manipur to the base of operations. The objective of this second invasion was to deal first with the Zomi tribe of the Sizang and the Kamhau groups, 'to march first against Koset [Khuasak], the head village of the Siyins [Sizang], and from the center to deal with the surrounding villages of the tribe' (ibid.: 27).

The Sizang, who had returned to normal life after they met Captain Raikes, were once again informed that the white men, *mirang/mikang*, had advanced with a larger force and were building a stockade at Phatzang, about ten miles from the boundary of Kale and Chinram. On hearing the news of the British advance, the Sizang council sent Khaikam of Khusak to Tiddim to meet Khawcin, the chief of the Kamhau. Chief Khawcin immediately rounded up 400 warriors and marched to Khuasak to stop the British from any further advance into the Chinram. Moreover, they sent a messenger to Tlaisun not only to inform but to ask for their assistance. The significance of this event, therefore, was that the Sizang of the Zomi tribe and the Tlaisun of the Laimi tribe, who had fought each other in a major tribal war as recently as the middle of the nineteenth century, 'were fighting side by side on this occasion' (ibid.).

Again, the British annexation of Chinram was an important mechanism by which the Chin were made aware of being one ethnic community, and standing no longer as different and divided peoples. Thus, the Sizang and Tlaisun, formerly rivals and bitter enemies, were now not just 'fighting side by side' but 'partaking' in a sacrificial ceremony called *Sathin-tuh*, the most important religious communion in terms of war. After they did this, they became a collective unity of one people; hence the revenge (*phuhlam*) for the soul of any casualty in this war must be taken not as an individual person but as a collective people. In this way, the Chin could put aside their inner feuds, overcome tribal differences and unite as one nation.

The morning after the *Sathin-tuh* sacrificial ceremony at Khuasak, a 'force consisting of 1,200 Sizang, four hundred Kamhau and thirty Suktes from Mualbem' marched to Phatzang (known in Chin as Leisan). The first encounter took place on 7 December, when the Chin attacked a working party making the road between the Kambale and Sizang Valleys. At Phatzang the British held their positions for some hours but were beaten, and retreated towards the plains. 'Khup Lian, the young chief of the Lophei, wrestled a rifle from a British soldier in a hand-to-hand fight' (Vumson 1986: 118). During that same battle, Lt Palmer and twelve Gurkha sepoys were killed, but there were no casualties on the Chin side.

On 10 December, a great Tlaisun force of 1,000 fighting men made a simultaneous attack on the camps and villages of Indin and Sihaung at 4 a.m. The same day the Sokte and Kamhau of the Zomi tribe attacked the military posts of the Kangye and Kale Valleys. Road-making parties were attacked every day from this time onward. The British troops found their routes blocked, and the Chin who held the stockades 'never ceased to ambush when an opportunity occurred, both day and night'. The British now recognized the Chin's tactical ability in war, as Carey and Tuck observed when they wrote:

> Whilst disputing every stage of our advance into the hills, the Chin showed considerable tactical ability by taking the offensive in the plans and attacking Shan villages and our posts in the rear of the advancing column (Carey and Tuck 1976: 27).

On 23 January 1889, Sir George White, General Faunce and Major Raikes reconnoitred to the summit of the Letha Range; Sizang challenged them all the way. On the 27th, fighting took place at Thayetpin (*Thing-u-nau* in Chin). Sir George White described the encounter to the Chief Commissioner of Burma as follows:

> Enemy yesterday attacked our working party on road above this and held our covering party, forty British and one hundred Gurkhas,

> from 9 till 2 when I arrived and ordered their positions to be charged. We carried on, driving them away ... Enemy is considerable number, using many rifles and plenty of ammunition. They fired at least a thousand rounds, standing resolutely until actually charged, even trying to outflank us. Their losses probably about eight or ten, but were carried down the khuds at once. Most difficult enemy to see or hit that I ever fought (ibid.: 28).

The result of this action was a serious blow to the Chin and they now realized that it was impossible to block the British forces and defend their homeland. The 'fight had taken place on one of their battlefields, for it was here that they had overthrown an army sent against them by the king of Burma in former days' (ibid.). In this battle Surgeon (Major) LeQuesne received the Victoria Cross, demonstrating how fierce the Chin resistance was that he had fought. The battlefield/stockade was named Fort White after General White, commanding officer of the forces of the second invasion of Chinram.

Between December 1888 and January 1889, 36 British were killed and 54 wounded. After they failed to defend at *Thing-u-nau*, the Chin had only a small chance of resistance. Thus, 'they did no more than fire a few shots and then busied themselves with carrying off household goods' and running away into the jungle. All the Sizang villages – more than 30 villages and 18 Kamhau (Tiddim) villages – were burnt to the ground. Despite heavy losses, the Sizang continuously attacked during the following rainy season. The Chin were, as Carey and Tuck wrote:

> [C]atlike people in their movements, [they] soon learnt that their power to annoy us lay in their skill in creeping inside the fort between sentries, and night after night the cattle-pens inside [and] the piglets ... were found to have been visited and stolen. On one occasion a whole herd of seventy heads was carried off ... Another time a *drabi* was shot and decapitated in the middle of the fort, the Chin escaping through the sentries. One night in June a determined attempt was made by the Chin to set fire to the roof of the house in which the *Myook* and the interpreters were living. The Chins crept into the post and set fire to the roof of the house; the damp thatch, however only smouldered and the Chin, being discovered, decamped after firing a volley through the house into the sepoys who were attracted by the shouts of the Burman. The intention was to shoot all the guides and interpreters and the plan was to reserve fire until the *Myook* and his people should rush out of the burning house, when by the light of the fire they could easily be picked off. The persistent annoyance did not end within the limits of the fort, for small parties of Chin hovered about the post and fired from ambush whenever occasion offered, and as there was much work going on inside the post and its approaches, the Chin had many chances of 'sniping' (ibid.: 32).

In August 1889, Sir Charles Crosthwaite, then Chief Commissioner, went up the Chindwin to meet Major Raikes, and together they planned the following season's punitive expeditions. Sir Charles first launched political warfare by proclaiming the following to the Sizang and Kamhau of the Zomi tribe and also to the Tlaisun of the Laimi tribe:

> A British army will march to the Thason (Tlaisun) Ywoma. The British government wishes to preserve your tribe and does not desire to punish you as it has punished the Kamhau and Siyin (Sizang) who have resisted the British forces.
>
> The British government desires from you only two things: first, that the captives (slaves) taken from Burman villages shall be released; secondly, that you shall in the future behave peacefully and cease to attack the subjects of the government.
>
> Therefore, the Chief Commissioner hereby declares and promises that you will be excused from punishment for the past if you comply with the following terms:
>
> 1. That you shall assist the British troops in their march through your country to your Ywoma, and that you will neither attack nor oppose them.;
> 2. That you shall to the utmost of your power compel the Siyin (Sizang) and Kamhow (Kamhau of Zomi) tribes to surrender their captives (slaves);
> 3. That the chief shall meet the officers in command of the British forces at the Ywoma and deliver up to him all the captives (slaves) in possession of your tribe and pay a fine of Rs. 1,000;
> 4. That you shall render annually a tribute of two elephant tusks and ten silk sheets to the British government.
>
> If you comply with these terms your lives and your property will be spared and the former order requiring you to deliver up the 'Shwegyobyu and other rebels will not be enforced. On the other hand, if you will not comply with these conditions, the Chief Commissioner will direct the troops to inflict on you the severest punishment. You will be attacked on all sides and your country will be laid waste. Choose now between the friendship and enmity of the British government'. (Quoted in Woodman 1962: 387)

The British knew very well from the start that the Chin would not comply with the terms. They therefore prepared to attack Chinram from all sides. A few weeks after the British authorities issued their demands, the Army headquarters in Burma and Chittagong were alerted with the following instructions:

1. to punitively visit certain tribes which had raided and committed depredation in British territory

2. to subjugate tribes as yet neutral, but now by force of circumstances brought within the sphere of British domination

3. to explore the unknown country lying between Burma and Chittagong and

4. to establish, if necessary, certain posts in the region visited, so as to ensure complete pacification and recognition of British power (ibid.: 388).

This was the preparation for the third and final invasion, known as the 'Chin-Lushai Expedition', which began in September 1889 and lasted until 1896 when the British finally promulgated the Chin Hills Regulation.

The Third and Final Invasion of Chinram (The Chin-Lushai Expedition)

The British government had thus adopted 'a policy of pacification through permanent occupation' for the whole of Chinram. The 'policy hitherto followed, of merely sending punitive expeditions to punish offending villages and then retiring from the hills, was now abandoned'. In dealing with the Chin, 'there were only two options', wrote Sir Charles Crosthwaite:

> [E]ither to make a well-prepared systematic advance into the Chin Hills and bring these people under British rule, or to retire altogether and leave an *enclave* of savagery between Burma and Bengal, trusting for the protection of the Burman villages to frontier posts and spasmodic expeditions. The long history of the dealings of the Bengal government with the Lushai and Nagas, very similar peoples, had proved the futility of the latter course. The inclination in Burma was all for the former, and this met with the thorough approval of the Supreme Government (1912: 310).

Thus, the Chin-Lushai Expedition was launched in September 1889. The expedition was entrusted with the implementation of this new policy of 'permanent occupation', which was to deal with the whole area inhabited by the Chin nation, and therefore the whole Chinram. As Crosthwaite had indicated in his ultimatums, the British invaded and attacked Chinram from three directions: along the eastern front from Burma, the western front from Assam (India) and the southern front from Chittagong (Bangladesh).[6]

The British forces on the eastern front were led by General W. P. Symon. He divided his troops into two columns. The northern column, led by Colonel Skene, was called the 'Fort White Column'. It consisted of 1,622 troops and operated from Fort White as a base against the Sizang and

Kamhau of the Zomi tribe and 'the tribes between the base and the Manipur River' (Carey and Tuck 1976: 33), i.e. the Laimi tribe. The second, southern column, led by General Symon himself, was called the 'Gangaw Column'. It consisted of 1,896 men strong and started from Gangaw as a base and advanced *via* Zokhua to Haka, capital of the Chin State in Burma. In addition to the above force, the whole length of the Kale and Yaw Valleys was guarded by military police posts, and the line of communication to Chinram was held by troops: the 10th, 33rd and 38th Bengal Infantry and the 2nd Madras Infantry supplying the garrisons. The Bengal Front, also known as the Chittagong Column, was put under the command of General V.W. Tregear. It consisted of 3,380 men, and it was to march from Rangamati (present district headquarters in Bangladesh) *via* Lungleh (Lunglei) to Haka, making a road as it advanced. The Chittagong Column was attacked by the Chin, led by Za Huat (Jahuta), a chief of Thlantlang (now a township headquarters in the Chin State, Burma), and Pu Huasata, chief of Lungthian (now in the Mizoram State, India). At the battle of Thangzang, the Chin forces, under the command of Lal Luai (Lalwe), younger brother of the Thlantlang chief Za Huat, annihilated a company under the command of Lt John Steward in a fierce battle. Lt. Steward himself was killed, but his head was recovered at Tao (Thau) when a company of the Chittagong Column under the command of Captain Rundall, and the Gankaw Column met on 26 February 1890.

The Assam Front, also called the Cachar Column, was put under the command of Colonel G. J. Skinner and consisted of 700 fighting men with the following assignment:

1. The construction of a mule path to Haka which would there unite with the one being constructed from Kan, and thus form a communication link between India and Burma.

2. The establishment of posts on the road so as to secure complete pacification and recognition of British power.

3. The subjugation of tribes as yet neutral, but now by force of circumstances brought within the sphere of British dominion (A. S. Reid 1893: 185–186).

Simultaneously, another column of military police under W.W. Daly was to be sent from Cachar in support of Col Skinner's forces, which also consisted of coolies, 2,511 Punjabis, 782 Manipuri, 2,196 mules and 71 elephants. The columns marched from one village to another, building roads and posts on the way. This column met with no opposition. Their work was mainly road making, reconnoitring and surveying – work of paramount importance in securing permanent peace. The end of 1890 saw the completion of a mule road from Cachar (Cilcar) via Aizwl to Lunglei,

Lunglei via Vanlaiphai (Fort Tregear) to Haka. At the same time, the Chittagong Column had constructed a road which connected Chittagong and Kale. They also established permanent posts at Aizawl, Lunglei, Vanlaiphai (Fort Tregear), Falam, Haka, Zokhua and Fort White.

During the Chin-Lushai Expedition, another expedition was sent against the Chin-boks, an Asho tribe, who had never moved over to the hill even after the original Chindwin Valley homeland was conquered by the Shan back in the twelfth century, but had remained in the Yaw and Gangaw Valleys of Upper Chindwin, west of Tilin. They surrendered on 16 December 1889.

Chin resistance lasted much longer than the British had expected. It was a difficult country to occupy since most of the area consisted of high, narrow mountain ridges and deep valleys running mainly north–south. Moreover, the Chins were ruled by the headmen of villages grouped tribally under chiefs. Thus, the occupation of Chinram could only be carried out by conquering one tribe after another. By the end of 1892, most of the chiefs in the Falam (Tlaisun), Haka and Lushai (Mizoram) areas had submitted to the British. The Sizang, Sukte and Kamhua of the Zomi tribe, and the Thlantlang of the Laimi tribe still opposed the British by force, and the Surkhua, Sakta (Thettas), Leitak and Aibur of the Laimi tribe refused to give up their slaves, thereby also refusing to submit to the British. The Cho tribe of Matu, Mindat and Kanpalet, and part of the Khuami tribe's area were left untouched by the British.

The Sizang, witnessing the surrender of the Tlaisun, decided to lay down their arms and give up their slaves in 1890. Even so, they revolted again later, as Mr. Carey himself reported:

> The Siyins (Sizang) have surrendered, but they are not afraid of us and this is due to the fact that we only travel by parts, that we do not know the country as we should, that we never night-march, and that we move about in large columns hampered by coolies and never surprise the Chins (Carey and Tuck 1976: 387).

The Sakta also surrendered once in 1891 when surrounded by two columns, one from Haka and another from Gangaw, but they refused to give up their slaves at that time and also revolted again openly in 1894. But when the Chin finally realized that their enemies were coming from three directions and their country was completely surrounded by a single power who deployed many different races and armies against them, 'they began to lose heart', as Sir Charles Crosthwaite put it. As the British had deployed 'Gurkhas, Afridis and Pathans who knew guerrilla tactics, they forecast that in one bout of cold weather the Chin would find themselves beaten at their own game of ambushing and would be stalked from khus and nullah and shot whilst eating or hunting or knifed whilst sleeping in their huts. When this takes place the Chins will grovel for mercy, [and will] even surrender their guns' (Woodman 1962: 395).

The Chin resistance was stopped in 1895–96, not by the British armed forces nor their colonial army who knew guerilla warfare, but by famine. An old Chin saying declares that 'an empty stomach cannot fight'. Famine occurred because the Chin had long been unable to cultivate the land properly because of the war. Moreover, the British armed forces burnt many villages to the ground together with a large quantity of grain in storehouses. As the British officers themselves admitted, the Chin were compelled to surrender by famine. In this way, the Chinram was occupied for the first time in its history by an outside power, namely the British. They were forcefully integrated as part of the British Empire in 1896, when the Chin Hills Regulation was officially promulgated.

THE PROMULGATION OF THE CHIN HILLS REGULATION IN 1896

The Significance of the Chin Hills Regulation

Chinram was occupied by the British as an independent and separate country that had its own boundaries for centuries. No Burman king or Bengali Indian king had ever occupied, ruled or propagandized the Chinram. The Chin people were administered by the headmen of villages grouped tribally under the chiefs (cf. Mang Tling 1998: 188–190). The British government had no choice but to rule Chinram separately and to apply an administrative system different from that of Assam, Bengal and Burma of British India.

Since the British Indian government had adopted the policy of 'permanent occupation' of Chinram in September 1889, they carefully studied a suitable administrative system for the Chin people. The guiding philosophy behind such an administrative system was as follows:

> They must be allowed to develop on their own lines and be protected against exploitation and subversion of their rules and customs by a different civilization which would be unsuited to them.[7]

On 1 October 1891, a proposal for the future administration of Chinram, what they then called Chin-Lushai Country, was presented to the Supreme Government of British India. The proposal stated that:

> It ... seems imperative that for a speedy and satisfactory settlement of the whole Chin-Lushai Country, it should be formed into a Commissionership or Chief Commissionership, the head of which should be responsible only to the (Supreme) Government of India. The officer appointed should have great personal knowledge of Hillmen and great experience in dealing with them.[8]

At this point, the Chin-Lushai Country was proposed to have the same political and administrative status as the governments of Assam, Bengal and Burma, each of which had a Chief Commissioner (later Governor) as the head of the government or administrator under the Supreme Government of British India.

This proposal was confirmed again by the conference that came to be known as the 'Chin-Lushai Conference of 1892', held at Calcutta on 29 January 1892. The conference passed a resolution, part of which stated:

> The majority of the Conference are of the opinion that it is very desirable that the whole tract of the country known as the 'Chin-Lushai Hills' should be brought under one administrative head as soon as this can be done.[9]

In line with this guiding principle, B. S. Carey drafted the Chin Hills Regulation, which was approved by the British government and officially promulgated on 13 August 1896. B. S. Carey, indeed, was very much aware of the ethnic origin of the Chin, although they were sometimes known by their various tribal names, such as Asho, Chó, Khuami, Laimi, Mizo (Lushai) and Zomi, instead of their national name of Chin. He therefore drafted the Chin Hills Regulation with reference to the common historical and cultural heritage of customary law, the political system and traditional ways of life, aiming to cover the whole territory of Chinram. He thus intended to integrate the whole Chin race under a single legal system. Under the protection of the new constitution provided by the Chin Hills Regulation, all the Chin tribes could now share their collective name 'Chin' legally, which in turn was strongly associated with the myth of their common descent and original homeland.

New Administrative Structures

The Chin Hills Regulation was the basic constitution for the entire administration of Chinram during the colonial period. According to this constitution, a Chief Commissioner was to be appointed for Chinram (then called the Chin-Lushai Country). For practical purposes, however, Chinram was administratively divided into three districts: the Chin Hills District (East Chinram, present Chin State in Burma); the Lushai Hills District (West Chinram, present Mizoram State in India); and a portion of southern Chinram which was attached to the Chittagong Hill Tracks of Bengal. In fact, the division of the Chinram administration into three units, or three districts, was based on the territory occupied from each military front during the third and final invasion. Until 1898, when the civilian administration became effective, each unit was put under the administration of their respective occupying military command.

The Chief Executive of each district was the Superintendent, under which the Deputy Commissioner, District Magistrate and Collector were appointed. The Superintendent and Assistant Superintendents were invested with all civil powers. As each district would constitute a Session Division (juridical unit), the Superintendent also served as the Session Judge. Subject to the control of the Chief Commissioner, the Superintendent could impose fines on the villagers or village community. He also had the power to prohibit the formation of new villages or the destruction of old villages, a power the Chin usually regarded as the supreme power of the tribal chief. He had the right to deport and detain villagers and confiscate their property.

The first Superintendent of the Chin Hills District (East Chinram) was B. S. Carey, author of the *Chin Hills Regulation* and co-author of *The Chin Hills,* Vol. 1 & 2. He was appointed first as Political Officer in 1890, and promoted to the latter post when Chinram was formerly annexed in 1896. B. S. Carey, however, left his job in 1898, 'when he came in conflict with the higher authorities' (Chatterjee 1990: 154–155). The conflict was over the division of Chinram into three separate administrative units; Carey strongly demanded the immediate appointment of a Chief Commissioner for the Chin-Lushai Country in accordance with the Chin Hills Regulation, and to cease the division of Chinram into three units according to the military administrative patterns. In contrast to B. S. Carey's suggestion, Sir Charles Crosthwaite proposed that instead of creating the post of Chief Commissioner for the Chin-Lushai country, it would be better to maintain the division so 'the Chin Hills can easily be assimilated into Burma' (Crosthwaite 1912: 29). The Lushai Hills could then be assimilated into Assam and the Chittagong into Bengal. Their main concern seemed to be the British lines of communication, as a British officer at Haka stated in his letter to the Chief Commissioner of Burma:

> The Tyo Var (Tio River), our boundary with the North Lushai Hills, as well as the Kaladyne River, our boundary with the South Lushai Hills, are for many months in the year impassable. Likewise, the Fort White charge is cut off from the rest of the hills by the Nankathe (Run or Manipur River) and even our main lines of communication with Burma can scarcely be used in the rainy season (cf. Chatterjee 1990: 155).

With the resignation of B. S. Carey, the military administrative pattern of three units remained virtually unchanged except for the transfer of South Lushai (present Lunglei District in Mizoram State of India) from the government of Bengal to the government of Assam in April 1898. Since South Lushai was transferred to Assam, it no longer formed a separate district for the Chin on the Bengal side, but was combined with the already existing Chittagong Hill Tracts District. In this way, Chinram was divided administratively, but not legally, into three different provinces of British India; the Chin Hills District of Burma, the Lushai Hills District of Assam and the Chittagong Hill Tracts of Bengal.

The Policy of 'Indirect Rule' and Its Effect

The first Superintendent of the Lushai Hills District (West Chinram) was Major J. Shakespear, author of *The Lushai Kuki Clan*, who was appointed in 1898. In B. S. Carey's place, Captain Dury was appointed as the Superintendent of the Chin Hills District. B. S. Carey, in fact, was the one who adopted the *Khua-bawi* system as the criteria for Chinram administration, but Shakespear implemented it officially. Soon after he came into power, Shakespear made land settlements of the entire districts and divided the whole area among the chiefs with well-defined boundaries for each of them. The chiefs – some of them were mere headmen, formerly appointed by *Ram-uk* – became the owners of the land for life. In 1901, the Circle Administration system was introduced in the Lushai Hills District, and applied also in the Chin Hills District in later years. Under the Circle Administration system, the Lushai Hills District was divided into eighteen circles, mainly based on sub-tribal or local-dialect groups, or what we call *Tual* community.[10]

Circle Inspectors were appointed for each circle, who usually supervised ten to twenty villages and worked in cooperation with the interpreters and chiefs. The interpreters, sometimes known as village writers, or '*Khawcchiar*', in Chin, were government employees but were absorbed in the traditional village administration under the chiefs.

According to the administration system known as 'indirect rule', the real function of the chiefs was to maintain law and order within their jurisdictions and to collect taxes on behalf of the British authority, though they enjoyed certain special privileges like feudal lords of the *Ram-uk*. The British authority, on the other hand, could effectively apply 'indirect rule' with minimum interference commensurate with the adequate protection of its economic interest, mainly because the Chin had traditionally a very reliable system of taxation. However, in contrast to the Chin tradition, the British imposed taxes based on individuals, not on households, which was traditionally unacceptable for the Chin. Thus, all the chiefs in Chinram collectively protested the new tax system, which caused many social problems for widows and elderly persons. As a result of this protest, the traditional system of taxation was reinstated in 1908.

According to the new administrative policy of 'indirect rule', the British recognized the existing system of chieftainship, especially the *Khua-bawi* system. As the British adopted this system as the criteria for the Chin Hills District administration, they abolished the Tlaisun Democratic Council and ignored the existence of the *Ram-uk* system until 1919. The main aim of the

Khua-bawi system was to accomplish the policy of 'indirect rule' and to maintain the *status quo*, i.e. wherever possible, village boundaries were recorded and treated as fixed.

The Chin had always taken village boundaries very seriously, because inter-village boundaries marked not just the territory of the chief but the sovereignty of their guardian god, *Khua-hrum*. Thus, the British policy of *status quo* strongly affected both their political system of tribal chieftainship, called *Ram-uk*, and the *Tlang Khua-hrum* worship. Since the British adopted the *Khua-bawi* system and fixed rigid boundary lines between the village; *Ram-uk Bawipa*, *Khua-bawi* and *Ram-uk Bawipa* appointed headmen (*Tlangpi* or *Upa*), all were turned into agents of the British government. The indirect rule policy was supposed to follow traditional political systems and customary law with minimum interference of the chief's powers, but the appointment of office bearers was always subject to approval by the British authorities. Thus the powers of the tribal chief, *Ram-uk Bawipa*, who usually ruled the whole tribe of his people, were reduced to a level of petty mediocrity. His political powers and social status became almost equal to his formerly appointed headmen. For the Chin, it was not just the reduction of *Ram-uk* powers but the humiliation of society itself which eventually led to such open rebellion as the Anglo–Chin War in 1917–19.

NOTES

1 The Burman King Bodawpaya had occupied Arakan in 1784, and King Bagyidaw had conquered Assam and Manipur in 1812.

2 After the British conquered Chinram in 1896, a part of Chinram was included into the Chittagong Hill Tracts District.

3 For a detailed account of the British occupation of Burma, see Crosthwaite 1968 and Ni Ni Myint 1983.

4 Raikes invited Cawn Bik not as a member of the Tlaisun Democratic Council, but as chief of Tlaisun. He seemed unaware of the fact that such a Democratic Council existed in Chinram.

5 Afterwards turned into the 2nd Burma Battalion

6 A detailed account of the Chin-Lushai Expedition is given in Carey and Tuck 1976, Chapter IV; Crosthwaite 1912, Chapter XXI; and A.S. Reid 1893.

7 This policy was reaffirmed by the Government of Assam in 1929. See Robert Reid 1941.

8 *Foreign Department Report on Chin-Lushai Hills* (September 1892), reprinted Aizawl: Tribal Research Institute, 1980, cited also in Lian Uk, ' Centennial Year of the Chin Hills Regulation' , in *Chin Journal*, (February, 1998), pp. 201–205.

9 *Foreign Department Report*, no. 32, p. 142.

10 Instead of using *Tual* for Circle Administration, the term *Peng* was used. The newly emerging Christian community did not like the term *Tual* because it was associated with their old religion, which they had already abandoned.

FIVE

The Coming of Christianity to East Chinram

British colonial power paved the way for the coming of Christianity to East Chinram. In this chapter, I shall explore the Chin encounter with Christianity and highlight how the Chin responded to the new religious challenge. In so doing, attention will be given to the formative process of the Chin Baptist Churches, from the coming of the pioneer American Baptist missionaries in 1899 to 1905, when the missionaries reaped the first fruit of their work in East Chinram.

I shall approach my study from a socio-cultural perspective, which views indigenous Christianity as an integral part of the socio-cultural history of the Chin people. Writing history from a socio-cultural perspective is very much about the history of mentality. It does not limit itself within the structural perspectives of social, economic and political history. While the aims of writing history from the socio-cultural perspective can be summarized as 'understanding of the past', the main task of the historian from this perspective is to discover 'mental and physical laws', which governed human action, rather than purely 'causal explanation', but without denying or excluding 'causal explanation' from an 'understanding of human behavior in its concrete manifestations' (Igger 1984: 43–45; see also Carr 1963: 164–65; Igger 1997: 51–61; Hunt 1989; Appleby, Hunt and Jacob 1994.)

The history of mentality, therefore, must be viewed as a history of consciousness, in which a social system is analysed – not only the ties between the social, economic and political structures but also the patterns of thought and behavior. The main purpose of the history of mentality therefore is to detect self-awareness and self-understanding, which indeed are the foundation for creating national identity. The history of mentality must be written from a local perspective, with the focus on the indigenous factors rather than purely foreign missionary factors. I shall, however, begin this chapter with a brief summary of early Burmese Church History and move on to developments in East Chinram in 1899–1905.

EARLY CATHOLIC VENTURES INTO BURMA

The official record of the Roman Catholic mission in Burma begins with the coming of two Jesuit missionaries, Pimenta and Boves, who were accompanied by the Portuguese adventurer and mercenary Philip de Brito y Nicote. According to Harvey, de Brito 'started his life as a cabin boy' (Harvey 1824: 185), then served many years as head of the Portuguese mercenaries employed by King Min Razagyi of Arakan. When Arakan conquered Syriam in 1599, the king appointed him 'to take charge of the custom house and control the Portuguese living there under their own laws' (Hall 1968: 375).

De Brito, however, quickly formed an ambitious plan to gain control over Syriam, expel the Arakanese governor and hold Syriam under the authority of the Viceroy of Goa, a Portuguese colony since the fifteenth century. At first, for de Brito, success followed success. He beat an Arakanese attack in league with the Toungoo Min, and in 1604 both rulers were forced to negotiate with de Brito. He also made a marriage alliance with the king of Mon, Binnya Dela, by which de Brito's son married a daughter of the king. In this way, de Brito became the unchallenged lord in Syriam and controlled most of lower Burma including Rangoon, present capital of Burma, which is located on the opposite side of the Rangoon River from Syriam.

Soon after he established a strong position in Syriam, de Brito, with the help of Jesuit priests, was 'pressing the mass conversion of Buddhists to Christianity, and plundered some Buddhist pagodas' (ibid.8: 376); 'he removed precious stones from the images, melted down the gold, beat it into leaf, and sold it' (Harvey 1924: 186). He even melted down the Dammazedi's great bell from the Shwedagon pagoda in Rangoon and other bronze bells of pagodas to save 'the expense of importing metal to make cannons' (Furnivall 1915: 53).

Like many other Portuguese adventurers and mercenaries of his time, de Brito seemed simply to have carried out the right of 'patronage' and brought the message of 'eternal redemption' to his newly formed semi-colony of Syriam in Burma. The so-called right of 'patronage' (*Patronato* in Spanish, *Padroado* in Portuguese) was granted by Pope Alexander VI in the Papal Bull *Inter Caetera Divinae* to the first two Catholic colonial powers, Spain and Portugal, in 1493. The bull divided the world outside of Europe between the kings of Spain and Portugal and granted them full authority over all the territories that had already been discovered as well as over those still to be discovered. It was based on the medieval Roman Catholic Church assumption that the Pope held supreme authority over the entire globe, including 'the pagan world', such as Burma and Chinram. Thus, de Brito assumed that he had the right to colonize and had full authority to convert the native Buddhists to Christianity, for 'colonialism and the mission were interdependent – the right to have colonies with the duty to Christianize the colonized' (Bosch 1993: 227).

De Brito and his Jesuit priests' work of Christianizing Burman Buddhists was said to have been quite successful. The most significant was the conversion of Nat Shin Naung, king of Taungoo and the 'most well known poet in the entire history of Burma' (Harvey 1924: 185). Indeed, the story of Nat Shin Naung, a young seventeenth-century Taungoo king, is a touching episode. He spent his young life as a soldier and poet. His poems, written away from the palace on the battlefield, were heartfelt rhymes to his loved one, Princess Da-Du-Kalya, seventeen years his senior. When he married her as king, she was a widow of forty-seven. This is considered by many to be the most touching love story in the history of Burma (ibid.: 186–189).

According to an Arakanese historian, Shwe Lu Maung, Nat Shin Naung embraced Christianity because he wanted to make friends with de Brito in order 'to revive the glory of Taungoo', the dynasty founded by King Ba Yin Naung and known as the 'Second Burman or Myanmar Empire' (Shwe Lu Maung 1989: 12). However, his daring act provoked the anger of the royal families all over the country and caused dissatisfaction among his people, for the conversion of Burman Buddhists to Christianity had always been viewed as an insult to Buddhism, the country and the people. Buddhism had made its way to Burma during the reign of King A'soka. According to Ceylonese chronicles, 'King A'soka sent a mission consisting of Theras Sona and Uttara to the "Land of Gold" (Suvannabhumi) to disseminate the Buddhist religion. This country can be identified as the country of the Mons' (cf. Bechert 1984: 147–158). Since the founding of the first Burman Kingdom of Pakan by King Annawrata in 1044, Buddhism had been adopted as the state religion, and the king was regarded not just as a ruler but as 'the defender of the faith' throughout his kingdom. Thus, Buddhism and Myanmar nationalism had been connected and thoroughly 'blended together and formed the same-whole', as the old Burmese saying put it: *Buddha-bata Myanmar Lu-myo* (to be a Myanmar is to be a Buddhist). For the *Buddha-bata Myanmar Lu-myo,* 'religion cannot be forsaken without giving up nationality' (Bigandet 1887: 11).

Taking the conversion of Nat Shin Naung as an insult to Buddhism, King Maha Dhamma Raja of Ava (also known as Anuakphet Lwun Min) marched to the south, and in 1613 crushed the Taungoo and destroyed Syriam. King Maha Dhamma Raja, an uncle of Nat Shin Naung who admired his poems and his bravery, asked him to return to Buddhism or face death. His nephew replied, 'I have done it with a blood oath; you may decide my fate as you wish' (cited in Shwe Lu Maung 1989: 13). According to Harvey, Nat Shin Naung said: 'Let me no longer be of his people; make me one of thy people (Portuguese)', and he received 'baptism from the white priest, de Brito standing sponsor' (Harvey 1967: 188;). Subsequently Nat Shin Naung and de Brito were crucified as heretics.

The conversion of Nat Shin Naung seems to have been the high noon of the Catholic mission in Burma. If the king had done well, then the whole country would have followed him and Burma might have become a Christian nation, for the Burman always regarded their king as 'the defender of the faith'. Unfortunately for the Roman Catholic mission in Burma, this was not to be the case. High noon was soon followed by the dark hours. Together with the crucifixion of Nat Shin Naung and de Brito, the Portuguese power in Burma was uprooted and its Catholic mission proved a failure. The remaining Christian followers were deported to a remote area between the Chindwin and the Mu Rivers in upper Burma. They and their descendents remained faithful to their religion, and in the 1780s Fr Sangermano saw them and reported that 'some 2,000 believers were in Burma' (Sangermano 1969: 19).

In 1720, one hundred years after the so-called de Brito episode, two Italian priests, Sigismund Calchi and Joseph Vittoni, were sent to Ava, then the capital of the Burman Kingdom. They were members of the company of five Barnabite fathers who accompanied the Patriarch of Alexandria, Monsigu Mazzabarba, to China. Pope Clement XI had sent Patriarch Mazzabarba as his ambassador to the Empire of China the previous year. When affairs there 'began to take a less favourable turn', Vittoni and Calchi were directed to establish a mission in Burma. The king of Ava, who seemed to have 'no memory of the de Brito episode, granted them permission to preach' (Trager 1966: 13). But when the Barnabite priests arrived on 'this unpromising scene, they had the additional handicap of being received badly by the resident Portuguese, who were undesired of competition' (ibid.). Perhaps this was the reason that the Barnabite priests had to move out of the capital, Ava, and re-establish themselves in Syriam, which at that time was under the control of the Mon king.

Calchi died after having served eight years in Burma and Vittoni carried on the missionary work alone. He made little progress and suffered many setbacks. Few local churches were founded although there was one in 1749 at Syriam 'which was built of brick with funds donated by a devoted and wealthy Armenian family' (ibid.). A boys' school and home for orphans, especially for girls, were established, but 'all were plundered and destroyed' when the Burman king Alaungpaya conquered Syriam and the whole kingdom of Mon in 1757. 'Members dispersed, supporters died out, and, subsequently, the foundation had literally to be laid anew' (Trager 1966.: 15).

After the Mon Kingdom was completely conquered by the Burman king Alaungpaya, the centuries-old rivalry and hundred-year war between Burman and Mon came to an end. The end of the war and subsequent political stability produced peace and tranquility in lower Burma. Political stability produced religious tolerance as well. When such an opportunity presented itself, Roman Catholic missionaries re-established themselves in Burma. They usually came from Italy, singly or in pairs, but never in large numbers. The most outstanding missionary among them was Father Sangermano, who came to Burma in 1783 and retired in 1806. He produced one of the earliest and best scholarly works on Burma *The Kingdom of the Burmese Empire,* published in 1893. He remarked that 'the Burmese government allowed' free exercise of their religion.

> But this tolerance arises more from political and religious motives ... Since the time that the Catholic missionaries have penetrated into these parts (lower Burma, formerly Mon country(there have indeed been some conversions but the number has not been so great as to excite the jealousy of the Talapoin or the government (Sangermano 1969: 111).

As mentioned already, for the Burman, conversion to another religion had always been a very sensitive matter not just for the community but even for the whole country. If a native Burman became a convert, he was called *Kala,* which means 'foreigner', as Paul Bigandit recorded:

> The few natives that became converts joined with the Christians .... They were called *Kalas* because in the opinion of the Burman, those who had embraced the religion of the *Kalas* had become *bona fide* strangers having lost their own nationality (Bigandit 1967: 4).

They not only called the native converts *Kalas,* the Roman Catholic Church itself was called, and is even today, *Bayingyi Bata,* practising the *Bayingyi* religion. The term *Bayingyi,* according to U Kaung, is derived from *feringhi,* a term corrupted from Persian which came to mean European and later to mean Indian-born Portuguese. As the term *Bayingyi* indicates, the Roman Catholic mission in the eighteenth century had few converts among the natives. The church members were mostly Portuguese descendents from previous centuries, other Europeans who had married native women, travelling merchants and some natives from the Malaban coast, which embraced Christianity.

THE PROTESTANT BACKGROUND IN BURMA

The London Missionary Society

In 1807, one year after the Roman Catholic missionary Father Sangermano left the country, the first Protestant missionaries came to Burma from British India. Their names were Marden and Charter, sent by the London Missionary Society. Felix Carey, son of the eminent English Baptist missionary to India, Dr William Carey, replaced Marden, who did not stay long, in 1808.

During the next four years, Charter and Carey occupied themselves chiefly with 'learning the Burmese language, translating extracts from the Old and New Testaments, and compiling a dictionary and grammar' (Maung Shwe Wa 1963: 2). Carey translated six chapters of the Gospel of

Matthew. The *Burma Baptist Chronicle* recorded that 'Felix Carey and Charter had proved anything but success', due to the fact that 'the Burmese people were not permitted to accept a new religion, and life in Rangoon was most precarious and difficult' (Maung Shwe Wa 1963: 2). Charter thus gave up his missionary work in Burma and went to Ceylon. According to Anderson, he gave up his missionary work because of the 'corruption of the government', whose king and local governors 'were absolute despots' (C. Anderson 1987: 141). The laws in Burma, he said, were 'the bloodiest on earth: the commonest punishments were beheading, crucifixion, and pouring melted lead in small quantities down the throat, and these were inflicted for such minor offenses as chewing opium or drinking spirits' (ibid.).

After Charter left, Felix Carey was the only missionary able to stay in Burma, probably because of his marriage to a Burmese woman of European extraction, and 'so enjoyed some small amount of official favor' (Trager 1966: 21). However, Felix Carey also had encountered many troubles during the 'dreadful event of the internal political state of affairs' in 1812. This event, according to William Carey, was caused by a Burmese commander who 'had ordered five hundred of his men to be buried alive merely because they were recruits sent to him by an officer whom he disliked' (C. Anderson 1987: 140). During that 'general massacre', Felix Carey and other Europeans had to take refuge for fifty days on a British frigate in Rangoon harbour. The governor of Rangoon demanded 'Felix, his wife, and the child of the captain, but the captain refused until he received a written guarantee that they would be protected'. William Carey wrote about this event to Dr Roger in Philadelphia: 'Dreadful ... confusion and hostility for a long time. My son was obliged to take refuge on board a ship for fifty days with his family' (cited in C. Anderson 1987: 143).

Despite such a dreadful state of internal political affairs, Felix Carey 'became greatly interested in the Burman' and he had good negotiations with King Bodawpaya in Ava, then the capital of the Burman Kingdom. Finally, he 'decided to give up his missionary career and enter the service of King Bodawpaya' (ibid.: 178). For this reason, the London Missionary Society's mission in Burma was handed over to the American Baptist mission in 1813.

The American Baptist Mission in Burma

The first American missionaries who came to Burma in 1813 were Ann and Adoniaram Judson, aged 24 and 25 respectively. They were the first Americans ever to leave their country as missionaries of the Protestant Church. Originally they were sent by the Congregationalist Church through the American Board of Commission for Foreign Mission, commonly known as the American Board on the conviction that distant Burma was 'composed of our Brethren, descended from the same common parents, involved in the consequences of the same fatal apostasy from God, and inhabiting the same world' (cited in W. Hutchinson 1987: 47). However, the Judsons and their colleague, Luther Rice, were converted to the Baptist religion on their voyage to the east and were baptized by Dr William Carey, a prominent English missionary in Serampore, India.

Unfortunately for the Judsons and their colleagues, the war, commonly known as the Second War of Independence or the War of 1812, was declared between Britain and America just the day after their arrival in Calcutta, on 17 June 1812. The British East India Company, which controlled the colonial authority in India, expelled the Judsons and all American missionaries on the grounds that 'they were suspected as American spies' (Maung Shwe Wa 1963: 2). Apart from the current war, the British East India Company had always been notorious for its policy on missionaries; it was, as Ann Judson recorded,

> hostile ... to missionary exertions and was 'opposed to every effort of evangelization of India ... determined by all means in their power to resist the introduction of Christianity among the native subjects of the British crown in Bengal' (cited in Maung Shwe Wa 1963: 2).

Indeed, the British East India Company which then enjoyed a monopoly of the British territory and trade, 'was hostile to the mission' (Latourette 1975: 1033), for the authorities of the company 'did not want the natives to learn Western ideas – particularly the subversive revolutionary ideas of the Gospel. They could lead to unrest, disturbances, even revolt' (C. Anderson 1987: 135). They thus expelled from the British colony of Calcutta even William Carey and his Baptist missionary colleagues in 1793, although they were British citizens by birth. William Carey and his colleagues, later known as the Serampore trio, finally settled in Serampore, a Danish colony. When the Judsons came to Calcutta in 1812, they found William Carey not just a Baptist missionary but also a professor of Oriental languages at the College of Fort William. Although the East India Company did not approve his mission, they recognized 'his attainments' and 'made him professor and gave him a good salary and a huge stone house'. He accepted the post 'in order to use his position to gain toleration for Serampore and his colleagues Marsham and Ward' (ibid.: 133).

William Carey, however, seemed unable to protect his fellow mission-aries from America when the deportation orders came from the British East India Company. The Judsons and their colleagues were ordered 'to return to America by the same ship on which they had come' (ibid.: 2). Finally, as an alternative, 'they were allowed to sail for Port Louis, four hundred miles away on the Isle of France [Mauritius], but they were unable to find there any true opportunity for missionary work' (ibid.: 3). Thus, the Judsons 'determined to go to Penang to settle', while 'Rice returned to America to seek aid for the fledgling American Baptist missions in the Far East' (Torbet 1987: 249).

In those days, there was no direct route from Port Louis to Penang across the Indian Ocean. Thus, the Judsons took their passage by way of India 'where they found that not a single ship was scheduled for Penang during the rainy season' (ibid.: 149). Still, they had no permit to remain in India, and the attitude of the East India Company was hardening. Their presence was promptly reported to the police, who forwarded an account of their arrival to the authorities. It would merely be a matter of time before an order came for their deportation. Thus, 'in their desperation they were willing to take passage on any ship heading in any direction rather than be forced to return to America. The only one they found due to sail was the schooner *Georgiana* bound for Rangoon'. When they reached Rangoon on 12 July 1813, the Judsons caught 'a breathtaking view through breaks in the rain of the great Shwedagon Pagoda' (Maung Shwe Wa 1963: 2). Ironically, the Shwedagon Pagoda was the symbol of Buddhism and its related civilization, which the Judsons had come to challenge in order to find converts.

The Burmans at that time, as Rev. Howard Malcom who visited the American Baptist mission in Burma in 1836 observed, were

> fully entitled to be called a civilized people. A regular government, a written language, an established literature, a settled abode, foreign commerce, respectable architecture, good roads and bridges, competent manufactures, adequate dress, gradations of rank, and the condition of women, conspire to establish their claim to be considered. Their exact place in the scale of civilization is not so easily settled. In intellect, morals, manners, and several of the points just named, they are certainly superior to any natives of this peninsula (Malcom 1839: 270; Cited also in Trager 1966: 115).

Moreover, as mentioned already, the people were molded by Buddhism and linked to life by institutions shaped by Buddhism. Buddhism provided, as Trager observes,

> a complete philosophical system and way of life. It had its scholars and learned books both in classical languages, Pali and Sanskrit, and in the major national languages, Burmese, Mon, and Shan. It provided an ever-renewing core of trained 'monks' (*pongyis*) who, as members of the 'clergy' were not only the custodians of this body of knowledge but also the active schoolteachers of the people. There were schools or *kyoungs* almost everywhere and travelers to Burma grudgingly or approvingly echoed the observation of Francis Buchanan, M.D., associate of Michael Symes, on the latter's first embassy to Ava, that 'there are but few of the Burman who do not read and write' (Trager 1966: 82–83).

In a nutshell, there was 'no religious vacuum' in the country where the Judsons had come to challenge people to become Christian converts. But being a good missionary, Dr Judson tried to see doctrinal insufficiency in Buddhism and wrote:

> Let those who plead the native innocence and purity of heathen nations visit Burma. The system of religion here has no power over the hearth or restraint on the passion. Though it forbids on pain of many years' suffering in hell, theft and falsehood, yet, I presume to say, there is not a single Burman in this country, who, if he had a good opportunity, without danger of detection, would hesitate to do either. Though the religion inculcates benevolence, tenderness, forgiveness of injuries, and love of enemies, though it forbids sensuality, love of pleasure, and attachment to worldly objects, yet it is destitute of the power to produce the former or subdue the latter in its votaries. In short, the Burman system of religion is like an alabaster image, perfect and beautiful in all its parts, but destitute of life. Besides being destitute of life, it provides no atonement for sin. Here also the Gospel triumphs over this and every other religion in the world (cited in Wayland 1853: 152).

He, however, seemed unable to see the doctrine of merit or *Kuhthu* in Buddhism, which is equivalent to the Christian doctrine of atonement for sin. As Trager points out, 'each Buddhist individual may wipe out his sins by acts of merit' (Trager 1966: 83). Thus, no Buddhist would abandon his religion because of any serious doctrinal insufficiency in Buddhism. When Ann Judson spoke to the women, they said, 'Your religion is good for you, ours for us. You will be rewarded for your good deeds in your ways – we in our ways' (C. Anderson 1987: 183).

Judson finally realized that here was no religious vacuum, thus people would not abandon their religion for the sake of the new foreign religion of Christianity. Moreover, conversion to Christianity would also mean forsaking their nationality, which no ordinary Burman would do without serious consideration. U Naw, the first convert among the Burman, was a case in point. Judson wrote:

> The first Burman I baptized had his attention drawn to Christianity by a tract that fell into his hands, and he kept reading and thinking to himself for two years before he called on me, and we had to converse and discuss every point of the Christian system and every doctrine of Buddhism a whole year before he could fully accept ... salvation; there has been no more faltering. He knows in whom he believes (cited in Mason 1860: 111).

It is true that the people attracted by early Christian teaching were what Judson called 'semi-atheists' and were no longer practising Buddhists, but believed in a diffuse sort of wisdom which took for them the place of God.

As the *Burma Baptist Chronicle* records, 'about half of the members of the early church' came out of this background.

> They were highly intelligent, very thoughtful, and most faithful disciples. The very fact that they had been skeptics showed their independence of character, mental alertness, and fearlessness. Though they often took a year or more to examine every phase of Christian thought, yet once they made up their minds, they became very staunch disciples. From such people the early church was formed (Maung Shwe Wa 1963: 40).

In 1822, the American Baptist mission in Burma received an opportunity to establish a church in Ava. The king of Ava, Bagyidawpaya, had heard of the skill of Dr Price, the first medical missionary to Burma, and had invited him and Judson to the capital. When the opportunity presented itself, Judson quickly took it. He even changed his missionary strategy altogether. Instead of pursuing stubborn individuals to convert, he now targeted the royal family first. The strategy was that if the king, the defender of the faith and head of the whole structure of society, converted, then the whole society would follow him and the country would become a Christian nation.

Judson's ambitious strategy seemed very well received at first. He explained to the king that to become a Christian did not necessarily mean forsaking his nationality, and that 'the Burmese Christians will retain the custom of the country and respect that national usage which was common to high and low, rich and poor, unless there were some special reason to change' (Wayland 1853: 387). The king asked what Judson had to say of Gautama Buddha. Judson, who spoke the highly respected Burmese royal court language which made a great impression on the king, replied that everyone knew that he was the son of King Thog-dau-dah-nah, that he regarded him as a wise man and a teacher, but did not call him God.

> 'That is right', exclaimed Atwinwun Maung Khah, Interior Minister, who proceeded to repeat what Judson had previously told him. Even Maung Zah, Minister of State, began to take the side of God before His Majesty, saying, 'Nearly all the world, Your Majesty, believes in an eternal God, all except Burma and Siam, these little spots!' His Majesty remained silent, and after some other desultory inquiries, he abruptly rose and departed (ibid.: 250).

The Royal family treated Judson very well. Atwinwun Maung Khah and Maung Tsoo, the Royal Secretary, were 'the most friendly of all the officials towards the missionaries', and Prince Mekara and his wife, who was the king's own sister, were 'deeply interested in discussing religion' (Maung Shwe Wa 1963: 44).

Unfortunately, all those opportunities and good relationships were ruined by the First Anglo–Burmese War. When the war broke out in 1824, Judson and all foreigners were arrested and put in jail until the end of the war. When the peace treaty was made between the British and Burman, Judson was assigned as translator at the peace talks. As a result, the Treaty of Yandabo was signed on 24 February 1826 and the Burman had to surrender Arakan, Tenasserim, Assam and Manipur. Judson was rewarded and allowed to work freely within the territory of the newly occupied British colony by Sir Archibald Campbell. He and his missionary colleagues thus established themselves again in Amherst, Tenasserim in 1827. The next year they moved to Moulmein, the present capital of the Mon State in Burma. He was still full of optimism for the future of the mission in Burma.

His optimism came mainly from his own understanding of the relationship of religion and state in Buddhist society. As hinted already, the royal patronage of the religious institution, particularly the monastic order, was a very important part of Buddhist society in Burma. Thus, Judson and his missionary colleagues hoped that the elimination of royal patronage in the newly occupied British colony would help to make individual conversion more easy. Buddhism, however, proved its power of resistance, as Howard Malcolm reported in 1836 (his book *Travel in Southeastern Asia* was published in 1839):

> In the British provinces, the national faith, being robbed of the support of the secular arms, seems to be cherished so much the more by national feeling. Expectancy that the religion of the new rulers may spread seems to awake greater vigilance than it may not. Pagodas, *kyoungs* (i.e. monasteries) and priests are well supported, and the clergy seem anxious to propitiate popular favour, to stand them in stead of government patronage (Malcolm 1839: 321).

Even after the British occupied lower Burma (including Rangoon) in 1852, and upper Burma (including Mandalay – the last capital of the Burman Kingdom) and dethroned King Thibaw Min in 1885, centuries-old identification with Buddhism was not to be denied. The people followed their old saying, *Buddha Bata Myanmar Lu-myo* (to be a Myanmar is to be a Buddhist), and prevented Buddhism from falling into decay as Judson and the Christian missionaries had hoped. Despite the fact that Judson produced some of the finest literary works – such as the translation of the Holy Bible, and a Burmese–English Dictionary which still is the most widely used in Burma – the American Baptist mission seems not to have been very successful among the Buddhist Burman as far as conversion is concerned.

Although most Burmans did not become Christians, three major ethnic groups within the present borders of Burma, the Chin, Kachin and Karen, did become predominantly Christian as a result of the work of Adoniaram Judson and his fellow American Baptist missionaries.[1]

A Survey of World Mission in 1961 paid a tribute to Judson as follows:

> Adoniaram Judson is surpassed by no missionary since the Apostle Paul in self devotion and scholarship, in labors and perils, in saintliness and humility, in the results of his toils on the future of an empire and its multitudes (cited by Caldwell 1961: 56–57).

THE AMERICAN BAPTIST CHIN MISSION

American Baptist Mission in a British Colony

The nature of the relationship between the colonial power and Christian mission in Chinram was quite different from that in India and Burma. It was due to the fact that during the course of a century or so, British colonial policy towards missionary work changed dramatically, as that policy was redirected from trading and commercial objectives to so-called 'benevolent colonialism'.

When the British East India Company was established in 1757, its sole interest was commerce and profit, and therefore it strictly maintained a policy of non-interference in the religious, cultural and social traditions of the Indians they ruled. But when the East India Company was brought under the direct control of the British government in 1783, some Parliament members, most notably William Wilberforce, the leader of the Clapham Sect, openly criticized the policy of non-interference, especially the policy that prohibited the introduction of Christianity into India. The Clapham Sect was formed by a group of MPs, bankers and lawyers who were the most ardent evangelicals and who used their wealth and influence to promote the Evangelical Revival and its causes. Moreover, they all took a leading part in the Antislavery Movement and missionary cause. As a result of their campaign, the British formally abolished the slave trade in 1807.

In 1813, a new charter for the East India Company, known as the 'Pious Clauses' was adopted by the British Parliament as a result of a long campaign launched by Wilberforce and his Clapham Sect. According to the Pious Clause the company would be required to adopt 'such measures ... for the interest and happiness of the native inhabitants of the British dominions in India', which would promote 'their advancement in useful knowledge and religious and moral improvement' (Christensen 1977: 335–342). This was the beginning of what came to be known as 'benevolent colonialism' which meant that 'the colonial power consciously took responsibility for the welfare of the inhabitants of its colonies'. It also meant that they had now taken on responsibility for sending out and maintaining schoolteachers (mostly missionaries!) to India and subsequently to other colonies. Thereafter, the missionaries were allowed to operate more or less freely anywhere in any British colony (Bosch 1991: 307).

By the time the British occupied Chinram, the Antislavery Movement had demonstrated its far-reaching consequences for the British missionary movement. Part of the 'benevolent colonialism' concept encompassed the principle that 'Britain as a Christian nation was bound to view its overseas possessions as a sacred trust' (Christensen 1977: 314). There was 'a growing consciousness among the colonial officials of the value and significance mission work had for the empire', as David Bosch has observed, 'particularly during the Victorian era' (Bosch 1991: 304). The spirit of this particular era can be seen also in a letter written by John Philip, Superintendent of the London Missionary Society at the Cape of Good Hope:

> While our missionaries ... are everywhere scattering the seeds of civilization, social order, happiness, they are, by the most unexceptionable means, extending British interests, British influence, and the British empire. Wherever the missionary places his standard among a savage tribe, their prejudices against the colonial government give way.
>
> Missionary stations are the most efficient agents which can be employed to promote the internal strength of our colonies, and the cheapest and best military posts that the wise government can employ to defend its frontier against the predatory incursion of savage tribes (cited in Bosch 1991: 305).

Chinram was occupied during the so-called 'high imperial era' (1880–1920) and Christian missionaries came in what Lattourette called 'the great century of the Christian mission' (i.e. 1814–1914). During that period, colonial expansion once again – as in the fifteenth to seventeenth centuries – acquired religious overtones and was also intimately linked with the mission. As James Scherer explains, 'since the time of Constantine, seldom have imperial policy and the interest of God's kingdom – Caesar and Christ – been in such close and comfortable convergence' (Scherer 1987: 14). Indeed, 'the missionaries were ideal allies' for the colonial authorities:

> They (the missionaries) lived among the local people, knew their languages, understood their customs. Who was better equipped than these missionaries to persuade the unwilling 'native' to submit to the *pax Britannica* or the *Teutonica*? (Bosch 1991: 303).

This was a time when colonial authorities enthusiastically welcomed missionaries into their territories. They found that the missionaries were useful in 'pacifying the hill tribes [including the Chin people] and providing education at a minimal cost to itself' (Downs 1992a: 30). Other contributions they made, for what was collectively known as 'social progress'; were explained by Eugene Smith as follows:

> The missionary movement made a prime contribution to the abolition of slavery, spread better methods of agriculture, established

> and maintained unnumbered schools, gave medical care to millions, elevated the status of women, created bonds between peoples of different countries which war could not sever, and trained a significant segment of the leadership of nations now newly independent (cited in Bosch 1991: 294).

It is therefore not surprising that during the peak of the 'high imperial era' the British colonial officers like B.S. Carey and Captain Dury, Superintendent of Chin Hills District, invited the American Baptist missionary couple Rev. Arthur Carson and Laura Carson to the newly occupied colony called Chinram in 1899.

The immediate concern for B.S. Carey and Captain Dury seems to be what they called 'the pacification' of the Chins. Even after the promulgation of the 'Chin Hills Regulation' in 1896, the Chin war of resistance against the British never died out but lasted until 1917–1919, when the Anglo–Chin War led to the defeat of the Chin, as I shall discuss in due course. The Chin, as an independent people, strongly opposed any foreign interference not only because of their political sovereignty but also because of their traditional religious beliefs and ritual practices.

Traditional Chin religion never allowed the presence of outsiders and foreigners at any kind of sacrificial ceremony to their *Khua-hrum*. Since their religion appeared to be the centre of their culture, the whole structure of Chin society was based on extreme exclusivism and isolationism. They believed that the presence of foreigners in their land would endanger their religion and society because tribal society could only be maintained through the traditional instruments of integration and fundamental isolation from other societies. To defend isolationism for them was to defend the fundamental doctrine of their religion and the foundation of their society on which they built their lives. They therefore defended isolationism as their own life and resisted any change in their social structures. After careful study of the Chin, the British realized that

> [i]f most of the Chin became Christians there would be a fairly good chance that they might welcome all other changes and the British considered it was wise to allow missionaries among the Chin (Thang Za Vung 1975: 3).

Thus, they invited the missionaries to come and work in their newly occupied colony, Chinram.

The Characteristic of the American Baptist Mission in Chinram

While the colonial power viewed the Christian mission as the best ally of *pax Britannica,* missionaries found the 'colonial power useful in underwriting, in many cases, their educational work and providing security for both themselves and their converts' (Downs 1992a: 30). Thus, it is appropriate to see how the American Protestant missions in general and the Baptist mission in particular viewed colonial power during the 'high imperial era'.

During this period (1880–1920), some of the old imperial powers such as Great Britain, France, Russia, Germany, and Belgium continued to enlarge their empires impressively, but others, like the Dutch, Spanish, and Portuguese stood still or found their colonies reduced. The United States was not involved in the scramble for colonies; the mission, however, provided Americans with an important 'moral equivalent' for imperialism (cf. W. Hutchinson 1987: 167–177).

The ideology of the American Protestant mission during the 'high imperial era' was the 'enthusiasm for the spread of Christian civilization'. The word 'civilization' in this case was defined by and as American Christendom. The concept of spreading Christian civilization was the 'most traditional and specific way of evangelizing the world', and theologically it was interpreted as 'the right of conquest of the world not only for Christ but for Christian civilization' (W. Hutchinson 1987: 168). Since it could justify 'the right of conquest of the world for Christian civilization' theologically, colonialism was morally correct as long as the gun was accompanied by the Bible, or 'gun and Bible go hand in hand' (Bosch 1991: 304). Thus, Robert Johnson sees the conquest of Chinram by the British colonial power as 'the divine plan of God':

> Who knows? Perhaps in the divine plan of God it was good for there to have been British rule for a time in the land of the Chins, to make the rough places smooth for the coming of the saving Gospel of Christ.

And he continues:

> I believe, without question, that the British rule was one of the best things that ever happened to the Chin people. The British opened the schools and hospitals and clinics, suppressed tribal fighting, made roads safe for travel, built mule paths where before only foot trails existed, built bridges, introduced mail and telegraph services, and made it possible for the mission to operate. They tried to introduce new crops and industries, though with limited success. The rise of the Chins out of animism coincided with the advent of the Westerners.
>
> The British courts became places where justice could be found for both high and low. For the chiefs and headmen it often meant being brought a bit low, but for the common people it meant rising up. Yes, the British rule was for the Chin a distinct blessing (Johnson 1988: 422).

This expansion of Christian civilization as the missionary motive also was closely linked with the 'assumption of the superiority of Western culture over all other cultures, but also [with] the conviction that God, in his providence, had chosen the Western nations because of their unique qualities to be the standard-bearers of his cause even to the uttermost ends of the world' (Bosch 1991: 298). This conviction was commonly referred to as 'manifest destiny' during the 'high imperial era'. It is therefore not surprising that when the first American Baptist missionaries came to Chinram in 1899, they 'looked down upon or abhorred the people they went out to serve'. Johnson, himself a former American Baptist missionary in Chinram, writes:

> As children of their age, they [the missionaries] did indeed feel that Western civilization was superior and that the civilization of the Chin was degraded and inferior. Certainly they felt keenly that the religion of the Chin was wrong and that the Christian faith could give not only eternal life to believers but also could raise the whole level of everyday life in education, in health, in family life, in morals, in governance, and even in agriculture and commerce. In short, the 'abundant life' which Jesus promises His followers would be received by the Chins if they believed and accepted the Gospel. No one can understand the motivation of all seven American couples who served the Chin as missionaries without understanding the power of this vision they had for what Laura Carson called 'the elevating, uplifting influence of the Gospel of Christ' (Johnson 1988: 48).

While the missionaries did many good things for the Chins, they also condemned traditional Chin religion and culture as 'pagan, heathen, demonic, evil', etc., and the traditional Chin way of life as 'sinful nature', and a 'life of permanent unrest and unhappiness'. While they called the Chin a 'wild tribe' and 'savage people', they quickly chose to title them-selves *Bawipa* and *Bawinu,* the titles for Chin chiefs and their paramount wives, to show that they were the new lords of the land like the colonial officers.

The main characteristic of the American Baptist mission in Chinram was the 'all-conquering approach' of the mission: conquering their political institution, changing their social structure, transforming their worldview and converting their religious beliefs and ritual systems. Chin scholar Za Tuah Ngur reappraised the 'all-conquering approach to missions' in her Master's thesis in 1993 and claimed that it was 'not necessarily culturally destructive' (1993: 6).

The Swedish–American Enterprise

Five out of seven of the early missionary families of the American Baptist Chin Mission – from Carson to Johnson – who served in East Chinram from 1899 to 1966 were American Baptists of Swedish descendents. It is therefore interesting to see how their Baptist background in Sweden related to the new enterprise they launched in East Chinram.

The Baptist churches in Sweden were founded by F. O. Nilsson in 1848. He was a sailor and converted to Christ when he visited, together with a fellow seaman named G. W. Schröder, the Baptist Mariner's Church in New York City in 1843. F. O. Nilsson did not become a Baptist at once, but returned to Sweden as a missionary to his fellow sailors. After a few years, he accepted the Baptist views and was baptized by immersion by Oncken, the well-known German Baptist pioneer, in Hamburg in 1847. In 1848 the Baptist church of Hamburg sent A. F. Forster, a Danish minister, to Sweden, and on 21 September he baptized Nilsson's wife, two of his bothers, and two other believers at a place called Vallersvik, near Gothenburg. That same night, Nilsson and the five who had received baptism by immersion were organized into the first Baptist church in Sweden (cf. Torbet 1987: 175–179; Johnson 1988: 609–610). Among them was August Swan Nilsson (who later changed his family name from Nilsson to Nelson when he applied for American citizenship). He was one of F. O. Nilsson's brothers and the grandfather of F. O. Nelson, who became a missionary to East Chinram.

In those days in Sweden, no religions denominations except the established Lutheran Church of Sweden were permitted to exist. Nilsson and his group were therefore severely persecuted. On New Year's Day 1850, Nilsson was arrested and thrown into prison for seven days for conducting illegal church services. He was hauled into court at Jönköping in February to answer charges of disseminating heretical doctrines contrary to 'the true evangelical faith'. Nilsson's clear statements of his faith in Christ and his dignified mien in court won him considerable sympathy; but under the existing law the court condemned him to perpetual banishment from his native country. He went into exile in Copenhagen in July 1851, where he served as pastor of a Baptist church until 1853. At that time he accepted the invitation of a band of twenty-four persecuted Swedish Baptists to lead them to the United States. He remained in America until 1860, when the King of Sweden annulled his banishment (cf. Johnson 1988: 176). During his stay in America, F. O. Nilsson and his followers established several Swedish Baptist churches, one of them being the First Swedish Baptist Church in Chicago, which sent Dr East, another Swedish–American Baptist, and his wife to Chinram as medical missionaries. Another was the First Swedish Baptist Church in St. Louis, which funded the first missionary hospital in Haka, East Chinram. I shall present biographical notes of other Swedish-American missionaries who served in Chimram in due course.

THE MISSIONARY FACTORS IN EAST CHINRAM

The Pioneers

The first regularly appointed missionaries to the Chin by the American Baptist Mission Society were Arthur and Laura Carson, a young couple from Columbus City, Iowa. The Carsons established and built two pioneer stations among the Chin in their twenty-one years of service together, the first in Thayetmyo for the Asho tribe and the second in Haka for the Laimi and Zomi tribes.

Before her marriage, Laura Hardin was appointed by the Women's Baptist Foreign Missionaries Society of the West in 1883 and was designated to work among the Karen at Bassien, Lower Burma in the Irrawaddy Delta, where she had charge of the Sgaw Karen mission high school. Meanwhile, Arthur Carson hurried to finish college and seminary work at Shurtleff College, Upper Alton, Illinois. In June 1886, he graduated from the Theological Department with a Bachelor of Divinity degree, 'having meantime carried his college work almost through the junior year' (A. Carson 1908: B–111). As H. H. Tilby wrote, 'He could have finished his college work and graduated with a B.A. degree the following year' (Tilby 1908: 1–3). However, he decided to come at once to Burma and join Laura, his fiancée. They had been classmates at Gibbon Seminary and she had preceded him to Burma by three years. They got married in December 1886, at the Sgaw Karen Mission Church in Bassien.

It was more than a year after their marriage before a location was decided upon and readied for establishing the new work. In the meantime, the Carsons, in the absence of other missionaries, looked after the Henzada Karen work and Prome Burman work. Finally, Thayetmyo was selected for the first station in the Chin mission and from 1888 to 1896 'Mr. and Mrs. Carson slowly but steadily and solidly built up a magnificent work and got ready an adequate plan for its conduct' (ibid.). The Thayetmyo station was established for the work among the Asho tribe of the Chin, also known as the 'Plain Chin', for they livid in the plains area of the Irrawaddy Delta and Chindwin Valley, stretching through the Minbu, Pandaung, Pakokko, Thayetmyo and Henzada Districts west of the Irrawaddy to the Prome District on the east, and the upper and lower Chindwin Valley. As mentioned already, the Asho were one of the Chin tribes who never moved over to the Hills even after the Shan conquered the original homeland of the Chin, the Chindwin Valley, in the thirteenth century.

As soon as the station was chosen, the Carsons 'cleared the ground, built a mission house, dug a well, erected the first schoolhouse for the Chins in the Irrawaddy Valley, learned the language,[2] and within a year laid a broad foundation for a prosperous mission and established a small but effective and self-supporting school from which converts were constantly won' (A. Carson 1908: B–111).[3] In eight years of service among the Asho tribe, there were 'several hundred baptisms, the school grew from four Chin boarders to eighty-six boarders besides a number of Burman and other day pupils, and every Chin that had been in the school as long as six months had been converted' (A. Carson 1908: B-111).[4] A number of young men were given such training as was possible and were doing excellent work as evangelists, while others were given better training in the Burman Seminary and in the

Baptist College, also known as Judson College, for more efficient service later on.

When Britain conquered Chinram and promulgated the Chin Hills Regulation in 1896, the British officer Captain Dury invited the Carsons to come to their newly occupied colony and work among the 'Hill Chins' in what is now the Chin State of Burma. Rev. Carson and the American Baptist Missionary Union viewed this as golden opportunity to evangelize the whole Chin race. Almost from the beginning of his work in Thayetmyo, Rev. Carson became intensely interested in the Hill Chins because of 'an ambition that quickly grew into a determined purpose to plant a mission in the Chin Hills, in the very midst of the untouched Chin masses of which those for whom he was working were but the Burmanized fringe' (A. Carson 1908: B–111). For the American missionaries of that time, there were only two categories of the Chin people: Plain Chins and Hill Chins. The 'Plain Chins' denoted the Asho tribe, and the 'Hill Chins' denoted the rest of the Chin tribes, particularly the Laimi and Zomi tribes.

In order to plan a new mission station at Haka, which had become the base of the new British administration, the Carsons left the Thayetmyo mission in 1899. After they left, the Chin mission among the Asho tribe at 'Thayetmyo ... suffered many setbacks'. Tilby described, however, that 'the good work so efficiently begun has never altogether ceased at any time and it has always taken on new life and given satisfactory results when receiving the care of good and capable missionaries' (Tilby 1908:1–3). Today, as a result of the good works of many capable missionaries who followed in the footsteps of the Carsons, the Asho Chin Baptist Convention has become one of the largest Baptist conventions in Burma. Beacuse I limit myself to the scope of the Christian movement in the present Chin State, however, I shall omit the continued development of the Asho Chin Baptist Convention in the present study.

The Opening of the Haka Mission Station

As they left Thayetmyo, the Carsons made their way to what they then called the Chin Hills and reached Haka on 15 March 1899. They were escorted by the British Indian sepoys, sent for their protection by Captain Dury, Superintendent of the Chin Hills, who not only invited them but arranged their journey. He also chose Haka as their missionary headquarters, where he himself had his headquarters. The Carsons were accompanied also by Karen preachers from Bassien, an Indian cook and his Burman wife and two Christian Chin from Thayetmyo.

As first missionaries at a new station, the first thing required, as Laura Carson explained, 'was to get a place of our own and to learn the language of our people' (L. Carson 1927: 166). Thus, soon after they arrived at Haka, they purchased thirty acres of 'the most beautiful land in the vicinity' from 'one big chief Lian Mo', and the cost was 'forty-five rupees' (ibid.). This land was retained as the Haka mission compound during the missionary era and was turned over to the Haka Baptist Association when the missionaries ultimately left in 1966. It is still the site of the Haka Baptist Church, the Haka Baptist Association headquarters and mission high school. The Burmese government nationalized the school in 1967.

Learning the language was a pressing necessity for pioneer missionaries, and there was no learning facility. Although the Lai dialect of the Chin language had been reduced to written form by the British officer Major Arthur Newland in 1893, that fact seemed unknown to the Carsons. Laura Carson therefore wrote: 'We had not one thing to help us, not even an alphabet, for the language was not yet reduced to writing' (ibid.). Moreover, they found it difficult to get a teacher, for the people were suspicious of them. They were seen as colonial agents who had come with the force of arms and established themselves as masters. Chin resistance against the foreign intruders was still fresh in people's minds, and anti-foreign sentiments were obvious. The Chin probably did not know at first the difference between English soldiers and American missionaries, and the Carsons were often mistaken for agents of the British working to capture young people as slaves. Laura Carson described this problem in her biography:

> They told us afterwards that they believed we were placed there by the government (British) to somehow work up some secret scheme to get hold of their children and send them to the British for slaves ... They could not believe it possible that we were there solely for their good (L. Carson 1927: 167).

The Carsons could not find a local teacher to teach them the language at first. It was an Indian called Mr Joseph who taught them the Lai dialect of the Chin language. Much later they got two Chin, Chia Khawm and Tum Kir, to become their teachers. In the 1901 Annual Report, after two years of learning the language, Carson admitted how little progress he had made:

> I wanted to preach but I couldn't make myself understood or make them accept my viewpoint. In fact, I do not know enough words in their dialect to express some delicate points in bringing them around to think in my way (A. Carson 1901: FM–182).

Another difficulty of learning the language were the many different Chin dialects. They had worked among the Asho tribe in Thayetmyo before, but regarding the Lai dialect in the Haka area 'we found it very different', as Carson noted, 'from that we had used in Thayetmyo ... hard for us to justify calling them dialects of the same language' (ibid.).

In order to establish the church firmly in East Chinram, they needed to learn the language and develop the new Chin script, invented by Major Arthur Newland and based on the Roman alphabet. Moreover, they wanted to translate the Gospel into the language of the people; Arthur Carson translated the Gospel of Matthew, many Sunday school lessons and a catechism into the Lai dialect. He also wrote the core of a Chin dictionary, which was completed by his wife Laura, ten years after his untimely death in 1908.

Soon after their arrival in Chinram, the pioneer missionaries also established schools, which they regarded as the most effective 'instruments of evangelism'. The Carsons had learnt the effectiveness of education from their own experiences at Thayetmyo. The good results they received from their work among the Asho tribe encouraged them. Thus, the first building they constructed in Haka was not their own residence but a schoolhouse, where the first teacher was the Karen preacher-teacher, San Win, who had accompanied the Carsons in 1899. Many more schools were opened in the years to come.

Unfortunately, the pioneer missionaries applied Burmese as a medium of instruction in the missionary schools, based on their experience at Thayetmyo. In Thayetmyo, Burmese was the *lingua franca* for both the Burman and Chin. Thus, Burmese as the medium of instruction at mission schools was more attractive for both the Chin and Burman pupils. But in the Chin heartland where Haka was located and the Chin were the only indigenous people, Burmese was simply a foreign language and, as such, not used. It was difficult to apply an experience from one place to another where the context was completely different. The use of Burmese as a medium of instruction at mission schools in Chinram caused many problems, not only for the growth of the church but also for the unity of the Chin people. I shall come back to the problem of standard language later.

The Beginning of the Medical Mission

In 1902, the first medical missionary, Dr East, arrived at Haka. The Carsons had requested the Home Board to send another missionary couple, particularly a medical missionary, for more effective mission work in Chinram. They felt quite correctly that 'a ministry to the pressing needs of the Chins for health and hygiene would break down barriers and commend the preaching of the word of God' (Johnson 1988: 68). Arthur Carson wrote to the Board and mentioned that 'every disease, and they are heir to them all, is assigned to the possession or influence of evil spirits, and sacrifice and feasting is the only remedy. We are sure that a medical missionary, beside the immense amount of suffering he could relieve, could unlock the heart of this simple people as no other could' (A. Carson 1899: FM–9).

In response to the urgent plea of the Carsons for a medical helper, the Home Board sent out a Swedish-American by the name of Erik Hjalmar Öster, who had changed his surname to East when he applied for U.S. citizenship. Born in Styrestad, four miles south of Norrköping in Östergötland on 2 February 1866, he graduated from the Southern Baptist Theological Seminary in Louisville, Kentucky in 1901 and received his

Doctor of Medicine from the Medical School of the University of Kentucky the same year. Soon after his graduation, he was appointed missionary of the American Baptist Missionary Union and sent to Chinram on 3 June 1901. He arrived in Haka on 21 March 1902. But just two months after his arrival he was forced to return to America for health reasons. He was 'taken with appendicitis and it became necessary to get him away as it would be perilous for him to remain where an operation would be impossible' (L. Carson 1927: 179). During his medical furlough, East married Miss Emily Johnson, who was born in Sundsvall, Sweden, on 28 January 1877. They were married at the First Swedish Baptist Church in Chicago on 5 August 1903. The Easts came back to Chinram as missionaries just two months after their wedding in November 1903.

Soon after their arrival in Haka, Dr East was running a hospital funded by the First Swedish Baptist Church in St. Louis and a gift of USD$ 1,500 from Dr. W. G. Tyzzer, a friend of Dr East, in memory of Dr. Tyzzer's wife, Emily, who had recently died. Thus, a twenty-bed mission hospital in Haka was named the Tyzzer Memorial Hospital and was completed in 1908. In his report for 1906, Dr East commented that he had treated 2,903 new patients and personally given 4,000 treatments (East 1907: FM–186). East was not just a good physician but a skilful carpenter as well. He and Carson built a home for the Easts in 1907, which is still used as the Haka Baptist Association headquarters. Moreover, he was 'large-hearted, efficient and sympathetic, and was rapidly winning a large place in the hearts of the people' (L. Carson 1927: 179).

Although Dr East was in charge of medical work as opposed to 'general work', he was keenly interested in the evangelization of the Chin. For that purpose he felt that it was important 'to tour the outlying villages'. When he did so, 'he also did a tremendous amount of medical practice in treating the numerous ailments of people who otherwise had absolutely no access to proper medical help' (Johnson 1988: 186). His letters constantly mentioned long hours of work treating cases in the villages. He also entered his experiences in his diary and later wrote his memoir in diary form, which was published only in 1984 by the Zomi Baptist Convention under the title *Burma Manuscript*.

During his years in Chinram, Dr East made several trips among the Laimi tribe in the Haka, Falam and Thlantlang areas, and even further north to the areas of Tiddim and Tawnzang where the Zomi tribe made their home. On 15 February 1904, he began a long tour to the Tiddim and Tawnzang areas, home of the most powerful chief among the Zomi tribe, Hau Cin Khup, who used to rule almost the whole Zomi tribe, except in the Sizang Valley. He evidently invited the missionary to come 'hoping to get a school established there' (ibid.: 194). Thus, Dr East arranged to establish two schools among the Zomi tribe, one in Tawnzang, the home of Hau Cin Khup, and another in Khuasak in the Sizang Valley. Two Karen preacher-teachers, Po Ko and Shwe Zan, were appointed as teachers in the respective villages.

East soon reaped good results from the Khuasak School. Through Shwe Zan's preaching, Pau Suan and his wife, Kham Ciang, Thuam Hang and his wife, Dim Khaw Cing, were converted and baptized by East on 11 May 1905 while the Carsons were in the United States on furlough. Thuam Hang was a 'remarkable man, heir to a tribal chieftainship', and they were 'the first fruits of our work in the Chin Hills' (L. Carson 1927: 104).

NOTES

1 On the American Baptist Kachin Mission, see Crider 1963: 367–382, and Tegenfeldt 1974. On the American Baptist work among the Karen, see Sowards and U Zan 1963: 302–326, and Hovermyr 1989.

2 Laura Carson learned Burmese and Karen when she was in Henzada. For her, only the Asho dialect of Chin was the new language she needed to learn. Arthur Carson, however, needed to learn both Chin and Burmese. Burmese was needed for them because it was the *lingua franca* in Thayetmyo where Burmese and the Asho tribe of Chin live side by side.

3 A sketch written in May 1908 by Miss F.A. Griffin of the Headquarters of the American Baptist Mission Society and sent to the Rev. G.G. Johnson, Buffalo, N.Y. and the Rev. Charles B. Stephens, Kearney, Neb. to be used at the memorial services of Rev. Arthur Carson, p. 3.

4 H. H. Tilby obituary article concerning Rev. A. Carson in *The Burma News* (Vol.XXI, No. 5), May 1908, pp.1–4; Carson also mentioned in his letter of 1.7.1892, 'all who have come to our school have been converted – and very few who have not', quoted by Johnson 1988: 15.

SIX

The Chin Response to the Christian Mission

Early conversions to Christianity among the Chin people were usually a product of a long-term process of change, which started before the turn of the twentieth century. When change was thrust upon the Chin by force, many types of social disintegration, disorientation and crisis occurred in Chin society because anything that affected one aspect of society affected its whole structure. Thus, political oppression of Chin society affected their socio-economic life and religious life as well. As long as the people remained unable to adjust to change or repel the change, there were many crises in life and the disintegration of society. These kinds of 'cumulative events and processes', as Lewis Rambo explains, 'are often crucial to conversion' (Rambo 1993: 46).

EARLY RESISTANCE AND THE FIRST CHIN CONVERTS

Occupation and Economic Crisis

The British occupation made deep inroads into the traditional social structure and caused many socio-political crises in Chin society. Life was not easy after the British launched their first annexation of Chinram in 1871, which was not completed until the final occupation in 1896. For the people whose economic life depended mostly on shifting agriculture, called *lothlawh,* which is the 'slash and burn' method of cultivation, every step required a perfect timetable. One had to start to fell trees in new sections called *lopil* in November so that they would be ready to burn in March or April. Planting crops of all kinds had to be finished before the end of May when the monsoon rain came. If one step of the cultivation procedures was missed, the result would be a poor harvest or perhaps even famine. During the wars it was impossible to follow such rigid seasonal patterns of cultivation. People were running and hiding for their lives in the jungle for many years without proper cultivation of the land. Thus, while many lost their lives in the battlefields, many more, especially children and women, perished because of famine and subsequent epidemics. It was the famine that forced the Chin to surrender and accept the Chin Hills Regulation in 1896.

Another factor which contributed to economic recession was the British policy of the abolishment of slavery. The British prohibited slave raids and abolished the whole system of slavery in Chinram. Carey and Tuck recorded that from 1892 to 1896, they released at least 5,000 recently captured Shan and Burman slaves from the Sizang, Kamhau, Tlaisun, Zokhua and Haka areas alone, not to mention the slaves released from the western Chinram of Mizoram (Lushai).[1]

Slavery had been a major economic institution for the rulers of peasant society in Chinram in the nineteenth century. When the British abolished their most reliable means of production, namely the institution of slavery, the central pillar of the tribal chiefs' economic power was broken as a result. The abolishment of slavery was the biggest assault ever on Chin society, especially the powers of the chiefs, *Ram-uk Bawipa,* which in turn affected every institution in life: political, cultural, economic, social and religious.

When the Chin lost the war and the British unilaterally promulgated the Chin Hills Regulation in 1896, the power of the traditional Chin chiefs was reduced to the level of mere headmen. The resulting political crisis affected the religious life as well. Since the chiefs' powers were reduced, they lost their hereditary right of taxation from their domain villages. Such taxes included the religious tax called *Khuaman,* which was used for the community sacrificial ceremony of *Tlang Khua-hrum* worship. When the chiefs lost their religious powers, it became very difficult to maintain traditional ways of life according to religious teaching and to continue costly sacrificial ceremonies which thus stopped in many *Tual* communities.

These crises affected all social strata in Chinram. Many chiefs viewed the political and religious crises as the result of their own inability to sacrifice to the guardian god *Khua-hrum* on behalf of the community and the Chin believed that the *Khua-hrum* had withdrawn all blessings, especially good harvests. According to Chin traditional religious understanding, famines, disease and epidemics resulted from an inability to offer the proper sacrifices to *Khua-hrum.* They believed that a proper ceremony would create abundant harvests and end famine. Moreover, an abundant harvest would prevent all kinds of diseases by providing proper sacrificial appeasement to the evil spirits, *Khua-chia,* which they believed caused their illness. The dilemma was obvious: they could not avoid famine because they could not carry out a proper sacrificial ceremony to *Khua-hrum*; they could not recover from their illness because they were unable to sacrifice for their illness; and worst of all, they could not perform proper funeral rites for their loved ones or communicate with the souls of their loved ones in *Mithi-khua* through *Khua-vang,* because it required costly sacrifice as well.

Soon after the British occupation, Chin society was miserable and chaotic. People were hungry, yet they could not offer sacrifice for a good harvest. When people were sick and dying, they could not offer any sacrifice to appease the evil spirit *Khua-chia,* which they believed caused their sickness. There was no remedy for the sick and hungry when the entire community suffered such poverty and hardship. It was at this time, when the Chin faced such profound social and religious crises, that they encountered Christianity. In a situation like this, crisis in life and conversion to Christianity, as we will see below, were more or less linked with each other.

Crisis and Conversion: The Story of Thuam Hang

There is abundant evidence among the first converts that they had suffered deep, prolonged and multi-faceted crises – spiritual, intellectual, and social – before they accepted baptism. A wide range of psychological, theological, philosophical and sociological factors were involved in religious transformation and conversion among the Chin. To illustrate my point, I would like to relate the story of Thuam Hang, the first convert among the Zomi tribe and the first to be baptized by Dr East, in 1906.

The crisis in Thuam Hang's life was about the poor health of his two sons, which caused him a great loss of economic and social status. He was bankrupt both economically and emotionally before he decided to convert. His eldest son, who would become the chief, for Thuam Hang was himself 'a local chief' (L. Carson 1927: 180), suffered from a disease which was not just physical but ritually unclean. Two of his teeth 'stuck right out and pierced the upper lip, so that about one-half inch of his teeth protruded from the lip' (Johnson 1988: 100). The Chin believed that any person who had such teeth was not an ordinary man but an incarnation of a tiger and called them *kei-thlak.* Since the Chin viewed the tiger as an unclean animal, such a person became ritually unclean. They could not participate in any sacrificial ceremonies to *khuahrum* or share sacrificial meals, nor were they allowed to participate in any kind of ritual dancing and singing. They were simply the outcasts of society. Thuam Hang and his family suffered both a physical and a social disease. These were cured by the missionary doctor, E. H. East. Johnson wrote:

> Dr East gave the boy chloroform and removed the teeth. Most of the villagers gathered to see the operation. When East placed the extracted teeth into the hand of the amazed father, all the people exclaimed, '*Aizaleh!*' – a term of wonderment. 'We have never seen such a thing before', they declared. When the boy awoke, the grateful father came over to the doctor and stroked his arm with his right hand as a signal of thankfulness (Johnson 1988: 100).

Thuam Hang had another son 'who had tuberculosis of the spine', which consumed most of his wealth as well as his emotional and psychological life. Johnson thus wrote:

> In supplication to the evil spirits the father had sacrificed many animals and had given all that he had, but the boy was no better. The newly arrived Karen preacher, Shwe Zan, asked the chief to worship the God of heaven who alone could give life and health and could save both him and his son. So Thuam Hang met with the preacher for prayer and finally came to the conclusion that, if it cost him his life, he would try to save his son's life. His fear was, of course, that the evil spirits would slay him for violation of the customary religion. The people believed that if anyone touched the skulls of sacrificial animals, that person would surely fall down dead.
>
> Shwe Zan told Thuam Hang that the whole sacrificial system was really the work of the devil and that if he wanted the God of heaven to bless him and his family, they would have to break with all the works of darkness, and if they did, God would make the son well (ibid.).

Dr East described the moment of decision in these words:

> Tum Harm (Thuam Hang) spent some days in thinking and was with the teacher who taught him how to pray to the God who made heaven and earth. One day when his whole being was in agony for his boy, he filled a corner of his blanket with stones and sat down before the heads and skulls and began to talk to them. He said, 'So if I touch you, you will kill me. So if I touch you, you will kill me!' He was agitated and his whole body had beads of perspiration all over it when he, full of fear, sprang to his feet, and taking stones from his blanket struck every skull with a stone. That done, he sat down to die! When death did not come he said, 'You are a lie; I will worship the God of heaven' (East 1984: 73).

Shwe Zan had been with Thuam Hang and his family through the whole struggle and 'rejoiced with him in the decision to embrace Christianity. A beautiful sequel to this story is that the son made a recovery from the tubercular process in his spine and lived on for many years' (Johnson 1988: 101). The Chin people now view such a story as the 'mighty work of God' in their history.

In present Chin society, where the social meanings of Christianity are widely shared and deeply felt, the conversion stories of Thuam Hang and others have become not just individual experience and the memory of inner transformation of the self, but a collective memory of communal experiences and a shared history of the transformation of society through collectively reproduced historical narratives. When individual memories of a conversion experience and inner transformation of the self become conceptualized and emotionalized by the entire society as a collective memory, socially constructed narrative practices produce a new meaning in life and a new identity for the people.

CONVERSION AND PERSECUTION

The Problem

Collectively acknowledged historical narratives of conversion events fostered a new identity for the Chin. The essence of the narrative plot in these stories is usually the conflict between old and new, and the most observable phenomenon of the conflict between them is persecution. The persecution of the new converts among the Chins therefore was the common and sometimes climactic manifestation of the conflict between the old traditional Chin ways of life, *Lai Phung*, and new Christian ways of life, *Krifa Phung*.

Not only the new converts but Christians in general regarded persecution as 'part of the plan and providence of God', or 'an integral consequence of following Jesus Christ' (Cunningham 1997: 14). As Jesus had warned his followers to expect persecution (Matt. 10: 17), the missionaries did the same to the new Chin converts. W.H.C. Frend wrote that the Christians regarded persecution as 'a personal witness to the truth of Christ's claim to be Messiah and a token of the closest possible identification with their Lord' (Frend in Eliade 1987: 251). Thus, the story of the persecution of the early converts contributes not only to the narrative account of shared experiences of the past, but also to the discourse of identity, just as identity is the discourse of history.

Although the stories of persecution among the Chin are trivial compared to the stories and experiences of the early Christians in the Mediterranean area and other parts of the world, they are still important shared experiences of the past, because both previous and new generations of Chin Christians have had to conceptualize and emotionalize such dramatic versions of conversion and persecution stories, which are repeatedly told and retold. In other words, the narrative of conversion and persecution creates a shared past that enlivens and personifies the understanding of self and community. It is therefore necessary to analyse scientifically the nature of persecution in Chin society and see how the conversions and persecutions challenged the traditional Chin ways of life and broke down centuries-old social structures in Chinram.

The persecution of new converts was very much related to the protection of the traditional Chin ways of life and social structure by the chiefs. It was not religious intolerance, as seen in other parts of the world, but the protection of long-standing isolationism in Chinram. Tribal society can only be maintained through traditional instruments of integration and isolation from other societies.

In order to maintain traditional instruments of integration, they protected the fundamental isolationism in their society and resisted any kind of change in their social structure. Even after the Chin Hills Regulation was promulgated, the Chin did not actually surrendered but fought against British colonial power whenever the opportunity presented itself, in order to regain isolation and maintain their social set-up through traditional instruments of integration.

The Chin gradually came to realize that they would not be able to expel the British by force. They thus applied diplomatic tactics instead of might to deal with the British. Their simple but important demand from the cultural point of view was not to allow the British sepoys to camp inside the village gate, especially during their religious ceremonies, because *Khua-hrum* sacrifice did not allow the presence of any stranger during the ceremony. Since the British officers always avoided such cultural confrontation in order to maintain law and order, they usually camped outside the village gate and even established their permanent stations outside the villages. For instance, the Haka station was outside the original village, but later became the center of town. So did the Tiddim and Falam stations.

In contrast to the British officers, however, the Christian missionaries came to challenge the very foundation of traditional Chin religion and the core of their culture. They persuaded the people to abandon their traditional religion and way of life altogether. They simply tried to be present at the center of society and live among the people. Their presence, especially when they made a convert, was a big crisis for the Chin and their society. The converts flatly refused to attend the *Tual Khua-hrum* worship ceremony. This meant that they had rejected the authorities of both *Khua-hrum* and *Khua-bawi.* As a result, they were not allowed to cultivate land and/or were expelled from the village. The Christians would quickly inform the missionaries and the missionaries would in turn take the case up with the British authorities. When the case reached the colonial office, the Chin chief would already be on the losing side. To illustrate this point, I shall tell the story of Thang Tsin of the Laimi tribe and the first convert from Lumbang (present Falam Township).

The Case of Thang Tsin

Thang Tsin was converted and baptized by Rev. Carson at Haka on 5 September 1906. The news of his conversion reached home ahead of him and he was promptly called before the chief to give an account of himself. When Thang Tsin tried to proclaim that 'the foreign God' whom he worshipped was 'a wonderful God – a God who loves us and helps us and saves us from our sin', the chief angrily stopped him and said:

'You need not tell me anything about this foreign God, and you cannot worship Him and live in this village. You know that if you do, our own gods will be angry, our crops will fail, our cattle will not reproduce, our children will die and all kinds of trouble will come upon us. You have either got to renounce this foreign religion or be driven from the village' (cited in L. Carson 1927: 184).

Thang Tsin consulted Rev. Carson, and in accordance with Carson's advice he met the Superintendent, the highest colonial officer in the region. The Superintendent consoled him and said:

'Well, Thang Tsin, go back to your village and as long as you live in accordance with the teaching of the missionaries, doing nothing wrong to anger the people, the British government will stand behind you. The chief cannot drive you out of your village simply because you worship the Christian God', and an order to this effect was sent to the chief (ibid.: 184).

Although the Chin chief could not overrule the order of the British officer, he exercised the power he still held, which was to take back the land, *lopil,* from Thang Tsin. According to Chin traditional *phunglam,* all lands were owned by the chiefs and were parceled out to the families and passed down from father to son, and the chief collected a certain percentage of the crops as a tax, called *longa.* However, the chief could take back land and give it to another if one abused customary law. The chief of Lunmbang, therefore, called in Thang Tsin and told him that 'unless he abandoned the worship of the foreign God, the land which he cultivated and which had been in the family for generations would be taken from him' (ibid.: 184). However, the chief gave him an alternative. If he dared to do so, he could cultivate the place called *ram-huai,* the 'sacred field', which the Chin believed to be the place or abode of the evil spirit *Khua-chia.* They also believed that 'anyone planting a seed in that sacred ground would drop dead on the spot'. In those days, no Chin dared to cut grass or trees, let alone cultivate in the *ram-huai,* which they believed would destroy *Khua-chia*'s abode.

However, Thang Tsin gladly accepted the offer. When he was 'digging the sacred field, people stood about in open-mouthed wonder expecting every moment to see him die. But he didn't die'. At the end of the season he went over to the mission station in Haka and told about farming the sacred field. He said:

Why, my corn was more than twice as tall as theirs (the land was rich, never having been tilled) and my pumpkins were huge, while theirs were not bigger than my double fists. The people tell me secretly that my God is greater than theirs because He gives me so much better crops than they get (ibid.: 186).

The missionaries and the newly converted Chin Christians capitalized on this event by using such stories to show the mighty work of God for the believers. Even today, the story has repeatedly been told and retold in narrative accounts: writings, songs, sermons and speeches on such occasions as Sunday services, feasts, Christmas, New Year and *civui,* which people commonly celebrate to recall the past.

The Case of Tsong Kham

The most repeated story of persecution among the Chin is the story of Tsong Kham (Cawng Kham), the first convert from the Bualkhua, and the Ngawn group in present Falam Township. His story is recorded by Laura Carson as follows:

> After a time a man named Tsong Kham in a nearby village got up the courage to publicly confess his faith in God.
>
> His chief was greatly excited over the matter and sent for the big tribal chief to decide what should be done with him. He came and Tsong Kham was summoned before him. Knowing of Thang Tsin's case, he began in a very conciliatory way of saying, 'We don't want to have any trouble over this matter. We don't want to go up to the English government about it, but you know if you worship the foreign God, all kinds of calamity will come upon your village. Just drop this foreign religion and I will give you thirty rupees (about ten dollars).
>
> 'But you do not understand', said Tsong Kham. 'I believe the Christian religion is true. I am a Christian and shall always be one.' For some time the chief continued trying to bribe, offering more and more, but when he found it was useless he became angry and said 'I have never in my life humiliated myself as I have in dealing with you today. You will either take what I have offered and give up the worship of this foreign God or you will take the worst beating any man ever had.' 'Then I will take the beating,' Tsong Kham replied. 'Bare your back', the chief commanded. Tsong Kham threw off his blanket. 'Put your hands on your knees', was the next order, and down went his hands to his knees.
>
> Then the chief called up three brutal men armed with bamboo and told them to give him fifteen strokes each and to lay them on hard. The first man finished his fiendish task and the second began when Tsong Kham fell to the earth. Raising his hand, he said, 'Wait a moment', and lifting his eyes to heaven, he prayed. He asked for strength to bear the torture, then added (like Stephen of old), 'Count it not against them, Father, for they do not understand – they don't understand.'
>
> Then he said, resuming his position with his hands on his knees, 'Come on, I am ready now. There is one to beat me still.' Filled with

> superstitious awe, the old chief slunk away and no man dared to strike again.

During his ordeal his fellow convert Thang Tsin was with him to support him in his faith all the time. Laura Carson continues:

> Thang Tsin lifted the poor man up, helped him to his own home and washed his terribly bruised and bleeding back. For five days he was unable to walk. When he got better they decided the best thing for him to do was to go home and go quietly about his work and await developments. This he did, and going out in his field to work one morning, he was met by the delegation, and the spokesman said, 'What are you doing here? Unless you are going to renounce the foreign religion, this is no longer your farm.' 'Then this is no longer my farm', said Tsong Kham. He gathered a load of firewood and went home. As he threw down the wood in the yard, some more men came saying, 'You need not be bringing wood here. This is no longer your home. The chief has already given it to another family. He has taken your wife for a slave, and unless you give up worshipping the foreign God you will have to be his slave, too.' 'Then I will be his slave', said Tsong Kham, and he actually went and served him for several months. He was so obedient and faithful that he won the confidence of the chief.
>
> As the time for the Association meeting (*civui*) at Haka drew near, Tsong Kham went to the chief and said, 'I have tried to do my work well. I have never asked for a day's leave. The Christians are going to have a big meeting at Haka and I very much want to go. Will you give me permission?' Permission was given, and when at the meeting a list of names of people applying for baptism and church membership was read, the name of Tsong Kham was on the list. When the question was asked if anybody knew any reason why he should not be received, Thang Tsin arose and very graphically told the story that I have told here, closing with this remark, 'I think when a man has given up his farm, his home, his wife and his liberty for Christ's sake, he ought to be baptized and numbered with Christians.' We thought so too! He was received and baptized and, needless to say, is now one of the staunchest Christians in all our hills.
>
> His case was taken up to the government by the missionaries, and the chief who ordered him beaten was fined and Tsong Kham's property, liberty and wife were restored (ibid.: 186–189).

Although the early converts endured local mistrust and persecution and were even seen as betrayers, the stories of their conversions and persecutions now have become a prime topic for ordinary conversation and ceremonial enactment, for example at the jubilee *civui*, which are the remembrance feasts for the pioneer Christians in Chinram. Every Christian in Chinram, from Sunday school children to the elders of the church, regard the first converts, such as Cawng Kham, Thang Cin, Pau Suan, Thuam Hang and Chia Khaw, as historic figures and heroes. Through these stories, the Chin people do 'identity work' together, for such stories hold history and identity together through socially reconstructed narrative practices.

THE FORMATION OF A NEW ECCLESIASTICAL STRUCTURE

The Educational Mission and the 'Magic of Letters'

As mentioned above, Chief Hau Cin Khup invited Dr East to establish a school in his village. Dr East and other missionaries mentioned in their reports and letters that Zokhua, Sakta, Thlantlang, Tiddim, Tawnzang, Khuasak and many other villages had asked for schools to be established. Indeed, the villagers themselves built the school buildings and provided housing and food for the teachers during the pioneer era. Almost all the Chin chiefs wanted to establish schools and have education for their children but without any attachment to the Christian religion. The Seipi chief, for instance, pointedly said:

> School I want, but the Heavenly Jesus I don't want. If we do not sacrifice chickens and pigs we die. Therefore, I don't want Jesus custom (East 1901: FM–186).

Thus, before I analyse the policy of American Baptist Foreign Mission Society's educational mission in Chinram, it is necessary to highlight the philosophical and theological reasons why the Chin wanted schools and why they called school work 'the magic of letters'. As mentioned, the Chin did not know the art of writing when the missionaries came to their homeland. But careful study has shown that the Chin had always considered 'the magic of letters' to be their long-lost treasure.

Almost all the Chin tribes had a legend about a written language. Although many different versions exists, all such legends mention that 'the Chin once possessed a script' which some versions called 'the magic of letters', and others called 'the leather book'.[2] As already told, the Chin ancestor lost his written language or leather book, which was given to him by God. Since then the Chin felt that they had become an inferior race compared to those who had a written language. They longed for a written medium of communication, a common written language.

The American Baptist's policy of educational mission, however, did not always correspond directly to the Chin concept, 'the magic of letters'. The Chin wanted 'the magic of letters' in order to overcome their own collective inferiority complex and for the sake of the unity of the Chin nation as a whole. In other words, they wanted a common language, and they knew that only 'the magic of letters' could produce it. Unfortunately, the American Baptist mission's philosophy of education was not concerned with the unity of the Chin or the creation of a common language for them but only with the conversion of the people to Christianity. The mission schools were therefore established as the ways and means of spreading the Gospel, not for the development of a common Chin language. They also rejected so-called 'secular education'. Thus, Dr East bluntly refused the request by the Seipi's chief to establish a school:

> I frankly told (them) that we would not for a moment consider any school without religious teaching, and our prime objective was and is to spread the Gospel, and while doing so we are willing to educate the people also (East 1902: FM–186).

One of the main weaknesses of the mission's philosophy of education was their use of the foreign language, Burmese, as the medium of teaching. Dr Cope, who came to Chinram in 1908, supported this philosophy in the mission schools and even suggested that the translation of the Bible into Chin dialects 'would be unnecessary since under the British administration the Burmese influence would gradually spread throughout the Chin Hills so that by the time the Chin could read such a translation, most of the Chin would be using Burmese' (Cung Lain Hup 1993: 118). Even Rev. Carson, who during the latter part of his life favoured the creation of a common Chin language, once remarked that 'the Chin had no word for hell, heaven, sin, forgiveness, repent, pray, and no word that really meant love' (A. Carson 1900: FM-9; Johnson 1988: 70). Since the missionaries viewed the Chin language as inadequate for proclaiming the Gospel, they chose a foreign language which they thought more adequate.

But imposing the foreign language was not always successful. Instead, it caused several problems and cultural confrontations. Ironically, the first such cultural confrontation came not from the Chin themselves but from the Burman. A Burman police surgeon appointed by the British government at Haka was surprised to see the Chin pupils being required to learn the Burmese language under the instruction of a Karen teacher at the mission school, but without providing food to the pupils. Since the pupils were learning Burmese, he might have thought that the mission school was a Buddhist monastery and the pupils were Buddhist novices. In a monastery school, pupils or novices were provided food and clothes, i.e. religious robes. So he simply suggested that the food and clothes be provided wherever the Burmese language was taught. Such a suggestion somehow seemed to be good news for the Chin, but for the missionaries it was 'a lie'. As Johnson wrote, 'he spread a lie that if the boys would go to a Buddhist or government school, they would be given food and clothes free'. He continued:

> The parents were thus incited to approach Carson with the story that they could no longer feed their boys if they attended the mission

> school. If the boys could not get free food and clothing, they would have to stop and go to the fields to farm their own food. Carson told the parents that he had no money for feeding pupils (Johnson 1988: 70).

Thus, the parents pulled their boys out of schools. Carson reported succinctly in June 1901: 'We have no school' (A. Carson 1901: FM–182).

But the 'food controversy', as Johnson put it, was not the end of the story of the mission school nor the end of requiring the Burmese language to be taught in East Chinram. The policy of using Burmese as the medium of instruction in mission schools lasted at least until the Anglo–Chin War of 1917–1919.

Despite many difficulties and cultural clashes, the educational mission proved to be the most effective instrument for evangelization of the Chin people, and the mission schools became the centres for conversion to Christianity. Although they did not produce the standard language 'magic of letters' that the Chin had expected, they certainly helped to create a new society in Chinram which had its roots in the Gospel of Love. Thus, in the new Chin Christian society, various Chin tribes could identify each other as brothers and sisters even without sharing a common language.

The Formation of the Chin Hills Baptist Association

Although all the Chin people professed the same religious doctrine and belief system before they became Christians, their ritual practices, which centred upon the sacrificial worship of *Khua-hrum*, divided them into many groups of clans and tribes. When they gradually became Christians, however, the Chin were provided with the same religious doctrine and belief system and also with a single ecclesiastical structure where all the tribes and clans of the Chin could join as members of a single community of believers. A most significant thing provided for the Chin people during the pioneer missionary era was the formation of the 'Chin Hills Baptist Association' (CHBA) at Haka in March 1907. Officially, the CHBA was formed by the election of its executive committee during a meeting of the representatives of the young churches and individual Christians, most of them new converts who attended the meeting in order to receive baptism. 'From a worldly standpoint', as Johnson observes, 'it was not a very important matter – just thirty-four people gathered to effect a little organization to bind scattered believers' in Chinram. But 'this was the tiny seed that later grew into today's Chin Baptist Convention, one of the largest church groups in Burma' (Johnson 1988: 145).

In 'Baptist polity', Johnson explains, 'an association brings together Baptist churches, not individuals, into an organization for fellowship. It is not an authority over churches, but provides a forum for the discussion of issues, for encouragement in the faith and acts as a body to advise churches on ordination' (ibid.). At that time in Chinram, however, there were very few organized churches and only a group of believers scattered in six or seven villages. The significance of the formation of the CHBA was that it was a 'binding medium' to these new and often weak Christians. Since the CHBA functioned as a binding medium throughout East Chinram, it could hold and unite all the Chin Christians who came from different tribes and clans within a single community of faith. It meant that, unlike their old religious institution of *Khua-hrum*, the CHBA could accept people from different tribes and clans as believers of the same faith within a single ecclesiastical structure. In joining the CHBA, the Chin people were encouraged to seek their self-awareness and national identity of Chin-ness through membership in a new Christian community. To put it another way, when they became Christians and joined the CHBA, the meanings and values of tribal and clan affiliation based on their traditional religion of *Khua-hrum* worship, which had been so effective in the Chin's resistance against the British, lost their strength. Thus, the national identity of Chin-ness gradually surpassed the meaning and effectiveness of clan and tribal identity.

From the moment of its formation, the CHBA challenged the notion of tribal identity, and 'racism begins to break down', as Johnson put it. As mentioned, no Chin would allow the presence of a stranger from another tribe at any kind of traditional religious ceremony. But the Christians in Haka now welcomed strangers from other tribes – especially the Zomi tribe from the Sizang and Kamhau groups in the Tiddim area – to their new religious ceremonies as their own brothers and sisters in Christ. This fact was very significant in Chin history; Johnson observes:

> The Hakas were used to calling the Sizang and Kamhau by the appellation 'Thaute', a derogatory term, and could not understand how Christians could accept these Thautes as brothers. The superstition that the Tiddim area people possessed the power of the 'evil eye' was still strong, and so the Hakas tended to shun them (ibid.).

Recognizing the effect of this event, the missionaries and Chin Christians did not only hold an Association Meeting as a recurring annual religious ceremony and festival; they also routinely shifted it from one place to another. In this way, the Association Meeting, also known in Chin as *Civui*, eventually became a major religious activity for the Chin Christians and the best known religious festival in Chinram. At the *Civui* festival, different tribes and clans of Chins who formerly had never shared a religious ceremony could now freely share their inner feelings and experiences in one faith, and also the same ritual practices, such as attending the same worship service, singing from the same hymn book, reading the same Bible, etc. And the *Civui* itself therefore constituted an identity-shaping source which provided for the individual's epistemological characteristics, i.e. his beliefs and basic values. It also catered to interpersonal relationships and brought about a sense of total solidarity between the individual and his fellow Christians who, in turn, shared his own ethnic nationality. Thus, the formation of the CHBA gradually provided the Chin people with a single ecclesiastical structure and its related religious festival called *Civui*; both constituted sources for the creation of their national identity in a new society, Christian Chinram. These new religious institutions transformed Chin society gradually from a 'tribal and clan-oriented' society to a 'community of faith' in Jesus Christ.

The Death of Arthur Carson

In East Chinram, one of the most important but unfortunate events was the death of Rev. Arthur Carson at the Haka mission station on 1 April 1908. He died of appendicitis at the age of forty-seven after serving among the Chin for more than twenty-one years, including his services among the Asho tribe in Thayetmyo. In his last report, written in February 1908, he mentioned the prospects for his mission and the new converts among the mission's students as follows:

> The last four, completing the fifty, were baptized about three weeks ago, two months before the end of the year, and there are others waiting. Many in the schools are only withheld from a public profession by opposition of their families. Furthermore four men from the tribes of the hills have been raised up to tell the gospel story to their people, who hear them with confidence and gladness.

He summarized the final year of his work:

> One long trip and four short ones, seventy days in all, constituted the whole of our touring. Besides the two months of training class, a lexicon of the language in the Roman characters carried almost to completion, a book of forty-two hymns ready for the printer, the translation each week of the Sunday school lesson, besides regular routine work that so often takes up all the time of a missionary, gave us all we could compass. The extent of territory over which we have to travel is about as large as all New England. This has to be done on foot or horseback, and I usually do it in the most of primitive fashion.

He closed his final report with these words:

> With courage unabated, faith strengthened and hope renewed for the harvests of the coming years (A. Carson 1908: FM–182).

At his memorial service, Rev. Arthur Carson was hailed as follows:

> Mr. Carson was a missionary pathfinder, a worthy exemplar of that pioneer spirit that has made American history great. He had the mark of all men of real power, the spirit which cannot be content merely to do an appointed work in an appointed way but that must attempt better methods and broader tasks. It was wisdom on the part of those in authority that designated the young man on his appointment in 1886 to open a new work, he being the first missionary sent out to labor solely among the Chins (A. Carson 1908: B–111).

ENLARGING THE MISSION IN THE FIRST PART OF SECOND ERA (1908–1919)

Arthur Carson's death ushered difficult new era, which brought much confusion and uncertainty to the work of the American Baptist mission in Chinram. The first part of the second missionary era – from 1908 to the end of the Anglo–Chin War in 1919 – can be called an 'era of confusion': the era when the ABCM suffered many setbacks and when they even considered closing the Haka mission station where the foundation of the Chin mission was laid down by the pioneer missionaries. However, during this period, the American Baptist Chin Mission was expanded to the northern part of the present Chin State in Burma.

Opening a New Mission Station at Tiddim

Dr East made many long tours to the northern part of the Chin State where the Zomi tribe of Chin lived. He was greatly impressed by the fact that the Zomi tribe, particularly the Sizang group, was 'more willing to open their hearts and accept the new way of life' (East 1904: FM–186), more willing to attend school and become Christians. His observation was correct. The first converts of the American Baptist Chin mission came from the Sizang group of the Zomi tribe, and the first mission school established after the 'food controversy' at Haka was at Khusak. Geographically, East found that Tiddim, the principal town among the Zomi tribe, was closer to Kalemyo, the plain area of Upper Chindwin. It was closer to Burma, and he therefore believed that

> [i]t was a mistake to have opened the mission station at Haka. He thought that Tidddim would have been a better site. This view was expressed gently while Arthur Carson lived, but after his death East became much more blunt in saying that Tiddim would have been a better choice and that the American Baptist Missionary Union ought to open a second station at Tiddim and take advantage of the northern Chin openness to change and conversion (Johnson 1988: 239).

In response to East's request, the 'missionary authorities agreed that another missionary couple was needed', and Rev. Joseph Herbert Cope and his wife Elizabeth were appointed for the work among the Zomi tribe and to open a new station at Tiddim.

J. Herbert Cope was born on 21 November 1882 in Germantown, Philadelphia, Pennsylvania. His mother, Elizabeth Danforth, was a daughter of missionaries to India: Rev. and Mrs Appleton Howe Danforth. His grandmother, born Frances Amelia Studley, returned to Assam after Rev. Danforth's death and married Rev. Miles Brownson, D.D., a pioneer and well known missionary in northeastern India. J. Herbert Cope entered the University of Pennsylvania in 1900. Four years later he received a Bachelor of Arts degree. After one year of service as missionary to the Mormons in Utah, he entered Rochester Theological Seminary and graduated in 1908 with a Bachelor of Divinity degree. He was appointed by the American Baptist Foreign Missionary Society[3] as missionary to the Chin in that same year. The Copes arrived in Haka on 12 December 1908, a few months after Rev. Arthur Carson had been buried.

When the Copes arrived they were asked to remain in Haka. This decision was made, according to Johnson, because of 'the grim financial situation of the Union' (Johnson 1988: 240). According to materials in the American Baptist Archive, however, this decision was made because of Laura Carson's disagreement with the idea of opening the new station at that time. She and her late husband Arthur Carson had a long-term plan for the Chin mission, that 'all the missionaries should stay in Haka', the centre of Chinram, so that the proper evangelistic tour could be made 'into every direction of the Hills in open season' (L. Carson 1908: FM-182). Their idea was that the missionaries would make evangelistic tours during the so-called open season, from the beginning of November to the end of May; during the rainy season all missionaries should stay in Haka and concentrate on teaching at the mission school, which was the most effective means of making converts. In this way, pupils from all parts of Chinram would come to Haka and learn the same dialect. She wrote: 'If they knew the common language the Gospel would be easier to spread' among the various tribes of the Chin. Thus, 'they need to learn only one dialect, not two' (ibid.). Her target seemed to be not just opening a new station and school but the creation of a common language for the Chin as their fellow Swedish-American Baptist missionary Olof Hanson had done for the Kachin people in northern Burma. She wrote in that same letter: 'There is no need to hurry to open a new station in Tiddim'. She believed that they needed to concentrate firmly on the already established missionary station at Haka.

Dr East on the other hand, was impatient and disappointed by this decision and wrote to the Home Board that 'if a station could not be opened at Tiddim, Haka could get along with a general missionary [Rev. Cope] and a medical missionary [himself] thus freeing Mrs. Carson to serve elsewhere in Burma' (East 1909: FM–186). 'Fortunately', comments Johnson on this letter, 'his idea did not prevail' (Johnson 1988: 240). Unfortunately for the Chin, however, such personal disputes and disagreements between the missionaries jeopardized the chance of establishing a common language. The struggle for a common language remains the most pressing need of the Chin.

While in Haka, Rev. Cope learned both the Lai and Kamhau dialects, the most commonly used among the Laimi and Zomi tribes respectively. Later, he also learned the Laizo dialect, one of the local dialects among the Laimi. In fact, he was the one who transcribed the Kamhau and Laizo dialects into written form. Finally, in 1911, the Tiddim station was opened and the Copes moved there and served continuously until 1938 when J. Herbert Cope died and his widow returned to the United States.

Issues over the Medical Mission

In 1910, the American Baptist Chin Mission suffered another major setback in its medical mission due to Dr East's poor health. He suffered a heart attack, which forced him to leave Chinram sooner than he had planned. On his last Sunday in Haka on 2 October 1910, he baptized two persons, and he commented on that day:

> So God allowed the baptismal water to be stirred once more. The day has been perfect, and so my work ends here, as I feel within me that my workday is passed in these Hills. I thank God that my best years and my best health have been given for the salvation of the heathen (East 1984: 18).

Because of his illness, such a highly respected and valuable medical missionary as Dr East and his family (three of their children were born in Haka) had to end their missionary careers and return to America. Dr East survived from the heart disease but was not allowed back to Haka. He passed away on 3 August 1939 at the age of 73.

In order to replace Dr. East at Haka mission hospital, the ABFMS appointed Dr John Gustav Woodin and his wife Bassie L. Woodin as medical missionaries to the Chin in that same year, 1910. John Woodin was born on 27 July 1883 in Iola, Kansas, the son of Swedish-American parents. He received his Bachelor of Arts degree in 1905 and M.D. in 1908 from the University of Kansas, but he had no seminary training. His wife, Bassie, was born on 20 February 1883 in Mendota, Illinois. She attended the University of Kansas for two years but did not continue for a degree.

They reached Haka just before Christmas in 1910. Bassie Woodin recorded in her diary their first day in Haka:

> Dec 23, 1910. Our first day in Haka. Mrs. Carson took us to the school, hospital, and our home, etc. The boys seem perfectly at home

> in her home. She is a mother to all and they love her. Their welcome is so genuine. The Burman told us of his escape from fire. Mrs. Carson is giving the Christians their Christmas feast (cited in Johnson 1988: 320).

As soon as their arrival at Haka, Dr Woodin started his work at the Emily Tyzzer Memorial Hospital and Bassie Woodin, like other missionaries' wives, started teaching at the Haka mission school. Dr Woodin found that, 'The Emily Tyzzer Memorial Hospital on the mission compound is fully operational, with beds, furniture, medicine, and instruments left by Dr. East' (J. Woodin 1911: FM–327).

The Woodins left Chinram after only five years. The intriguing thing about the Woodin years in Haka is not what they did for the Chins but why they left Chinram so soon. There are two possible reasons: either John Woodin's negative attitude toward the medical mission in Chinram, or Mrs Woodin's poor health, perhaps the decisive factor.

Unlike Rev. Carson and Dr East, Dr Woodin could not see the advantage of a medical mission in Chinram. In 1914, he wrote to the Home Board that 'medical work' is not necessary. And he described the state of the medical work as follows:

> I want to urge strongly to the Board that this is not a field for medical work. I feel sure that medicine is more of a harm than help in the work. There is no prejudice against missionaries in evangelistic work ... The greater part of the medicine we give is for intestinal parasites, itch, and the like and is merely thrown away, as they will take no precautions and are sure to get reinfected in a very short time ... The hospital is well equipped for the work here but is almost unused. It is unusual to have even one patient, and he will stay only a few days. A Chin is prejudiced against a man who claims to be a medicine chief and can't cure all his chronic troubles with a pill or by rubbing something on two or three times. I can't see any excuse for putting more money into medical work here. At least I hope no medical man will be sent out without knowing the conditions (J. Woodin 1914: FM–327).

This letter contains many interesting facts such as why the Chins were unwilling to come to the hospital and why they failed to stay long enough to get proper treatment. This is not just human behaviour in general but a cultural issue.

When Woodin's letter reached the United States, it caused a 'hot' debate on the question of discontinuation of the medical mission in Chinram. The retired missionary, Dr East took a strong stand on the issue in his letter to Dr Haggard of the Mission Society in Boston. He wrote:

> I have heard that Dr. Woodin has recommended the discontinuation of the medical work in Haka. This recommendation is positive proof

> that he is a failure and his recommendation should not be taken seriously. The medical work as carried on by us was a means in God's hands for the opening of the work in the Chin Hills. By relieving their suffering they were willing to listen to us and we gained their confidence and everywhere they received me gladly after they knew my mission. It was the medical work that opened the Tiddim field and work in the Falam District, but this work must go hand in hand with the evangelistic work, and so the medical missionary should also be an evangelist.
>
> People object to remaining in hospitals in America and the same feeling dominates the Chins; we have to train the people here for operations and hospital care and so we must do in the Chin Hills. It is a matter of training and growth, and the man that thinks the people will come in large numbers without getting their confidence through patient and constant training makes a blunder. I had an average of 10,000 patients each year in my touring and in the hospital and I say that is not a failure. And I further insist upon that it was money well spent and brought a greater return than any other money expended. Mr. Carson was not a fool and it was he who asked for the medical work, and I am still to prove from my own experience that he was right. The young men you have up there now did not see the travail of our souls in the beginning of the work. They stepped into the work after the ice was broken and when the harvest had already begun and a sentiment in our favor had permeated the masses, and so they are not ripe to sit in judgement on our work ...
>
> The medical mission should not be discontinued but carried out in a practical manner and in the Spirit of Christ who loved us and gave Himself for us (East 1914: FM–186).

The second reason was Mrs Woodin's poor health. John Woodin's many letters confirmed that Mrs Woodin had suffered from headaches and dizzy spells ever since they came to Haka. On one occasion he wrote:

> We went down [to Rangoon] to the Judson meetings, partly to see if she would be relieved while lower down. She was quite well the month spent below, except for a go with ptomaine. As soon as we reached the hills, her heart went bad; I put her on a stimulant and we got along fairly well. One day she had to lie by the road till we could get the effect of a stimulant [to begin]. She had a sick spell that kept her in bed two weeks; at times I was very anxious as to how it would come out. I thought then that as soon as she got stronger I would get her to a lower place, but she doesn't want to leave me alone (J. Woodin 1914: FM–327).

The Woodins applied for a transfer to a lower place and left for Bhamo the following year. Strangely enough, Rev. Cope supported Dr Woodin's idea of the discontinuation of the medical mission in Haka. This affected the decision of the Mission board. Thus, it was very unfortunate for the Chin people that Dr Woodin became the second and last medical missionary sent by the American Foreign Mission Society in the entire history of the American Baptist mission in Chinram.

The Medical Mission and the Secularization of the Chin Worldviews

Although the medical mission had a short life in Chinram, from 1902 to 1915, it made a great impact on Chin society and had far-reaching consequences in the changing of the Chin traditional worldview.

The Chin sacrificial system can be categorized into the sacrificial ceremonies related to *Khua-hrum,* which are a communion between the guardian god *Khua-hrum* and man, and the sacrificial ceremonies for illness to appease the evil spirit, *Khua-chia.* However, it is difficult to differentiate between the religious purpose and the medical purpose of sacrifice since religious and secular aspects of life were closely intertwined. Thus, from the traditional Chin religious point of view, it was impossible to see Christianity and Western medicine differently. It meant that one could not accept Western medicine without forsaking one's own traditional way of life and religion.

I have mentioned in a previous section that Dr East treated a tremendous amount of various illnesses during his tenure as missionary doctor in Chinram. But Dr Woodin correctly pointed out that such treatment was only 'for intestinal parasites, itch, and the like' (J. Woodin 1914: FM–327). He was right, because the Chin would accept treatment for diseases, which they believed were caused not by the *Khua-chia* but by other factors. Had their illnesses been caused by the *Khua-chia,* they would not have accepted any treatment from an outsider. They would have had to perform a sacrificial ceremony in order to appease the *Khua-chia,* and this was not just a matter of health care but the basic doctrine of their religion, which they could not easily forsake without forsaking the roots of their religion.

John Woodin once complained that although the mission hospital in Haka was well equipped, the Chin 'didn't like to stay in the hospital', and 'it is unusual to have even one patient, and he will stay [just] a few days' (ibid.). Missionary doctors never asked why the Chin did not like to stay in the hospital long enough to recover from their illness. For the Chin, their main concern was not just recovery from illness but preparation for life in *Mithi-khua.* The Chin wanted to recover at home or die at home where one sacrificed to one's own guardian god, called *Zing,* the family guardian god, called *Chung-um,* and the *Khua-hrum.* A man's *Khua-hrum* knew him in illness and even in death. His *Zing* would take care of him as long as he was alive, and if he died his family *Khua-hrum* would accompany him and inform all his ancestors' souls who died before him. In the Chin context, a matter of life and death was a matter for the home, not something that should happen outside the house or in a hospital. No Chin who practised traditional religion would leave home during such a critical period as a serious illness. That is the reason why the Chin did not like to stay at the hospital for any prolonged period of time.

The theological concept behind such religious understanding and practice was based on the traditional Chin religious doctrine of eschatology, which I discussed in Chapter 2. According to traditional Chin religious thought, there are two places for the human soul in the life after death, namely *Pulthi-khua* and *Sarthi-khua*. *Pulthi-khua* is a place for those who died of natural causes at home. The *Sarthi-khua* was a place for the souls of people who had died an accidental death or outside the village or the compound of their own house. The only way of redeeming the soul from *Sarthi-khua*, was in the case of violence. If the deceased died as a victim of war or by the hand of somebody else, and if his relatives could avenge his death, then his soul would be redeemed from *Sarthi-khua* and could be transferred to *Pulthi-khua*. However, there is no way to redeem a soul who died outside his house because of carelessness. If somebody died such a careless death, he would not be buried in the family grave inside the compound of the house. So no Chin wanted to die outside his house, and no families and relatives would ever let their loved one die outside the gate of his house or in the hospital.

To illustrate my point, I quote Bassie Woodin's letter about Lian Zam, a young chief from Haka who was brought into the mission hospital for pneumonia, a common illness which, according to Chin understanding, was caused by the *Khua-chia*. He was brought into the hospital not by his own will or the will of his relatives but by order of Mr Fisher, the local British administrator. Mrs Woodin wrote:

> John had an awful time to get him here. He fixed a stretcher but they wouldn't use it because it is Chin custom to carry only dead people on a stretcher. So they carried the young man on their backs as they do their babies which was a very bad thing for his lungs. He was rather heavy and the climb was very steep, so they changed off and they kept making excuses to stop – to fix his hair, etc, etc.
>
> Finally, they arrived. When Mrs. Carson and I heard of it, we went to the hospital. Well, no sooner had the people gotten him here than they began begging to go back because it wouldn't be Chin custom for him to die out of his house. If he did, his spirit would wander. John told them that if they took him he would die before they reached the village. He was very weak and exhausted. When we told them over and over that he might live if left in the hospital and would die if they went home, they wanted to know if they might bring some of the meat and sacrifice here. We said yes. Poor people, they believe just as much in their religion as we do ours. But they didn't bring meat. I think they made up their minds they would take him, whether or not. They came up about 2 o'clock. We managed to keep him until 4:30 or 4:45. His heart was getting a little stronger. It was

> one of the hottest days we've had, and with the room crowded with Chins it was not comfortable! John finally managed to get all but his mother, brother, and sisters out, which left only twelve. Then amidst the weeping and wailing and begging, John managed to give the hypodermic, which strengthened his heart. Chins lined each side of the bed and constantly talked to him, held his hands, feet, etc. John kept him pretty well until just a little while before they left.
>
> When he told them he was a little better, they seemed to go stark crazy – the whole outfit. They probably thought that was the chance to get him home before he died; it never seemed to occur to them that he might get well. Because we couldn't tell them with absolute certainty that he would not die, nothing else would do but take him home.
>
> The young chief, Lian Zam, died on the way home as we thought he would. I do not know whether he can be buried in the village or not. Being a chief, they might get him in, but usually people who die outside have to be buried outside.
>
> Protherto, the Superintendent, is still here; he says that the brother who carried him away from the hospital and who is the eldest cannot be chief. It will be a punishment, but I do hope Prothero will stick to it. He appoints the chiefs. It will be better for us if the government upholds us and the fellow will be told it is because he shortened his brother's life that he cannot be chief (Woodin 1911: FM–327).

Mrs Woodin's letter vividly shows that the traditional Chin concept of life and death was not only the fundamental doctrine of their religion but the cornerstone of their customs and at the heart of the people's behaviour in times of sadness and happiness, it manifested itself in the way they expressed love and sympathy. If the missionaries had shown a better understanding of the traditional concept of life and death, one could expect the church growth and social changes to have taken place more quickly than with only the social aspects of the mission. But without the social aspect of the mission, such as hospitals and schools, evangelism alone could not properly relieve human suffering. Evangelization needed to target the concepts of life, death and salvation, and the missions needed to emphasize the social aspects of human suffering, liberty, needing each other, etc.

The American Baptist Chin Mission applied what used to be called the 'holistic approach of mission', emphasizing both evangelization and the social development of the people to whom they proclaimed the Gospel. The pioneer missionaries were aware of the Chin context and thus they very much emphasized the medical and educational mission. During the eras of Dr Woodin and Rev. Cope, however, the emphasis was shifted too much towards the educational mission where results could quickly be seen, while completely neglecting the medical mission, which the Chin were reluctant to accept anyway. And when neglecting the medical mission, they neglected the whole concept of traditional Chin religion and its related social structure and customs.

Although the Chin were reluctant to accept Western medicine, their traditional understanding of illness and the practice of costly sacrifices to appease the *Khua-chia* in reality were burdens for the majority of the people. Ordinary and poor people who possessed no animal to be sacrificed in time of illness, poor families who could not perform the funeral rites called *lung-donh le thlan-ser* for their loved ones who had passed away many years before, a poor widow who was unable to contact her late husband in *Mithi-khua* through *Khua-vang* – these were not just burdens but misery for those involved and even for the entire society. But if a missionary had emphasized the medical mission, a small clinic could have helped the entire community tremendously. A small, relatively inexpensive Western tablet – called *Si* or *damdawih,* meaning the magic of health – could have substituted for a tremendous amount of Chin resources, wealth and labour which they would have spent as sacrificial offerings in times of illness. If the American Baptist mission could have spent more resources and personnel in the Chin fields, as the Welsh Presbyterian and English Baptist missions did among the Mizo tribe of the Chin people in West Chinram, the mass conversion probably would have occurred already by the 1910s as it happened there. While the American Baptists sent no more than two missionary families at one time to work among two tribes, the Laimi and Zomi, at least twelve missionary families, including doctors and nurses, worked among the Mizo, just one tribe of the Chin.

Regrettably, the American Baptist Chin Mission abandoned their medical mission in 1915 when the last medical missionary, Dr Woodin, left the field. However short-lived, it still had a great impact on Chin society. One way or another, the Chin came to realize that illness could be cured without sacrificial appeasement to *Khua-chia.* This was a big step for the secularization of the traditional Chin worldview. This process of secularization was strengthened when the British government established a civil hospital at Falam in the early 1920s and later also at Haka and Tiddim.

The Dark Hours of the Mission in East Chinram

History has proved Laura Carson right; she wanted to concentrate firmly on the first mission station and disagreed with the idea of hurrying to open another one. Soon after the new station was opened at Tiddim, the American Baptist Chin mission suffered many setbacks. When Dr Woodin left in 1915, only Laura Carson herself was working at the Haka station, and an ordained minister, Rev. Cope, was working at the Tiddim station, 180 miles (290 km, or at least ten days' journey on foot) away from Haka. To make matters worse, the political situation during that period was very unstable and characterized by turmoil. The Anglo–Chin War, which I shall discuss below, broke out in 1917 and lasted for two years. It was soon followed by epidemics and famine, which caused a great deal of trouble for both the missionaries and the Chin. During the war years, Laura Carson was evacuated from Haka and Rev. Cope was the only missionary who remained, trying but unable to cover the vast field of two stations. Moreover, it was a time that Robert Johnson called a 'great controversy' among the missionaries themselves regarding the question: 'Should the missionary effort at Haka be continued?' (Johnson 1988: 425).

After Dr Woodin left his work in Chinram, Laura Carson wrote several letters to Boston, the headquarters of the ABFMS, and requested that they recruit another missionary family for the work in Chinram. In order to see to the situation personally, Dr J.C. Robbin, the Mission Secretary, went to Burma and interviewed Rev. Cope during the missionary conference at Bassein in 1917, just before the Anglo–Chin War broke out. Unfortunately, Rev. Cope presented a very different and 'more pessimistic view of the Chin work than did Laura Carson', as Robert Johnson put it (1988: 427). He told Dr Robbin that the 'Haka work should not be strengthened; may be it should even be abandoned, and the American Baptist effort should be transferred to the northern part of the Chin Hills [that is, the Tiddim and Tawnzang areas where the Zomi tribe is living], and Haka might well be turned over to another missionary agency' (Cope 1918: FM–184). Moreover, he accused Laura Carson of being the one whose 'life was wrapped up in the Haka work and could not bear to see it slighted, whereas he felt that there were other more needy fields' (Johnson 1988: 427).

When Rev. Cope returned from the conference and told Laura Carson that 'he had withdrawn her request for another missionary family, she gave him a good scolding and told him that there was no more needy field than the Chin Hills in the whole of Burma' (ibid.: 427). Mrs Carson thus quickly went down to Rangoon and, on her request, the Reference Committee of the American Foreign Mission Society in Rangoon permitted her to address the committee on the matter. She did so and strongly made such a fervent plea for continuing the Haka work that 'the committee was quite convinced', wrote Rev. Cope to Dr Robbin. 'What she said I do not know, but at the end both Dr McGuire and Dr Seagraves said they would never vote for the closing of Haka' (Cope 1918: FM–184). Although the Reference Committee did not recommend the abandonment of the Haka station and the Mission Board in Boston took no action, such a personal dispute between the missionaries themselves had a bad impact on the Chin mission. Rev. Cope, for instance, blocked Mrs Carson's attempt to send a medical missionary to Haka after her retirement. Perhaps it was because of personal antagonism between the missionaries themselves.[4]

While the American Baptist Chin Mission was facing such problems, the Anglo–Chin War broke out in 1917.

NOTES

1 Carey and Tuck (1976) recorded that they had released at least 5,000 slaves from Chinram between 1888 and 1892.

2 According to the Zomi tradition, mentioned in 'A Brief History of the Chin' by Gin Za Tuang, the Chin carried the leather book from the Kale Valley, their original homeland. 'That book, being of leather, was eaten up by a hungry dog.' Cited by Za Tuah Ngur (1993), p. 48.

3 The name 'American Baptist Mission Union' (ABMU) was changed to the 'American Baptist Foreign Mission Society' in 1906.

4 Laura Carson and her new husband Judge J. H. Merriam (married in 1924) raised enough funds for Haka Hospital from their own personal resources, income from the sale of Laura Carson's book, and from friends and churches. They wrote a series of letters to the Mission Board in New York insisting that the missionary society send a medical doctor to Haka to take up the work begun by East and Woodin. The correspondence stretched out from 1924 to 1931. However, Rev. Cope strongly rejected the idea of sending a medical doctor to Haka. In 1926, for instance, he wrote a four-page letter and argued that 'I really do not believe a doctor from America should be sent out...I believe too we should emphasize the school and literary work.' See Cope's letter dated on 4 September 1926.

SEVEN

The Church after the Anglo–Chin War

The outbreak of the Anglo–Chin War of 1917–1919 immensely influenced both missionary activities and Chin politics. From the political point of view, it eventually ended an open conflict and launched a new relationship between the Chin and the British. Although the Chin admitted defeat, the British authorities were forced to reconsider their policies, especially their administrative and educational policies. In order to adjust their colonial policy to the new situation after the war, the British amended the Chin Hills Regulation in 1919, and restored the traditional political system called *Ram-uk*. As a result, Chin tribal and local chiefs regained their political power in their traditional domains. Thus, from the Chin's point of view, something was gained from the war, which they had fought so hard. Moreover, in order to appease the defeated Chin, the British government decided to adopt the Chin language as the medium of instruction in schools all over eastern Chinram.

The adoption of the Chin language in the schools resulted in far-reaching consequences for the Christian movement in Chinram. First, the missionary schooling became 'much more attractive' (Za Tuah Ngur 1993: 51), and almost all of the students converted at the schools. Many students who attended school by a government order of 1924 were the sons of ruling tribal and local chiefs. Thus, the conversion of this new generation of the ruling class spearheaded the growth of the church after the war and a change of society itself. Second, a new arrangement of cooperation between the mission and the colonial power in the field of education developed after the war. Rev. J. H. Cope accepted the government post of inspector of the schools in 1924, and all the teachers formerly employed by the American Baptist Chin mission were accepted as government employees. While 'this new arrangement relieved the mission's heavy expenses', as Johnson observes, 'it did not materially affect the Christian witness of the schools since most of the teachers were Christians' (Johnson 1963: 389). Third, the emergence of a Chin elite based on professional soldiers and teacher-cum-preachers also contributed many ways and means for church growth after the war.

When Chin resistance had been defeated and there was a new emphasis on education, the Christian missions had a strong modernizing effect and the number of converts increased tremendously. This also reinforced the new national self-awareness of the Chin.

THE ANGLO–CHIN WAR (*LAI RAL*), 1917–1919

The Last and Final Resistance

As Dawt Sung, a Chin scholar, correctly observed, the Anglo–Chin War from 1917 to 1919 was the last and final resistance against British imperialism in Chinram (Dawt Sung 1976). After that war, there was no other major conflict between the Chin and British until the end of the colonial period in 1948. Although the war was fought two decades after the Chin Hills Regulation of 1896, we cannot see, as Dawt Sung claims, the Anglo–Chin War as a separate historical event beginning as late as 1917, but as a series of resistance wars which began in 1871. It was resistance not only to physical invasion, but a struggle to maintain their traditional ways of life and social structure based on traditional religion, belief systems and ritual practices. The war was not only against the British occupation of Chinram but also against the 'strange new laws and customs that were molding society in ways alarming to them', as the American Baptist missionary Johnson put it. The war was caused by 'the perceived loss of power on the part of the chiefs' (Johnson 1988: 410), which itself was the result of British occupation and the abolition of traditional Chin tribal chieftainship and its political system, *Ram-uk.*

At one time or another, from the first invasion in 1871 to the final war in 1919, every branch, clan and tribe of the Chin was involved in the resistance. But final wai was fought by the Laimi tribe, particularly from Haka and its surrounding areas such as Zokhua, Senthang, Zophei, Lautu and Mara. Therefore, the war was commonly known in Chin as *Lai Ral.*

The Immediate Cause of the War

The immediate cause of the war was related to World War I. During the war, Great Britain mobilized all its human resources and recruited soldiers and non-combatants from all its colonies. In 1916 about a million soldiers and half a million non-combatants from British India and Burma were sent to Mesopotamia, Iran, Turkey and France.

The British demanded a fixed number of young, able-bodied men from the Chin for the French labour corps. The Mizo (Lushai) tribe joined the labour forces in 1916, but in 1917 more men were needed, and altogether 1,000 men joined from the administrative subdivisions of Haka, Falam (both Laimi tribes) and Tiddim (the Zomi tribe). The Zomi tribe joined the forces in 1917, but the Laimi tribe from the Haka subdivision rebelled against the British order.

As indicated, the main cause of the war was political, which in turn was closely associated with religious matters. Furthermore, isolationism and exclusivism in traditional Chin religion implied both exclusion of strangers from their society and isolation from other societies, and their confinement to the sovereignty of their *Khua-hrum,* i.e. within the boundary of the power of their chief. They simply refused to go beyond their boundaries and the sovereignty of their *Khua-hrum.* They would say: *Mi ram dang ah cun kan khua-hrum pei a phu a nem cang cu!* 'The power of our *Khua-hrum* is already softened in other countries'. This meant that their *Khua-hrum* would not be able to protect them in any other country. This religious concept developed into a strong conviction as a result of experiences already in 1890 and 1891. For them, it was not a superstitious belief but the reality of their own life.

During the invasion, the British combined military and political warfare against the Chin. For example, eighteen Chin chiefs from Haka, Thlantlang (Klang Klang), Sakta (Thetta), Khuapi (Kapi) and Zokhua (Yokwa) were persuaded to go to Rangoon for a demonstration of British power in September 1890. It was hoped that the visit would show the Chin 'how small and insignificant their country was and give them an idea of the power and resources of the British'. 'They were shown', wrote Carey and Tuck, 'the garrison and batteries at Rangoon, and parades were held for the edification at Myinyan and Mandalay' (Carey and Tuck 1976: 60).

The result of the enforced visit, however, was a disaster for both sides. During their stay, the bubonic plague struck Rangoon and environs and killed six of the visiting Chin chiefs. The Chin simply thought that they were traveling too far and had reached beyond the sovereignty of *Khua-hrum* where they had no divine power of protection. They were convinced more than before of their traditional belief *Mi khua le Mi ram ah cun kan Khua-hrum pei a phu a nem cang cu* [Our *Khua-hrum* cannot protect us in a foreign country]. Thus, when they returned home, they kept silent about their journey and mourned instead for their lost friends. Carey and Tuck concluded: 'They were too uncivilized to understand all they had seen ... but put down and silent about their travel tales' (ibid.).

This incident was still fresh in their memories when news came that the British government was attempting to raise members for the labour corps. This caused great excitement and fear that their *Khua-hrum* would not be able to protect them in such a far-away country as France. They 'absolutely refused to go to France', Laura Carson wrote. 'They said they would commit suicide rather than go' (L. Carson 1927: 227). The phrase 'I will commit suicide rather than ...' (*Kaa that deuh lai*) implies not killing oneself but rather 'I will never surrender but fight until I die'.

Inspired by their religious convictions and the politics of what can be called 'ethno-nationalism', the Chin prepared to fight back rather than surrender and obey the orders of alien intruders. Thus, in 1917, some 5,000 men from the Haka, Zokhua, Thlantlang, Senthang, Zophei, Miram, Lautu and Zotung areas of the Laimi tribe united and took a sacred oath, called *Sathintuh,* at Sakta to make war against the British. Unfortunately, Sakta, where they had taken the sacred oath, happened to be a village of Chia Khaw (Shia Kaw), a former slave of the Sakta chief and the first Christian convert among the Laimi tribe. Chia Khaw betrayed his fellow Chin and leaked the war plans against the British to Laura Carson. She graphically described the event in her book as follows:

> One evening Shia Kaw (Chia Khaw), our first Haka Christian convert, who was teaching school in the large village of Sakta twenty miles distant (from Haka), appeared at the mission closely followed by two Sakta Chins. He made me understand that he wanted to speak to me in private. The men were watching his every move. After casually talking to him for a moment I stepped into my storeroom and called him to come and get some rice I had for him. Understanding the ruse he came quickly and saying as he took the rice, 'I will come tonight. I have something to tell you', and hastened back, meeting the two Sakta men following him into the storeroom. It was very evident they did not intend he should have any private conversation with me. After the Chins had gone to bed, Shia Kaw slipped softly in at my back door, saw that the blinds were closely drawn, and then, shaking with excitement, said that if the Chins knew what he was going to tell me they would certainly kill him.
>
> He said that thirty villages had united, taking the sacred oath that they would attack Haka, kill the sepoys, take their guns and with them clear the Hills of the British and resume their government. He said that the men of Sakta had secreted their women and children in the jungle and carried out six months provisions for them and that they were spending their time day and night making ammunition; that two days before, six hundred armed men had congregated less than three miles from Haka with plans all made for attack but that he and Maung Lun, a Karen preacher in the employ of the mission, had told them of the strength of the British and of their own certain ultimate defeat and had persuaded them to disband and return to their villages. But upon their return they were so derided by their friends over the outcome of the wonderful things they had boasted they would do that they planned to gather in larger numbers and be met in the north by an equally large force. 'And if they do, God pity us, for we will every one be killed', he said, tragically striking his breast with both hands. Maung Lun was not permitted to come to Haka lest he give information (ibid.: 227).

Laura Carson seemed reluctant to forward the information that she had received from Chia Khaw to the British authorities. However, the next day, to Mrs Carson's relief (but very unfortunately for the Chin), Rev. Cope returned from Rangoon after his meeting with Dr. Robin. Rev. Cope quickly passed the information from Chia Khaw to the British officers. Thus, when the Chin attacked Haka, the colonial army was ready and armed. The missionaries and all foreigners also had to take refuge inside the fort of the garrison, or behind the 'police lines' as Mrs. Carson put it, which were surrounded by the outer lines of a trench. 'In this trench', wrote Laura Carson, 'were placed our sepoys with their guns'. She also mentioned that 'our Karen teachers and a few Christian Chins who had guns joined them (the sepoys)' (*ibid.*: 231).

As the British sepoys took their positions, the Chin could not easily overrun Haka, which was now ruled by the British, as they had expected. However, they besieged the fort of Haka for 'twenty-two days'. Mysteriously, they did not storm the fort itself. There would have been 'some reluctance to storm the fort', as Robert Johnson argues, 'since the missionaries were inside the garrison' (Johnson 1988: 422). If that was the case, then it was because of Laura Carson, who had won the hearts of the Chin and created a good relationship with them. One of her greatest abilities, her knowledge of midwifery, showed them her tender love of the poor. It was said that she assisted almost all the pregnant women in Haka during their labour and attended many of the name-giving ceremonies with presents. Even the day before Haka was besieged, she assisted a woman who gave birth at Laura's own house, described as follows:

> During the night, amid of mob of men, women and children, a little child was born to the wife of one of our teachers. He is called *Ral Zam* (Fleeing from the War).
>
> Before we had half finished breakfast there came a messenger from the Assistant Superintendent saying that we must hasten to the police lines, that an army of five thousand were advancing and would reach Haka within an hour or two. Mr. Cope immediately hastened from the table to look after the Christians. He had barely gone when a second messenger came saying that it was the order of the Assistant Superintendent that 'The Boinu' (I am the Boinu) 'hasten to the police line'. What was I to do? The poor woman with the two-hour old baby could not be left. Her husband had gone to their home on some errand and there seemed to be no help available. All was confusion and excitement. A third messenger came with peremptory orders. 'The Bawinu *must* go to the police lines immediately!' My cook came around at last, a big, strong man. He put his hands on his knees and, well wrapped in a blanket, we got the new mother on his back and started up the hill. I called a passing schoolboy, wrapped little Ral Zam in another blanket, gave him to the boy and sent him to follow the mother. Then I went to my storeroom and collected such food as we could carry. Mr. Cope, having done all that he could for the Christians, came and we slowly climbed the steep hill, hearing several shots fired as we went (L. Carson 1927: 230).

Laura Carson knew that the people trusted her. She wrote with confidence that even during the siege, 'several of the leading chiefs sent me word that I need not fear, that I had always been their friend and they would not harm me. And I really believe that they would not have intentionally done so' (ibid.: 233).

After nearly a month, a 'relief column' from the British Burma Army arrived at Haka. On their way, the relief column was attacked by the Chin and, according to Laura Carson, 'there were thirty to forty casualties' before they reached Haka. Despite the fact that she had received protection from the British authorities, Laura Carson strongly condemned the brutality of the sepoys:

> The relief column remained behind to do a shameless thing. They went to Hniarlawn, which has been the leading village in the rebellion, and shot the cattle and hogs, confiscated the grain and fowls and *burned the village* – a thoroughly German performance and quite unworthy of the British government. The poor women and the children will have to sleep in the jungle and we are having freezing weather. I do not see what is to prevent many of them starving to death. It is brutal and inhuman and I resent it with all the strength of my soul (L. Carson 1917: FM–182).

Perhaps because of her outspoken criticism of the British government, she was evacuated almost by force from Haka. She wrote:

> The Superintendent told me that as soon as the roads were cleared I must leave the Hills until things were in a more settled condition. I protested, saying that my people had never needed me so much as at that time, that I was not afraid and that I would feel like a deserter to leave them. But it was no use. He was obdurate. He said he had not sufficient force to protect me and at the same time settle the surrounding country, that if anything happened to me he would be responsible to the governments of both my country and his and that he asked *as a personal favor* that I go without further protest (L. Carson 1927: 134).

Laura Carson was unable to witness further brutality of the British sepoys towards the Chin people. They burnt to the ground all of the villages in the Senthang and Zophei areas, at least forty villages, and many more villages in other areas as well. They burnt the villages together with all kinds of livestock and destroyed grain in the fields. Despite such brutalities, the war lasted for two years. However, due to the unequal armed forces and a famine, which hit severely because of the war, the Chin surrendered in 1919.

The Chin leaders were brought to trial in 1919 and 61 were sentenced for their part in the resistance. Among them three were sentenced to death, but before the sentence could be carried out two of these committed suicide. Ral Chum, the headman of Hniarlawn, was executed at Mingkin. The others were imprisoned outside Chinram in Mingkin, in lowland Burma, and Lashio and Taunggyi in the Shan State.[1]

But the British did not only imprison all the chiefs and headmen who had fought that war; also fines amounting to over Rs. 13,000 were lavied on them. The war created many thousands of refugees, especially from the Senthang area, who fled to the Gankaw Valley and never returned home. The war was also followed by a prolonged famine and influenza epidemic which claimed many more lives in Chinram.

After the Anglo–Chin War from 1917 to 1919, people from East Chinram were divided into two groups: those who had gone to Europe as part of the labour corps and those who had fought the war at home. The following sections describe how this transition effected Chin history.

THE IMPACT OF THE WAR

The Anglo–Chin War was a landmark in Chin history and had a strong impact on Chin society for many years to come. Politically, it was the end of Chin resistance and the beginning of a new relationship between the Chin people and the British authorities. One of the positive results of the war in terms of politics was that in order to appease the defeated Chin chiefs, the British government amended the Chin Hills Regulation in 1919 restored the *Ram-uk* system and the powers of the tribal and local chiefs who had ruled at least two villages. For the Chin, restoration of the *Ram-uk* system was a partial victory; it was something they had fought hard for. Afterwards, the Chin faithfully served the British until the end of the colonial period.

From the ecclesiastical point of view, it was also a new beginning of what Johnson called 'the edification of churches' in Chinram. Church growth during this period was mainly the result of the fact that the Zomi tribe had gone to Europe as part of the labour corps instead of fighting in Chinram. There was also the effect of the changing educational policy, which adopted the Chin language as the medium of instruction in schools, and a new cooperation between the colonial power and the Christian mission. Since the impact of the war and thereby church development among the Zomi and Laimi tribes were quite different, I shall focus in this section on the continued development of the Chin Baptist Association. This involved the struggle over the continued differentiation of the Chin community: a breakthrough for the Zomi and slowed development for the Laimi tribe in the Haka area.

The Breakthrough among the Zomi Tribe

The effect of the war on the Zomi tribe was positive and rewarding. As mentioned, the Zomi from the Tiddim and Tawnzang areas followed the orders of the British colonial power and sent men to France during World War I. According to Vumson, about 4,000 young Zomi men went to Europe.

> Their journey from Chittagong, Akyab and Rangoon took them to France where they evacuated the wounded and loaded and unloaded military supplies going to the front. Each of them, except for those who died on the ship or in Europe, brought enough money to home to pay for any bride they chose to marry (Vumson 1986: 134).

As Vumson observes, on their return they were apparently better off than before they left for Europe, at least in economic terms. Some of them had learned a trade, which was still novel among the Chin. Moreover, most of them were recruited as professional soldiers after the war. Not only did the army provide a rare opportunity for a young man to venture into the world outside Chinram; it also let him earn money, which was not possible in Chinram at that time. The Chin labour corps during World War I contributed much to change Chin society and their worldview. Vumson explains the development as follows:

> The experiences of those who went to Europe were not easily forgotten and in some cases *changed beliefs*. The war machinery of the Europeans impressed them, as the planes, ships and guns were immense developments for the boys from Zo country ... Before they left for Europe they had been certain that the sun rose out of the mountain ranges, but when they returned they were convinced the sun rose from the ocean (ibid.).

Their journey to Europe also made them realize that one could travel beyond the sovereignty of their *Khua-hrum*. Before then, they were confined within the boundaries of the *Khua-hrum*'s sovereignty, which limited their worldview and their identity. Many of them, as Vumson observes, changed their beliefs. They were convinced that the European God, that is, the Christian God, was more powerful than their local *Khua-hrum*. Although the American Baptist missionaries had been proclaiming the Gospel for some time at home, many of these young men had been reluctant to accept it. But now, after they had seen Europe with their own eyes, they were prepared to accept the powerful Christian God who, in their view, made Europeans stronger than themselves. Since their worldview had already changed, nothing was left to worry about, they could even abandon their traditional religious practices and belief systems in *Khua-hrum*. It is therefore not surprising that the first group of mass conversions occurred among those who had gone to Europe. To support my point, I would like to present some figures concerning Church membership before and soon after the Anglo–Chin War.

In 1918, according to Rev. Cope's report, there were only 'five hundred Christians at the Association' (Cope 1918: FM–184). At that time, the 'Chin

Hills Baptist Association' included both the Zomi tribe in the Tiddim–Tawnzang areas and the Laimi tribe in the Haka–Falam areas. Thus, the membership of 500 Christians included both the Zomi and Laimi tribes. The Government Census of 1921, however, showed that there were 788 Christians in the Chin Hills District – the same place but different name for those areas. However, in 1926, according to the *Mission Census and Christian Progress in Burma*, a report written by A. McLeish in the Survey of the National Christian Council of India, Burma, and Ceylon, 'The Christian community among the Chin amounted to 4,046' (McLeish 1929: 19). The report also said that 'around Haka, the work has been difficult and most of the converts come from Tiddim and the North' (ibid.: 54). Tiddim and the northern part of the Chin State are the home of the Zomi tribe where most of the 'Labour Corps-turned-Army' converted to Christianity after they returned from Europe.

The most far-reaching consequence of the war was the emergence of a Chin elite made up primarily of professional soldiers. The British had recruited them to form the First Chin Hills Battalion soon after the war.[2] These professional soldiers not only converted *en masse* to Christianity; they were apparently better off economically than their friends who had stayed behind on the traditional farmland. Thus, this new elite of professional soldiers – now Christians – became role models for people in the new Chin society. After a course of fifty or so years, their professional army (which in normal times had nothing to do with political power) was empowered by General Ne Win's military regime, and from 1962 to 1988 (from Major Son Kho Lian in 1962 to Lt Col E.K. Kim Ngin in 1988) the rulers of the Chin State came from army officers of the First Chin Hills Battalion, the Zomi tribe in general and Sizang in particular.[3] They controlled the Chin Baptist Church as well, especially the Zomi Baptist Convention.

Slowed Development among the Laimi Tribe in the Haka Area

McLeish's report mentioned that missionary work among the Laimi tribe in the Haka area was more difficult, especially during the Anglo–Chin War. They rejected British orders and fought the war instead of going to Europe as part of a labour corps, and they suffered many casualties and lost many lives after the war because of famine, influenza and other epidemics.

The famine was caused by the Chin's inability to cultivate land properly during the war and the savage actions of the British sepoys who 'burned several villages along with large quantities of food' (Johnson 1988: 428). Thus, fire consumed almost all of their life-saving grain. One fire took seven days to destroy the house of Chief Mang Hnin of Leitak, because the house was big and his people had stored much livestock and grain inside. The famine was also caused by the use of forced labour, which the British had imposed upon the Chin. As Johnson pointed out, 'prisoners of war were put to work making roads and constructing the artificial lake at Haka at the very time they should have been in their fields. The result was a terrible shortage of food' (ibid.).

In 1918, the flu epidemic followed the war. The Chin believed that the British had used chemical weapons, *Mirang nih thih si an thlah* (The white men spread the poison gas), but it was the pandemic Spanish influenza and it hit the Chinram hard, especially in the Haka area 'because of famine conditions weakening the people' (ibid.). The reason was not only the food shortage; they did not have proper shelters because the British had burnt their villages to the ground. In January 1919, the 'sickness had worked havoc for the Chins', and Cope reported that some villages lost an average of one person per family, and in others, 'the entire family had been wiped out', and death 'will come to way over a thousand, I believe, and this in a population of less than 115,000' (Cope 1918: FM–184). According to Johnson, Cope 'had to revise this figure to more than 5,000 deaths' (Johnson 1988: 429). But my field research has shown that the death tolls could have been as high as 10,000. The missionary's report did not mention the figures from the Senthang and Zophei areas where most of the deaths occurred. Aibur in the Zophei area, for instance, which had more than 300 households before the war, was reduced to just seventy-five households, by the time of the 1921 census. Most lives were lost because of famine and influenza epidemic.

Because of all this, many Chin refugees, especially from the Senthang area, left their homes for the Gankaw Valley. According to Rev. Cope's report, the Anglo–Chin War had created at least 4,000 refugees by 1918, even before the war was over. It was a depressing figure for Rev. Cope. He thought that the Chin who 'moved to' the Kale-Gankaw Valley would be 'assimilated by the Burman' (Cope 1918: FM–184). He also saw not only the lack of response in Haka but even the resentment of the local people, who burnt the mission school after the war. He seemed aware that the people's resentment was directed at him, for he had informed the British about the secret plans behind the war. He suggested closing the Haka station again in 1918 in a long letter to Dr. Robin. Part of his letter read as follows:

> Perhaps you think I have changed my mind. Not in the slightest. I stand where I stood at Bassein. I believe our work in Burma is in a bad way and something desperate must be done. I believe some stations must be closed or turned over to people who will work them while we concentrate on stations in need. I believe from the little I know from the work that the Chin Hills can be as easily as any other, if not more so, turned over to the Wesleyans since the field touches theirs and, in fact, they lie between us and the other Baptist missions (ibid.).

Unlike J. H. Cope, Laura Carson stood firm for the continuation of the Haka station; she also helped the people during the famine and flu epidemic. During the war, she was evacuated but returned to Haka before the end of 1919 and saw that the people were suffering severely. She sent men down to the Kale Valley to buy and carry rice to Haka with her own money. She had trouble finding men strong enough to make the trip but she finally got together a band of 150 men with one person to do the buying and superintending. For three days the men walked down to the Kale Valley, spent a day resting and buying and then started back, each with a seventy pound load of rice. It took them five days to make the trip back up the steep mountain trails. Laura Carson supervised the distribution of rice in Haka – first to the sick and to those with little children. She gave the food free to the widows and orphans but made those pay who could afford to do so. After the porters rested, she sent them for more rice (cf. Johnson 1988: 428).

It was a time when Laura Carson won what she herself once called the 'stone-heart' of the Haka men, not just the womenfolk, and the Chin still remember her as 'Our beloved lady who stands with us in the critical times of life and death': *Thihnak le nunnak kar ah a kan dir pi mi, Kan dawtmi kan Bawinu* (from a popular Christian song the Chin composed about her). For new Christian converts taught from childhood to make sacrifices for good health, it was a period of special testing, and Laura Carson and Rev. Cope were glad to report that, by and large, 'the Christians remained faithful to their newfound religion' (ibid.).

The Departure of Laura Carson

After the Anglo–Chin War, one of the most significant events in the American Baptist Chin mission was the departure of Laura Carson, the most remarkable and courageous woman. She stood firmly during the period that we call 'the era of confusion', not only for the course of the mission itself but for the Chin people with whom she always identified herself. Her departure was a loss for the Chin and the mission.

Despite the sorrowful days which followed her husband's death, Laura Carson remained as a missionary in Haka until her retirement in 1920. She, indeed, was one of the greatest missionaries among the Chin, 'ready to surrender all her personal interests to the task of winning people to Christ'[4], even leaving behind her children in America for all those years. She was a devoted Christian who could say 'every breath was a prayer' (L. Carson 1927: 154). Her main task was teaching in the newly established Haka mission school which, as Arthur Carson reported, 'under her care has grown in numbers and effectiveness and is much more promising as an agent of our ultimate objective: evangelization of these people' (A. Carson 1908: FM–182). Indeed, the first convert among the Laimi tribe, namely Chia Khaw, was a student of Laura's at the mission school in Haka. In addition to her main task as schoolteacher and midwife, Laura compiled a Chin primer and wrote the story of Jesus Christ in the Lai dialect. She also translated the Gospels of Luke and John, and Acts, together with 126 hymns in the same dialect. Moreover, she completed a Chin–English dictionary, which was started but left unfinished by her late husband.

For the American Baptist Chin mission, one of her greatest achievements was her ability to persuade four Karen preachers, most of them former students, to come with her from the former mission field of Bassien and Henzada. This legacy bore fruit for many years to come, and many Karen came to help the Chin mission work and had 'a definite Christian impact on their schools and villages and were evangelists as well as teachers'. Dr Cope gave this generous tribute to those Karen preachers-cum-teachers:

> We owe everything to the Karens. We do not know what we would do without them. When Mr. and Mrs. Carson first came up they brought three or four Karens with them and from that time on, with few exceptions, they have proven splendid men on whom one could place no end of responsibility. For a long time they were the only evangelists here. They went out to strange villages where no preparations had been made for them and where they were threatened direly. The first Chin Christian came seven-days journey from Haka where a Henzada Karen, Thra Shwe Zan, worked alone, seeing the missionaries only once a year. The Chin preachers were trained by them. They learned the language, learned the ways of the people, and won their confidence. In the first literary work I did, it was the Karens who helped me. In the schoolwork as well we have Karen headmasters, and they proved as valuable there as in the evangelistic work (Maung Shwe Wa 1962: 388).

After her retirement in 1920, Laura Carson married Judge J. H. Merriam, her former classmate at the Baptist Seminary in Gibbon, Nebraska. With the help of her new husband, she raised funds for the missionary work in Chinram, particularly for the mission hospital in Haka. She also published her memoir, *Pioneer Trails, Trials and Triumph,* in 1927. She died on 19 July 1942. At her memorial service, Mrs Selder W. Cummings paid her homage on behalf of The Women's Society and said:

> In Haka she left churches, schools, a hospital, the scriptures in their own language, and loveable Christian communities... She ended a life that was proof beyond doubt of the power of Christianity to redeem, instruct, and transform even the most degraded (Johnson 1988: 460).

NEW COOPERATION BETWEEN THE COLONIAL POWER AND THE MISSION

Changing of Colonial Policy on Education

Although the Chins were defeated in the Anglo–Chin War, the British authorities in Chinram were forced to reconsider the colonial policy toward the Chin people. The British government had no choice but to amend the Chin Hills Regulation in 1919 and restored *Ram-uk* in East Chinram. British government also decided to adopt the Chin language as the medium of instruction in schools all over East Chinram because they wanted to prevent further rebellion and they recognized Chin antagonism toward learning Burmese, a foreign language, in the schools. Chin patriots who had just returned from serving their prison terms for their part during the war in foreign lands, burned the mission school in Haka in 1922. Their action challenged to the British and destroyed the symbol of enslavement, the place where they were forced to learn Burmese. Johnson noted that Rev. Cope 'knew who the arsonist was' (Johnson 1988: 460), and he probably informed the British authorities. The British, however, put aside the case because they feared that the Chin might revolt again if they took action on such challenging behavior. Instead, they changed their own government educational policy in order to appease the defeated Chin chiefs.

In 1923, the British authorities decided at the Maymyo Education Conference that 'education should be carried on in the vernacular, using but one dialect' (Johnson 1988: 460). The language chosen for the whole Chin race was the Laizo dialect, originally spoken by the Laimi tribe in the Falam area. At that conference, Major Arthur Newland, who invented the first Chin script based on the Roman alphabet, proposed that the Lai dialect from the Haka area (also known as Haka dialect) be chosen as the standard language. Major Burne, Superintendent of the Chin Hills, however, rejected Newland's proposal on the grounds of his personal antagonism towards the Laimi tribe from the Haka, Senthang and Zophei areas, who had fought the British in the Anglo–Chin War. Major Burne and his troops had been ambushed by the Chin patriotic army led by Chief Mang Hnin of Leitak, between Dawngva and Leitak. In that battle, Major Burne's personal assistant, Captain Kennedy, had been killed and he himself seriously wounded.[5] Major Burne took this event personally and therefore rejected the idea of adopting his enemy's language as the standard language for the whole area of his administration.[6]

As a result of new cooperation between the colonial power and the Christian mission on education, the Laizo dialect did not become the standard language for the Chin either. Rev. Cope reversed the British policy when he became Inspector of Schools in 1924.

New Cooperation and the Problem of a Standard Language

'Christian education' was what concerned J. H. Cope most when he accepted the post of the Inspector of Schools. In other words, he was worried about the spread of secular education in Chinram. As he knew, if the British government stressed education after the Anglo–Chin War, it would become impossible for the mission to compete with the government's secular schools, which apparently would be more secure financially. From the British government's point of view, J. H. Cope was the right person at the right time. He was 'a very intelligent and hard-working man, completely absorbed in services to the people' (Johnson 1988: 463). His loyalty to the colonial power during the Anglo–Chin War was unquestioned by the British officers, especially Deputy Commissioner Major Burne, who offered the post of colonial officer to Rev. Cope as a reward for his loyalty.

When he accepted the post, J. H. Cope became the most powerful person in terms of policy-making on education for both the Christian mission and the colonial power. Unfortunately, he abused his power. He insisted on the use of three dialects, or what came to be known as the three standard dialects: Kamhau dialect, Lai (Haka) dialect and Laizo (Falam) dialect. He may have been biased toward these dialects by his own expertise of the Kamhau dialect. When he reversed the government decision in 1924 and created three standard languages, the Chin lost for the second time a golden opportunity to establish a common language.

According to Robert Johnson, Rev. Cope's main concern was to create three standard languages, produce 'school textbooks' and spread the Gospel to various tribes and dialect groups of the Chin. The creation of a common language for the Chin was neither his concern nor part of the American Baptist mission's educational policy. Thus, Chin scholars like Za Tuah Ngur criticize the policy as 'overemphasizing the immediate needs while neglecting the long-term development of the people' (1993: 53). Unlike Za Tuah Ngur, however, an older generation of Chin scholars, like Hau Go, who received their education under the instruction of missionaries themselves, still argue for the sake of Rev. Cope's policy of three standard languages. Hau Go thus says:

> We must not criticize Rev. Cope for not imposing one *lingua franca* for all the people in the Chin Hills, since it would have been absolutely impossible at that time. He should rather be given credit for his achievement in reducing over forty dialects into three *lingua franca* (Hau Go 1986: 13).

The new generation of Chin scholars, like Cung Lian Hup, compares the context with the Kachin, where a Swedish-American Baptist missionary, Olof Hanson, created a common language for all the Kachin tribes. They reject the above argument completely. Cung Lian Hup says:

> If the missionaries to the Kachin could produce a common language, i.e. Jingphaw, for all the Kachin tribes, who had many local dialects just as the Chin did, there is no doubt that the missionaries to the Chin Hills could do it also for the Chin.

He also points out a historical fact in the Chin context and argues:

> And if the Tiddim and Falam students at Strait's Bible school in Haka could write their examination papers in Haka (Lai) dialect before they had been in Haka for six months, it is strongly believable that for the Chin there would have been no big problem in learning another dialect if that language was prescribed as a common language for the Chin (Cung Lian Hup 1993: 180).

Indeed, most Chin people regret the fact that in 1924 they lost an opportunity to establish a common language, and they are still unable to produce one today. The creation of three standard languages has 'made it difficult today for the Chins to agree upon one common language or script' (Za Tuah Ngur 1993: 53). It is mainly because 'the Chins have not been encouraged to forsake their localism, but instead stick to their own dialects, creating a barrier to their unity and integration for the future', said Ngur. She continues:

> This (three standard languages) has caused great difficulties in the production of Bibles, hymnbooks and other religious and secular literature. Creating autographies for each tribal group was a serious mistake on the part of missionaries. Some rationalization in this area is urgently required to facilitate the evangelization of each tribal group, but in the long term, the churches and the whole of the Chin people have suffered from the mistakes of missionaries (1993: 53).

The creation of three standard dialects was a real disappointment for the Chin. Although Christianity provided the means of preserving their national identity across tribal borders, the creation of three standard languages also produced a countermeasure for reviving old tribal and local identities, which became a stumbling block for unity. In terms of disunity, the Chin people suffered from Zomi tribalism defined by the disparity between the Kamhau dialect and two localisms, namely Haka and Falam, which were based on two different standard dialects which the missionaries created for just one and exactly the same tribe of Laimi.

Despite the fact that the missionaries created three standard dialects for the Chins and no common language, the change-over from Burmese to three Chin dialects as the medium of instruction made 'schooling more attractive', and the 'value of the change over from Burmese to Chin as the medium of instruction', as Robert Johnson observes, 'was incalculable'. He said:

> Up to that point, all schoolwork was done in what really was a foreign language – Burmese. The reader must realize that unlike people in many other parts of Burma, the Chin were and remain still a very homogenous group of people. Burmans did not live among them; there was a sharp line of demarcation between them. Burmese was indeed a foreign language, hard to learn and easy to forget. But now,

> with schooling in their own tribal languages, the way was paved for rapid progress – as soon as the textbooks could be produced! (Johnson 1988: 463).

As Johnson observes correctly, the mission schools proved themselves the centres of conversion. In Chin *Cathiam le Krifa* was a phenomenon in those days: 'To be a learned person is to be a Christian'. Almost all the students became Christians.

A New Cooperation to Self-supporting Churches

In 1924 when Dr Cope became the Inspector of Schools, the British government took over all the mission schools, and almost all of the Karen preacher-teachers who were employed by the American Baptist Chin mission accepted government employment. Rev. Cope viewed this new arrangement of cooperation between the mission and colonial power as the 'opportunity of a lifetime'. He thus wrote:

> We have the opportunity of a lifetime here in the Hills and I want to take advantage of it. It means also advance in our work even if there were no more missionaries or helpers. At first the large majority of teachers in the Haka and Falam subdivisions will be Christians and everyone will be an evangelist. It means also that one language will slowly come to dominate the lower two-thirds of the Hills and one in the Tiddim subdivision, thus doing away with the most exasperating obstacle to the progress of work here in this field. It will mean more solidarity in the work and in the people. All will study this one dialect, or two (Cope 1924: FM–236).

As far as the growth of the church was concerned, this new arrangement of cooperation between the colonial power and the Christian mission was very effective. While 'this arrangement relieved the mission's heavy expenses', as Johnson observes, 'it did not materially affect the Christian witness of the schools since most of the teachers were Christians' (Johnson 1963: 389). Moreover, since the Karen preacher-teachers received their salaries from the British government, 'the tithes and offerings of the Chin Christians reached the point of paying for their Chin pastors, and thus we can say that the churches had achieved self-support' (ibid.). This was a landmark in the emergence of the Chin churches, which also ushered in a new era of Christian movement into East Chinram.

When the British government took over the mission schools, there were only eleven primary schools and three middle schools at Haka, Falam and Tiddim. As the British government supported education with strong budgets, the number of schools and pupils increased rapidly. The statistics of 1929 and 1930 show that there were 25 primary schools and three middle schools with some 1,500 pupils, compared to a mere 75 enrollments in

1924. As the number of students increased, the number of converts also increased rapidly during this period. Although it has been difficult to find the statistics for 1930, church membership figures for 1939 show that there were some 12,000 Christians (5,000 from the Tiddim, Tawnzang and Sizang areas, 4,000 from the Haka area, and 3,000 from the Falam area), an increase from a base of less than 5,000 Christians in 1926 (cf. Strait 1939: FM–303). Rev. Strait's report also mentioned that during the 1930s the churches in the Haka area alone admitted newly baptized converts at a rate of 'from three to five hundred a year!'

As already implied, the motor for church growth during this period was cooperation between the colonial power and the Christian mission, with a strong emphasis on modern education. Although all mission schools became government schools, Christian influence continued through the presence of Christian teachers and, more importantly, the influence of Rev. Cope himself who, as Inspector of the Schools, produced all the curricula for the schools all over East Chinram. In his position as the Inspector of the Schools, Cope produced 35 textbooks for primary schools in the Lai and Kamhau dialects. As a missionary, he translated a complete New Testament into the Kamhau dialect in 1936 and produced a hymn book including nearly 300 hymns. He also translated John Bunyan's *The Pilgrim's Progress* into the same dialect. Rev. Cope was 'given a generous travel allowance to enable him to visit all the schools', which he always combined with an evangelical tour, 'the care of the church and the propagation of the Gospel' (Cope 1938: B–201). This 'travel allowance enabled him, after paying actual expenses, to run a monthly magazine for fifteen years' (Johnson 1988: 462).

THE EDIFICATION OF THE CHIN CHURCHES AFTER THE ANGLO–CHIN WAR

The Mission and an Emerging Chin Elite

As the colonial administration created the opportunity for many young Chin men to enlist as professional soldiers in the newly formed Chin Hills Battalion of the British India Army, the Christian mission also created a new society where two different kinds of Chin elite, namely professional teachers and preachers, began to emerge.

This was closely associated with the changing of the British government's policy on education, which was carried out by the Christian mission. In addition to the change over of the language of instruction in schools from Burmese to Chin, the British government issued a decree in 1924 which stated that there would be no more appointments of headmen who could not read or write. Thus, all the hereditary Chin chiefs had no choice but to send their sons to mission school in order to maintain their political positions. But even so, almost all of those students converted to Christianity.

In this way, a new cooperation between the colonial power and the Christian mission accelerated social and religious change in Chin society.

The main actor to contribute to the emergence of a Chin teacher-preacher elite was Chester Strait, who came to Chinram as a missionary in 1926. Dr Cope also played an important role, but he was unable to open a Bible school because so much of his many duties. Chester U. Strait was born on 21 December 1893 in Glenwood City, Wisconsin, a small town about fifty miles east of Minneapolis, Minnesota. As a young man he served in the American Army during World War I, taking part in the war effort in France. After the war he entered the University of Minnesota and graduated in 1923 with a Bachelor of Arts degree. He then went to California and received a Bachelor of Divinity degree from Berkeley Baptist Divinity School in 1925. He later earned his Masters of Theology in 1932 and Doctor of Theology in 1933; his dissertation was on traditional Chin religion. This was the first major work on the subject. He received the first doctoral degree awarded by the Department of the History of Religion, Berkeley Baptist Divinity School while on his first furlough. Both degrees were from the same institute. While he was still in college in Minneapolis, Chester married Florence Talbot on 31 July 1920. Florence Strait received her education in Minnesota, and he secured an R. N. (Registered Nurse) certificate at the St. Paul City and County Hospital in 1919. The couple were appointed for the work in Chinram on 2 June 1925 by ABFMS but did not reach Haka until 5 April 1926.

The Bible school at Haka was the most important contribution that Chester Strait made to the Chin. More importantly, the medium of instruction for this Bible school was the Chin language. As mentioned already, the earliest preachers to the Chin were the Karen, brought up by the Carsons and their successors. Gradually, as the Chin converted and volunteered for Christian service, some of the young Chins needed to be trained at the Bible school at Insein near Rangoon. Instruction there was either in the Pow or Sgow Karen languages, which the Chin could not use, or in Burmese. A higher level of theological instruction was in English. But when Strait came to Chinram, he realized that the time had come to train the Chin in their own language, in their own country, instead of sending them to Burma. He argued, among other reasons, that expenses were much greater if students went to Rangoon to study. Thus, he opened the Chin Hills Bible School at Haka on 1 May 1928 with thirteen students – four from the Tiddim area of the Zomi tribe, four from Falam and five from the Haka areas of the Laimi tribe.

Although the teaching was in the Lai dialect, it was no problem at all, as far as language was concerned, even for Zomi students. Strait reported that 'one of them quickly picked up the language but the other three were slower. But all the students were able to write their final exam in Lai before they had been in Haka for six months' (Johnson 1988: 510–512). Being a missionary himself, Strait said that he experienced little difficulty, and by the end of six weeks could teach with comprehensive ease. Strait's teaching method, as Johnson observes, was 'slow and tedious'. In order for the students to have something to take home with them, he wrote everything on the blackboard to be copied. But such a 'tedious and slow' teaching system had its own value. First of all, the students were able to correct his Chin. Second, the students went away with what was in fact a written commentary on the Scriptures. Third, the teacher was able to turn these written lessons into published books for the benefit of all (ibid.: 512).

In this way, during the first three years of the course, Strait produced the 'Old Testament stories in chronological order, from Genesis right up to the time of Jesus, and the New Testament explanatory notes which may serve as a commentary, of which we are much in need' (Strait 1930: FM–303). He also produced many Sunday school lessons and corrected the Chin hymn-book with the aid of his students and a Haka preacher, Rev. Sang Ling; he also added 100 new songs to the 125 hymns already translated and published with the catechism in 1937. Perhaps his most important work was the translation of the whole New Testament with the help of Rev. Sang Ling into the Lai dialect, which was published by the Mission Press at Rangoon in 1940.

After the first batch graduated from the Bible School in 1931, Rev. Strait did not continue the Bible school because the American Baptist Chin mission could hire only six of the graduates for pastoral work. But it was a blessing in disguise. Although the Bible school was discontinued, it was transformed into a teachers' training school where the Bible was still taught. In fact, the two types of school were almost identical in all but name: the students were taught the same lesson by the same teacher in the same building, but for a slightly different purpose. At the teachers' training school, students were trained to become both preachers and teachers. They were even taught first aid, an elementary knowledge of medicine and health care. The advantage of the training school for the Chin was that while it remained a training ground for Christian workers, it also produced a greater number of Chin elite than a Bible school could (whose only purpose was to produce preachers).

When Chin students graduated from the teachers' training school at Haka, they gradually replaced the Karen preacher-teachers, and finally in 1934 all the Karen preacher-teachers were sent home. Following the model of the Karen, this newly emerging Chin elite took over as preacher-teachers in the new Chin society. On weekdays they were teachers at the schools which were run jointly by the colonial government and the Christian mission (the British government paid the bills and the missionaries controlled the educational system, since Dr Cope was the Inspector of the Schools). On Sundays they were the preachers at the newly established churches in Chinram. In this way, this newly emerging Chin elite not only contributed to the rapid growth of the church, but also became the main force for the transformation of the Chin social structure into a new form of 'community of faith', where the church itself stood as a central pillar of society.

Church Growth and the Indigenous Chin Mission

The major factors which contributed to rapid church growth in Chinram after the Anglo–Chin War were the adoption of the Chin language as the medium of instruction in the schools in 1923 and the emergence of Chin professional elite, especially teachers and preachers, who considered themselves Christian workers. Thus Dr. Strait was glad to report:

> Think of a field that was recommended to be abandoned less than fifteen years ago now having over a hundred Christian groups with about fifty organized churches, reporting three to five hundred baptisms a year! This is the record of the Haka field, and the whole of the Chin Hills is just as promising (Strait 1939: FM–303).

From the beginning church growth in Chinram, especially in the Haka area, was intertwined with the indigenous Chin mission to the south. In fact, students who graduated from the mission school in Haka under Dr. Strait's training, and became 'teachers-cum-preachers' in their respective local communities, were in a sense indigenous missionaries who made many converts. But I would like to differentiate between teachers-cum-preachers who worked among their local people and those who were sent to other places than their own local communities and tribes.

The first indigenous mission field to the south was Matupi, the principal village of the Cho tribe of the Ngalang group, and the first indigenous missionaries who worked as teachers-cum-preacher were Sakhong of Aibur from the Zophei area[7] and Lian Kar of Khuapi from the Senthang area. Both were sent to school by their parents in order to become chiefs in their respective villages. However, instead of becoming tribal chiefs, these two young men converted to Christianity while attending the school at Haka under the instruction of Strait, and they finally became teachers and missionaries in the south.[8] They were sent to Matupi, established a school there in 1933, and used the Lai dialect as the medium of instruction. Since then, the Lai dialect has become the *lingua franca* for the whole area of Matupi and the Cho tribe.

At first, according to Sakhong, the school was unattractive for the local people. In those days, Matupi was very isolated and even the British sepoys had no regular station there yet. And like other Chin tribes, the Cho did not welcome strangers. Moreover, as discussed in Part One, warfare between neighbouring villages and tribes was almost a way of life – the mainspring of many political, social and cultural institutions. Although the British brought it to an end, they still had some suspicion and animosity toward strangers. Thus, one can imagine the difficulties that the pioneer teacher-cum-missionary confronted in order to open a new field. Sakhong recorded an experience where he attracted the local boys with his Winchester rifle and the promise of a hunting expedition, for he himself was a good hunter. He let the boys fire the gun and join his hunting expedition if they promised in return to attend his school. Most of the students who enrolled at the newly established school therefore were all 'grown boys who [were] old enough to get married'.[9]

By the summer of 1935, the American Baptist mission had withdrawn completely from the general education mission, except for Dr Cope, who continued as the Inspector of Schools until his death in 1938.This took place because the British government had issued a new order, which prohibited daily prayers and chapel services at government schools. Although all mission schools were turned over to the government in 1924, the schools in many parts of Chinram functioned, as did the buildings themselves, as both schools and churches. This new order was a big blow for newly established churches, especially in a new field like Matupi. The prohibition of daily prayers and chapel services at school made it very difficult to make converts, and the schools subsequently ceased to be the centre of conversion. Personally, it was a big disappointment for Sakhong and Lian Kar. After Dr Cope passed away, Sakhong resigned from his teaching post: he always regarded himself as a missionary, not a government employee, although he received his salary from the government.

After he returned home, Sakhong tried hard to recruit a new missionary to be sent to Matupi. With the help of Rev. Heng Cin, one of the most prominent pastors in the Zophei area, he recruited two men for purely missionary purposes. The recruits were That Dun and Pa Hrek, both from Tluangram of the Zophei area, but they were sponsored by the Bualpeng Area Baptist Church and the Haka Baptist Association. They went to Matupi in 1944 during World War II.

The significance of the indigenous mission, unlike the foreign mission, was that it provided a new relationship between different Chin tribes who in their previous history had made war with each other as part of their way of life. The Christians now emphasized love of neighbour and enemy alike, which provided a fresh ideology and identity for the Chin. In their new society, various Chin tribes could identify themselves as being part of a single people. Not only did the old society and tribal barriers of the religion of *Khua-hrum* break down; a self-awareness of national identity based on their new Christian religion emerged.

NOTES

1 On both sides, my grandfathers were imprisoned in Mingkin and Lashio. That was why my mother got her Burmese name 'Men Tang' (May Than), in memory of

'Mintaingpin', the prison at Mingkin, where her father Mang Hnin, chief of Leitak, served a political prison term for six years. One of my aunts is called 'La-la', derived from La-shio, where her father served his prison terms.

2 The 1st Chin Hills Battalion consisted exclusively of the Zomi tribe who went to France, and the 2nd Chin Hills Battalion, which was formed later. was made of members from the rest of the Chin tribes in Burma.

3 Lt Col Kap Cung Nung, who served as Chairman of Chin State Council in 1974–78, was the only non-Zomi who ruled Chin State during Ne Win's regime.

4 'Memorial Address in Honor of Mrs. Laura H. Carson' given at the request of 'The Woman's Society of the First Baptist Church of Pasadena, California,' on 17 September 1942, by Mrs. Selder W. Cummings.

5 The place is still known as *Zipengkah-nak*, 'where the superintendent was shot'.

6 Interview with Za Peng Sakhong, a former lecturer in History at Mandalay University and the author of 'Lai History'. He is now a political exile in the USA. I interviewed him during the 'Chin Seminar' in Ottawa, Canada, in May 1998.

7 Sakhong is my father and I am able to draw on his diary.

8 Lian Kar remained as a school teacher throughout his long life, but Sakhong resigned in 1938 when Dr Cope died and the American Baptist Chin Mission withdrew completely from the education mission. He became a politician.

9 Pu Sakhong, 'Matupi Saza Rian Ttuannak Dairy' (unpublished personal memoir).

PART THREE

THE CHIN IN A NEW CONTEXT OF INDEPENDENT BURMA

EIGHT

The Burma Act of 1935 and World War II

Vital political events occurred during the colonial period without the Chin being aware of them: the twin promulgation of the India Act of 1935 and the Burma Act of 1935. These arrangements officially separated Burma from the Province of British India, and provided a government structure for Burma Proper with a Burman prime minister and cabinet. Subsequently, Burma Proper was called 'Ministerial Burma'. Although the government structure was created for 'Ministerial Burma', the essential power remained firmly in the hands of the British governor and Westminster. Moreover, Chinram and other non-Burman nationality areas were excluded from Ministerial Burma because they were conquered separately by the British as independent countries. As they were under the direct administration of the British governor, the Chin, Kachin, Shan and Karrenni were not represented in the Legislative Council of Ministerial Burma.

Although Chinram was excluded from Ministerial Burma, Assam (India) and Bengal (Bangladesh), the promulgation of both the Burma Act and the India Act of 1935 strongly affected the division of Chinram into three countries. Before these acts, the division of Chinram into three administrative units under a single authority of the Supreme Government of British India did not affect the separation of Chinram. In other words, the Chin people in the three different provinces of British India could still enjoy the same legal protection and judicial system under the constitution provided by the Chin Hills Regulation. Legally, it was not even a political separation but a mere division for the convenience of administration and communication.

When the Burma Act and the India Act of 1935 were officially implemented in 1937, however, the separation of the two countries strongly affected the division of Chinram. As Burma and India became separate countries under separate viceroys or governors, a boundary demarcation was needed between the two countries. Unfortunately for the Chin, this boundary followed administrative boundaries within Chinram. As a result, East Chinram, or the Chin Hills District, became the Frontier Areas of Burma and West Chinram of present Mizoram became the Lushai Hills

District of Assam State in India. Chittagong Hill Tracts remained in Bengal, or present Bangladesh. Since the Frontier Areas, including East Chimram, were excluded from so-called Ministerial Burma, the term 'Excluded Areas' was also used to denote the Frontier Areas.

In this chapter, I shall explore two different responses to the Burma Act of 1935, one by the Chin and the other by the Burman. I shall also study the parallel but different developments of the Chin Christian movement and the Burmese nationalist movement prior to World War II. Although the Chin were little aware of its political implications, the so-called Crown Colonial Scheme of 1941 should be viewed as the Chin response (or more correctly, the British response on behalf of the Chin) to the Burma Act of 1935. For the Chin, the separation of Chinram because of the Burma Act became a major political issue only after World War II, but the demarcation of borderlines within Chinram already affected the missionary work in 1937, when it was officially implemented. One such example was the dispute between the American Baptist Chin Mission (ABCM) and the Lakher Pioneer Mission (LPM) in 1938. Even more important, however, were the differences between the Chin and the Burman responses to the Japanese occupation in the 1940s.

THE CHIN MISSION AFTER THE BURMA ACT OF 1935

The Demarcation of the New Border and the Dispute between the American Baptist Chin Mission and the Lakher Pioneer Mission

The immediate effect of the Burma Act of 1935 was felt more by the Christian missions than by the political system; it caused a dispute between the American Baptist Chin Mission (ABCM) and the Lakher Pioneer Mission (LPM) when implemented in 1937.

The LPM was founded by Rev. and Mrs Reginald A. Lorrain. Mr Lorrain was a brother of James Herbert Lorrain, a pioneer missionary of the Baptist Missionary Society of London (BMSL) working at Lunglei, in the southern part of Mizoram. R. A. Lorrain did not secure support from the BMSL, so he and his wife left England without any institutional support to establish a mission, and depended upon the contributions of friends and individual supporters. They began the Lakher Pioneer Mision in 1907, and settled at Sherkor among the Lakher or Mara group of Chin in the adjoining area of southernmost Mizoram and the western part of the Chin State. The only long-term missionaries who served in the LPM were Rev. and Mrs. Reginald Lorrain, their only daughter, Louise, and her husband, Albert B. Lorrain-Foxhall. It was a one-family enterprise, which effectively evangelized among the Mara.

Within five years, Lorrain reduced the Mara dialect into writing, compiled two sets of dictionaries containing 7,000 to 8,000 words each, made a complete grammar of the language and prepared a primer for use in the schools. The primer was printed by the Government of Assam free of charge. He also prepared a hymn book with 68 selected songs in Mara, a catechism containing questions and answers using scriptural texts ranging from Genesis to Revelation, and he translated the Gospel of John. In addition, he and his wife established a school for 22 pupils, which enabled some 50 Mara to read and write their language. Eventually, most of the students were converted (cf. Kipgen 1996: 203–295). In 1938, J. H. Cope wrote, 'They have two ordained workers, six unordained, five schools aside from headquarters where there is a boys' and a girls' school' (Cope 1938: FM–283). Cope also mentioned the LPM's literature in his correspondence:

> The New Testament is soon coming out in a second edition, the first having been completed 4 years before the Kamhau [that is, in about 1928]. Genesis is now going through the press and they have other books in the process of translation. They have a radio station. They claim 600 Christians, some of whom are in the Chin Hills, and 4,000 in the Christian community (Cope 1938: FM–283).

As Cope's letter indicated, the LPM's work covered both sides of the Lushai Hills and Chin Hills Districts. The Mara group of the Laimi tribe are spread along both the eastern and western banks of the Bawinu River, which became the natural boundary between India and Burma. Thus, when the borderline was officially demarcated, the LPM became an 'unauthorized mission' in the Chin Hills District. Moreover, the ABCM accused them of 'encroachment of preachers into territory not their own' (Johnson 1988: 590). In this way, the separation of Chinram by alien law aggravated a competitive problem that used to be avoided by so-called mission comity.

To settle the dispute regarding the territory served by the ABCM and the LPM, Mr Naylor, the Deputy Commissioner of the Chin Hills District, invited the missionaries from both parties to a meeting held at Pako, a small village on the border, on 24 May 1938. Cope and Strait represented the ABCM and Rev. Lorrain and Mr Lorrain-Foxhall represented the LPM. Naylor chaired the meeting. Cope mentioned in his letter that Lorrain was two years his senior, good-looking, grey-haired, and he 'was especially pleasant to me. He had long looked forward to meeting me. Lorrain's brother, who was at Lungleh, has retired and is finishing a Lushai dictionary at home' (Cope 1938: FM–283). While Lorrain pleasantly expressed his concern for the Mara people as a whole, regardless of the artificial boundary or foreign law, Cope on the other hand seemed to act like an over zealous missionary-cum-colonial-officer, saying,

> I then summed up, giving four reasons why I believed they should not come into the Chin Hills at all. They were shocked when I declared I was against their coming at all. They made no summing up and neither did Strait. The chiefs in this area were called in and

all said they preferred having the Haka mission work in their country (Cope 1938: FM–283).

Unfortunately, nowhere in his letter did Cope spell out the details of the agreement or the decisions they made at the meeting, nor did Strait.[1]

What we know from the present situation, however, is that Rev. Lorrain and his LPM work were forced to confine themselves to the western (Indian) bank of the Bawinu River, and the Mara people on the eastern (Burmese) side were forced to join the ABCM. This is one illustration of how the Burma Act and the India Act of 1935 forced the separation of the Chin people into two countries.

The Death of Rev. Cope

After the meeting at Pako, Strait went to the Matupi area to inspect the schools but Cope and Naylor returned directly home. On their way back, Dr Cope suffered a severe attack of malaria. When they reached Haka, his malaria was followed by dysentery. Weakened by sickness, he could not continue his journey to Tiddim where his family awaited his return. On 5 June, Dr Strait returned from his tour, and found that his colleague was very weak and sick. He immediately sent a telegram to Falam and requested the Civil Surgeon to come to Haka. He also sent a telegram to Mrs Cope to tell her that her husband was dangerously ill and that she should come, but before she could reach Haka, Rev. Cope passed away on 11 June 1938 in his 55th year.

As Cope was the longest serving missionary in Chinram, his contribution to the Chin mission was most varied. His work was recognized by many different groups and peoples, from the colonial power to the Chin natives and his fellow Americans. In 1927, Rev. Cope was awarded the KIH (Kaiser-I-Hind) medal, one of the most prestigious medals for civil servants of the British Colonial Crown, in 'recognition of unselfish service, which he had rendered the Chin people'.[2] In 1934 when he was in America on furlough, Colgate University awarded him the degree of Doctor of Divinity.

His contribution to the Chin mission was enormous, and Dr Strait paid warm of words of tribute to him as follows:

> His work among the Chin had affected every phase of their existence. He established and built churches, he administered to the spiritual needs of the people, teaching and preaching, and at all times was keenly interested in their health, economic life and mental development, rendering continual service (Cope 1938: B–201).

Robert Johnson also commented on Dr Cope's contribution to the American Baptist Chin Mission as follows:

> Dr and Mrs Cope came to the Chin Hills in 1908 when the Christian work was still in its infancy. When he died, the churches were well

> established and viable. It is quite possible that even if the American Baptist Mission had been forced then, in 1938, to stop its work entirely, the churches would have been rooted strongly enough to survive and grow (Johnson 1988: 602).

J. Herbert Cope was buried in the Haka mission compound where Arthur Carson had been buried some thirty years earlier. An open chapel was built over their graves by the Haka Baptist Church in 1984, where Chia Khaw, the first convert from the Haka was also buried, as a memorial to these pioneer Christian missionaries and workers. In Chinram, two churches and a publishing house were named in his honour: the Cope Memorial Baptist Church at Tiddim and Tawnzang, and the Cope Memorial Press at Tiddim.

The Baptist Mission on the Eve of World War II

In August 1938, just two months after Cope's death, Strait wrote a long and impassioned letter to Dr Robbin and strongly demanded that the ABFMS send more missionaries for the work in Chinram. In this letter, he reminded Robbin that 'when Dr Cope took over the position as Honorary Inspector of Schools fifteen years earlier, he (Cope) felt that he should stick with the job until the government schools were established'. That task was now accomplished. Cope had been forced to give so much time to the schools that evangelism had lagged in the Tiddim area, and now it was time to press on with full-time 'evangelism and reap the harvest of past seeds sown by the Chin Hills missionaries'. In the same letter, Strait also requested the opening of a new station at Falam. He said that the Haka and Tiddim fields had about one hundred Christian groups each, 'but the Falam district fell behind because of the lack of supervision. The work is too great, too big, for even two missionary men; a third must be sent out' (Strait 1938: FM–303).

In 1939, he wrote another long and eloquent letter to Robbin in which he compared the missionary work in the eastern and western parts of Chinram, then called the Chin Hills and Lushai Hills. The western part of Chinram, was the missionary field of the English Baptist and Scottish Presbyterian Mission Societies. Strait pointed out that among the Mizo (Lushai), there were 60,000 Christians, while there were only 5,000 among the Zomi and Laimi tribes in the Chin State. The 'Lushai work began about the same time as the Chin work, but the difference lay in the fact that among the former there were 10–12 missionaries working at all times. If the Baptists could supply more missionaries, the same result could happen among the Chins,' he said. And he continued:

> I am not asking for ten. I am only asking for three missionary families for the Chin Hills, and we shall, with three well-trained men and their wives, in the condition of the field now, step up with the Lushais

in almost a decade. After initial expenditure for buildings, I am asking for $ 7,000 a year to minister to 150,000 people. What city in America of that size begrudges $ 7,000 spent by its people for the cause and benefits of religion? What church spending a budget of $ 7,000 a year baptizes a thousand persons a year and reaches into every phase of the people to whom they administered their religious life in all its phases, their social life, their economic life and agricultural life, and their future planning? What church spending $ 7,000 a year is expecting to see a harvest of twenty thousand souls in the next decade or little more?

My wife and I have been willing to bury ourselves in this far-away place, which is a modified form of solitary confinement, that we may be a blessing to a wretched and ill-favored group through the presentation of the Gospel, which lifts a people like this more than even imagination can picture to the average Christian mind in America. If a missionary has trained himself well and is willing to make sacrifices, he should not be forgotten. Think of a field that was recommend to be abandoned less than fifteen years ago now having over a hundred Christian groups with about fifty organized churches, reporting from three to five hundred baptisms a year! This is the record of the Haka field, and the whole of the Chin Hills is just as promising (Strait 1939: FM–303).

In response to Dr Strait's request, the Mission Society in New York appointed two missionary couples for the work in Chinram. The first couple, Franklin and Phileda Nelson, were sent to the Tiddim station in Chinram in November 1939 for the replacement of Dr Cope. The second couple, Robert and Elizabeth Johnson, was appointed and scheduled to sail in 1941, but their travel plans were disrupted by World War II.

Franklin Nelson was born on 30 November 1909 in Minneapolis, Minnesota, and came from the family of F. O. Nilsson, the first Swedish Baptist. He was baptized in March 1921 at the Swedish Baptist Church in Turlock, California, where the family lived for a couple of years. He received his Bachelor of Theology degree from Bethel Seminary in St.Paul, Minnesota, in 1934 and graduated with a Bachelor of Arts at the University of Minnesota in 1938. During the war years, he studied at the Kennedy School of Mission, Hartford, Connecticut, and got a Master's degree in 1944. After his second term he returned to the seminary again and earned a Bachelor of Divinity degree in 1952. Phileda Mai Ogren also came from a Swedish-American family and was born on 21 December 1916. She received her education at St. Cloud Teacher's College. During the war years, after returning from their first term in Chinram, she attended the Hartford School of Religious Education and received a Bachelor of Religious Education (BRE) degree in 1944. Phileda and Franklin were married on 7 June 1939 and appointed as missionaries by ABFMS and designated to Tiddim. Phidela wrote about their first experience of Tiddim:

> Suddenly, on the spur of the mountain in front of us lay the place where God had called us, the village of Tiddim. As we viewed it from across the valley it was flawless, so beautiful as the Sun seems to warm it and make it glow. This was the field where we would serve. It was the place where we would establish our first home. I stopped, fell silent, and there was a lump in my throat. All along God has said to us as He did to the Psalmist, 'I will instruct thee and teach thee in the way which thou shalt go; I will guide thee with mine eyes' (Ps 32: 8). He had done that, and here we were viewing the spot to which He had led us (Nelson 1940: FM–343).

Their willing service as missionaries in Chinram was interrupted by World War II, and the Nelsons left Chinram in 1942.

Strait's Final Report

On 3 January 1941, just a year before Burma became a battlefield, the Straits left Chinram for good. At first it was supposed to be a furlough but the war prevented their return. During the war Chester strait served as an army chaplain. After the war, although he seemed willing to return, he was not sent back to Haka.

Just three days before he left Haka for the last time, Strait wrote his annual report for the year 1940. This report – rather like a summary of his entire work as a missionary in Chinram – gives an excellent view of the missionary's own assessment of his service and the Christian movement in Eastern Chinram during the last decade before the war. He wrote:

> The close of this year also terminates the missionary's second term of service. Less than twenty years ago the Haka subdivision was always looked upon as an unproductive field that would never show any results, and recommended to be abandoned. My wife and I came into this field in 1925 when there were some sixty Christians and one or two weak, struggling churches. There are now a hundred churches in the section and something like 6,000 baptized and unbaptized Christians. The Haka field is sufficiently manned and the work is well organized with local and district associations and Bible assemblies. This year there may be as many as five hundreds baptisms in this section that two decades ago was so unpromising. Our whole Chin Hills area reports an annual number of well on to a thousand baptisms a year.
>
> Some of the best preachers we have are those trained in the Haka Bible School, which was conducted for four years, without outside help. The salary budget for the entire Chin Hills is about forty-four hundred rupees a year, of which amount the Chin (25 for the whole of the Chin Hills)[3] of the Haka area, there are perhaps sixty other men duly appointed by the local association to act as lay pastors of their respective local churches.

He described his literary achievements as follows:

> These Christians now have a hymnbook of over three hundred hymns with a catechism in the back, the New Testament, and a Christian manual and Sunday School lessons of over three hundred pages. All this translation was done without any extra appropriation. Usually when one does translating he is given a native helper, but since finances were so low, all the translation which has been accomplished in the Haka dialect has been done without any additional expense to the Society.
>
> The Haka Chins now with the New Testament in their own dialect may be expected to gain in strength and wisdom. The New Testament has a much further radius than merely the Haka subdivision, as one finds these Bibles in all parts of the Chin Hills. Perhaps that is one of the growing features of the work, but change from debauched drunkenness to industriousness and thrift is a transformation no less marvelous, which has occurred time and again all over the subdivision. Imagine a group where only one or two can read, and then within a few months find all of the young people reading the hymnbook and Bible, although the majority of them never went a day to school in their lives. This may be witnessed in church after church in the various out-villages.
>
> Such then, in brief, are the results at the end of this year and the second term of service. It was accomplished by patience and a faith in the Living God (Strait 1940: FM–341).

As already implied, the Chin themselves and the leaders of the Chin Hills Baptist Association were not primarily concerned with the promulgation of the Burma Act of 1935. Strait's report is the most reliable source for our information about the stages of Christianity among the Chin up to World War II.

THE CROWN COLONIAL SCHEME

In 1937, when the Burma Act of 1935 was officially implemented and Burma was separated from the Province of India, Sir Robert Reid, the Governor of Assam, strongly protested against the policy of the Supreme British India government, which had adopted the administrative boundaries within Chinram as the boundary between Burma and India, and later between India and Bangladesh as well. He thus wrote to the government in London that 'the separation of Burma from India on lines of the present frontiers will permanently divorce portions of tribes, which naturally should comprise a single unit' (R. Reid 1942: 6).

From the very beginning, Sir Robert Reid, like B. S. Carey, was in favour of forming a united Chinram and strongly opposed to the idea of a Chinram divided into three administrative units and thereby three countries. As he could foresee the dangerous consequences of dividing Chinram, he updated the provision of the Inner Line Regulation of 1873 in 1935 in order to protect the Chin and other tribal groups in what is now Northeast India from the exploitation of outsiders. The British policy of inner line regulation was designed to prohibit the permanent residence in the area by persons not native to it, and it therefore became known as Excluded Areas administration. The guiding philosophy of Excluded Areas administration was based on the independent status of Chinram before the colonial period, and also on the traditional Chin religious concept of exclusivism, in which outsiders were excluded during religious sacrificial ceremonies.

In 1941, Sir Robert Reid made his strongest proposal, known as the Crown Colonial Scheme of 1941. He declared that 'they (the Chin) are not Indian in any sense of the word, neither in origin nor in outlook, and it is a historical accident that they have been taken into an Indian province' (R. Reid 1942: 6). He thus proposed to form a separate colonial province, what he called the Chin-Lushai Province, which would have had its own governorship and the same political and administrative status as the governments of Assam, Bengal and Burma. His proposal was accepted by the Conservative Party in London led by Sir Winston Churchill. Unfortunately, World War II prevented the implementation of Sir Robert Reid's proposal of the Crown Colonial scheme in 1941.[4]

I shall come back to later developments in the next chapter. The attention will now be turned to the critical response of Burmese nationalists to the Burma Act of 1935.

THE RESPONSE OF BURMESE NATIONALISTS

The Rise of Traditional Burmese Nationalism

Until 1938, when Aung San became the leader of the Dobama Asi-Azone (We the Burman Organization), which was founded by Ba Tawng and his fellow nationalists on 30 May 1930, Buddhism was an extremely important factor in the development of the Burman national movement. In terms of 'traditional national identity and the choice of nationalist issues', the history of the Burman had been built upon a foundation of 'Buddhism as religion and political ideology'(Schecter 1967: 106). In other words, the tradition of the Burman national movement until that time was more or less a revival of Buddhism as a political ideology, and most of the nationalists who played a leading role in that movement were either monks or lay religious leaders. Thus, 'traditional nationalism' among the Burmans, as Smith argues, was not the product of British rule, as in India, but a 'clearly indigenous, traditionalist pattern of nationalism' (D. E. Smith 1965: 113), in which national and religious symbols were fused together. Indeed, traditional nationalism among the Burmans was 'not simply anti-British sentiment and a movement for freedom from foreign rule'. Instead it was based, among other things, 'on a common race, language, and religion' (ibid.: 112). The notion of '*Buddha-Bata Myanmar Lu-myo:* To be a Burman is to be a Buddhist' can therefore be seen as an integral part of the Buddhism-oriented national movements.

As mentioned briefly in Chapter 5, the notion of *Buddha-Bata Myanmar Lu-myo* had begun already in 1044, when King Annawratha founded the first Myanmar (Burman) Kingdom of the Pagan Dynasty. Since then, Burman 'history has been built on Buddhism as religion and political ideology', and the kings were regarded as 'the defenders of faith, the promoters of Buddhism, builders of pagodas, and the patrons of the sangha' (Schecter 1967: 106). Thus, Buddhism was a traditional Burman national symbol, and 'Buddha was an important figure in the national heritage, and many Burman villagers refused (and still refuse) to believe that he was an Indian. The Dhamma was ideally a national way of life; the Sanga was the visible symbol of national religion; the Faithful Ruler (Buddhist King) became a chief defender of faith' (D. E. Smith 1965: 113). In this way, Buddhism (Buddha, Dhamma and Sanga) provided a vital component of the corporate identity of traditional nationalism in which religion (Buddhism) and state (the Myanmar Kingdom) were inseparably intertwined.

This traditional pattern of relationship between state and religion came to an end in 1886, when the British deposed King Thibaw. The extinction of the monarchy left Buddhism, a religion of the nation, without a defender of the faith – a king who appoints and rules the Buddhist hierarchy. Donald E. Smith suggests that the British did not systematically suppress Buddhism, but the decline of Buddhism in Burma under the British was 'more the result of the social, economic, political and intellectual upheaval brought about by the sudden confrontation with the West, than of government policies dealing specially with religion' (ibid.: 80). He also observed the link between religion and nationalism:

> The British had destroyed the traditional monarchy with one stroke, but by 1920 it was clear that foreign rule was also rapidly eroding the status and prestige of the monk. There was no place for him in the new western-oriented social hierarchy, his educational functions were assumed by other agencies, an unknown foreign language prevented him from understanding what was going on, and westernized Burmese laymen increasingly regarded him as irrelevant to modern life. Of all sections of Burmese society, the *pongyis* had the strongest reasons for hating the British and became the most uncompromising nationalists (D. E. Smith 1965: 93).

It was not surprising that the first anti-British traditional nationalist movements were organized and led by either the monks themselves or religiously oriented organizations, such as the Young Men's Buddhist Association (YMBA, f.1908) and the General Council of Buddhist Associations (GCBA, f.1911), which took an active part in demonstrations in the 1920s against the British. The first open resistance was sparked by 'the practice of Europeans in Burma wearing shoes while walking on the pagoda premises', contrary to the Buddhist custom of removing all footwear before entering the hallowed pagoda areas. As a religious issue, the 'footwear controversy' became the 'first nationalist issue, so a religious leader, a monk U Ottama, became the first popular nationalist hero and martyr' (ibid.: 95).

U Ottama lived in India for some time as a student at Calcutta University. He traveled widely to such Asian countries as Japan, Korea, Manchuria and China. When in India, he served as president of the Hindu Mahasabha, and was intimately acquainted with activists and principles of the Indian National Congress. He returned to Burma in 1921 bringing with him the inspiration and techniques of Gandhi's non-violence movement and organized *Wunthanu Athin* ('the nationalist association'). The supporters of *Wunthanu Athin,* like Gandhi's followers in India, boycotted foreign goods and wore homespun cloths. His political movement got inspiration from the idea that 'Buddhism was threatened with eclipse by an alien government'. He also insisted that 'the *pongyis* themselves must leave the monasteries to defend the faith' (Cady 1958: 231). He argued that 'if the Buddha were alive he would not preach on political topics but on Nirvana. However, in light of the Burman enslavement to the British, the people should not ask for Nirvana yet'. 'When Lord Buddha was alive', he said, 'man had a predilection for Nirvana. There is nothing left now. The reason why it is so is because the government is English' (ibid.: 232). Nirvana is the release from cosmic slavery (the wheel of rebirth), but it cannot be attained by those who do not even enjoy earthly life. '*Pongyi* pray for Nirvana but slaves can never obtain it, therefore they must pray for release from slavery in this life' (cited in E. Smith 1965: 96). U Ottama was arrested in 1921 on a charge of incitement to sedition. He was released in 1922, but jailed again from 1924 to 1928, and yet again later in 1928 until his death in 1939, when he became a martyr.

When the non-violence method of U Ottama failed under the strong pressure of the colonial power, the religiously oriented traditional nationalist movement prompted an open rebellion, led by Saya San in 1930–31. Saya San himself was an ex-monk, a quack doctor and formerly a member of the monk-influenced GCBA's So Thein faction. He proclaimed himself king with the title of *Thupannak Galon Raja* and declared: 'In the name of our Lord Buddha and for the Sangha's glory, I, *Thupannaka Galon Raja,* declare war upon the heathen English who enslaved us ... Burma is meant only for Burma ... the heretics removed the King by force. They have ruined our race and religion. The heathen English are rebels' (cited in

Gravers 1996a: 297). Saya San prayed and asked help from 'Guardian spirits of Gotama Buddha's religion, together with the inside 37 *nats*, outside 37 *nats* and *Weik-za*, I entreat you to help my soldiers to victory and protect them from any harm'. The symbol of his army was the *galon* (or *garuda*) of Hindu mythology, a bird which attacked and destroyed the *naga* (snake). The aim of the Saya San rebellion, as Gravers observes, was 'to challenge the British with a legitimate political authority based on the symbolic power of Buddhism and its cosmology' (ibid.: 299).

The Saya San rebellion spread widely, largely through the activities of itinerant *pongyis* in the village nationalist associations, and its complete suppression by the government took eighteen months. The colonial government deployed 10,000 soldiers from the British India Army, including the Karen and Chin Hills battalions, to suppress 'poorly-armed peasants who relied on magic tattoos for protection against bullets' (Gravers 1996a: 303). On many occasions, as the government's forces resorted to brutal suppression, entire villages were burnt to the ground, 'suspected rebels were decapitated and their severed heads displayed as warning to others' (Kin Oung 1993: 20). In one particularly gruesome incident, fifteen severed heads were displayed in front of the Deputy Commissioner's office in Prome. Photographs of the gory exhibition appeared in the *Sun* newspaper, owned and edited by U Saw, and copies were sent to the British parliament. In the end, 'more than 3,000 rebels were killed or wounded, and 9,000 were interned. 1,389 were given prison sentences or deported, and 128 were hanged, including Saya San' (Gravers 1996a: 303).

As Gravers observes, the Saya San rebellion had a 'deepening ... religious and ethnic opposition character' made up of traditionalist Burman nationalism, which opposed not only the British colonial power but also non-Burman and non-Buddhist nationalities like the Chin, Kachin and Karen, because 1,600 Christian Karen and Chin soldiers were deployed by the British government 'in hunting down the rebels'.[5]

That was the reason why the BIA (Burma Independence Army) took revenge on the Karen during the Japanese occupation, but thanks to the 'Chin Levies' the BIA could not extended their vengeful military rule to the Chinram, as we will see below.

Aung San's Radical Option

When Aung San became the leader of Dobama Asi-Azone in 1938, the nationalist movement became the flag bearer of a purely 'secular nationalism'. They regarded 'political independence for Burma as an end in itself and saw no need to resort to religious appeals to fight their cause' (U Maung Maung 1980: 115).

Aung San entered the University of Rangoon the year after the Saya San uprising had been suppressed and its leaders executed. For Aung San, it seemed that the decapitation of Saya San's rebels refreshed the memory of the fate of his own uncle, U Min Yawng, who led 'one of the earliest resistance groups against the British until he was captured and beheaded' (Aung San Suu Kyi 1984: 4). Although he 'did not find Saya San himself an attractive figure', Aung San and many of the young intelligentsia in Rangoon and elsewhere, 'were stirred by the courage and the national spirit of the rebels and moved to pity by harsh reprisals taken by the government, and it aroused greater sympathy than might have appeared from the degree of active involvement' (ibid.: 6). Thus, as his life proved from the very beginning of his university days to the end at the age of 32, the total picture of Aung San was 'one of the young men of great integrity and strong character who led his country to independence with single-mindedness and a high sense of purpose' (ibid.: 36).

At Rangoon University, Aung San immediately proved to be an outstanding student, and he was avidly involved in student politics. The 1930s was an era when new nationalist leaders began to emerge from the youthful ranks of Rangoon University, working hand-in-hand with other nationalists off the campus under the banner of Dobama Asi-Azone (also known as the *Thakin* movement, led by such prominent figures as Ba Thawng and Thakin Kodaw Hmaing, the latter known as the Grand Old Man of Letters and Politics). In 1935, as the editor of the Rangoon University Student Union's Magazine, *O-way*, Aung San published an article, 'Hell Hound at Large', which indirectly helped to trigger off the far-reaching university strike, very well known in Burmese history as the 'University Strike of 1936'. The article was taken to be 'a scurrilous attack on a university official, and the authorities demanded the name of the author from Aung San, who refused on the grounds that it would be against journalistic ethics. This provided the administration with an excuse to serve the editor with an expulsion notice' (ibid.: 9).

Shortly afterwards, U Nu, the president of the Student Union, was also expelled because of his critical speeches. The Student Union responded with a nation-wide strike. As a result, the 'government was forced to consider the grievances of the strikers seriously, and the eventual outcome was the retirement of the high-handed principal, and the forming of a committee to look into the amendments the students wanted to be incorporated into the University Act' (ibid.: 9). Eventually, Aung San and U Nu were readmitted. Not only did the 1936 strike make Aung San a widely known student leader; it was also an important landmark in the political development of young nationalists, such as U Nu, Hla Pe (later known as Let Ya), Thein Pe and Kyaw Nyein, who were to become figures in the Burma independence movement and leaders of independent Burma. As his prestige grew steadily, Aung San became president of the Rangoon University Student Union and the All Burma Student Union in the following academic years.

In order to devote himself to full-time national politics, Aung San left the university in 1938 and became Secretary General of Dobama Asi-Azone. In that same year, he drafted a manifesto of Dobama Asi-Azone in which he clearly opted for a more radical, non-racial and non-religious approach to the independence movement. Aung San stated in this manifesto that the Dobama Asi-Azone with its Thakhin movement was 'the only non-racial, non-religious, and impersonal movement that has ever existed in Burma. Formerly, and still now among a certain section of the Burmese public, nationalism was conceived in terms of race and religion' (Kyaw Yin 1969: 115). In his concept of a nation, Aung San explicitly rejected the centrality of race and religion. In a manifesto of Dobama Asi-Azone, which he repeatedly quoted in his speeches delivered in the 1940s, Aung San declared:

> A nation is a collective term applied to a people, irrespective of their ethnic origin, living in close contact with one another and having common interests and sharing joys and sorrows together for such a historic period as to have acquired a sense of oneness. Though race, religion, and language are important factors it is only their traditional desire and will to live in unity through weal and woe that binds the people together and makes them a nation and their spirit a patriotism (ibid.: 115).

Although Aung San claimed that the Dobama Asi-Azone was the 'only non-racial, non-religious movement that ever existed in Burma', some elements of traditional nationalism, which blended the nationalist movement with the revival of Theravada Buddhism, still existed, as it was the founding principle of the organization when it was established in 1930. This Buddhism oriented traditional nationalism was represented by such prominent figures as Tun Ok and Ba Sein. Thus, while Aung San's policy, defined by an inclusive radical secular approach, allowed a certain level of inclusiveness towards the non-Burman nationalities in the Burma independence movement, this same policy caused a split of the Dobama Asi-Azone into two factions in March 1938. A group who opposed Aung San's policy of inclusivism and secularism was led by Tun Ok and Ba Sein, and thereby known as the 'Tun Ok – Ba Sein' faction. The remaining majority faction was led by Thakin Kodaw Hmaing and Aung San. Although each claimed to be the Dobama Asi-Azone, 'they were in reality two separate parties' (U Maung Maung 1989: 20).

While Kodaw Hmaing and Aung San opted for a 'non-racial, non-religious secular approach', Tun Ok and Ba Sein centred their political conviction on race and religion, namely the Myanmar race and the religion of Buddhism. While Kodaw Hmaing and Aung San advocated democracy and a federal union, Ba Sein and Tun Ok were in 'favor of a totalitarian form of national organization'. They both admired Hitler and Mussolini, believed in Fascism and declared that 'totalitarianism would benefit Burma'. They also 'favored restoration of the monarchy' (ibid.: 21), which was inseparably associated with the state religion of Buddhism. Buddhism for them was not just a religion but a political ideology as well, and they could not conceive of religion without a defender of the faith, i.e. the 'king who appointed and ruled the Buddhist hierarchy' (cf. Schector 1967: 105–127). They proposed the revival of monarchy as the best means of independence.

As Tun Ok and Ba Sein had opted for exclusivism for Buddhism and the Burman race under such slogans as 'one race, one blood, one voice,' and 'a purer race, a purer religion and a purer language' (Khin Yi 1988: 255), they excluded non-Burman nationalities, such as the Chin, Kachin and Shan, and they even ignored the existence of these nationalities and peoples, as we shall see below. Aung San's policy of radical secularism was the reason why he was assassinated in 1947, and why confessional Buddhists in Burmese politics had rejected his policy of a secular state even before Burma gained her independence.

WORLD WAR II AND THE JAPANESE INVASION

Alliance with the Japanese

In 1939, soon after the outbreak of war in Europe, the Burman nationalists founded the 'Freedom Bloc', an alliance of Aung San's radical wing of Dobama Asi-Azone, Ba Maw's Sinyetha Party, students and some individual politicians. Dr Ba Maw was an older generation politician and a Law-at-Bar who became the first Prime Minister of Burma according to the Burma Act of 1935. Dr Ba Maw was chosen as the chairman while Aung San became the general secretary of the organization. The 'Freedom Bloc' sent their message to the British authorities stating that 'they should support the British effort only if they were promised independence at the end of the war; if the British government was not prepared to make such a declaration, the people should oppose the war effort strenuously' (Aung San Suu Kyi 1991: 12). They took 'colonialism's difficulty' as a chance for 'freedom's opportunity'.[6] The colonial authorities responded to the nationalist movement by making large-scale arrests.

By the time of the Japanese invasion of Southeast Asia, Ba Maw and many nationalists were arrested but Aung San and his comrade Hla Myaing (later known as Yan Aung) fled to Amoy in China. At first they intended to establish contact with Chinese communists but this did not materialize. Instead, 'they were approached by Japanese agents and flown out to Tokyo to meet Colonel Keiji Suzuki, an officer of the Japanese army who was to become famous as the head of the Minamikan, a secret organization whose task would be to aid Burmese independence and to close the Burma Road' (ibid.: 13–14). In this way, Aung San and his legendary 'thirty comrades' received military training from the Japanese army and formed the Burma Independence Army (BIA) in 1941 at a time when the Japanese occupied Bangkok. Colonel Suzuki acted as commander-in-chief of the BIA. He was given a Burmese name, Bo Moe Gyo, meaning 'Thunderbolt', when he assumed the command of the BIA.

Burma became a battlefield of World War II when the Japanese bombed Rangoon on Christmas Eve, 1941. Meanwhile, the BIA began their military operations and marched alongside the invading Japanese ground forces. Thousands of other young Burman nationalists joined the BIA. Since the Japanese Imperial Army was marching with the BIA, they easily eliminated the British Burma Army with the support of the Burman nationalist movement and captured Rangoon on 7 March 1942. In May 1942, the governor of Burma fled to Simla and established the Burmese government-in-exile there. Having successfully driven the British into India, the Japanese army and the BIA occupied Burma proper.

When the BIA set up its military administration in the Irrawaddy delta area, where the majority of the population was Karen, a violent racial conflict took place between Burmans and Karen, especially in the Basein and Myangmya areas. The cause of this racial conflict, as the Swedish journalist Bertil Lintner observed, was that 'the Burman took revenge' (1989: 36), because the British authority had deployed the Karen and the Chin battalions of the British Burma Army to crush the Burman nationalist movement of the Saza San rebellion in the early 1930s. During the colonial period, the British recruited for the British Burma Army among non-Burman nationalities. At the close of the war it was 22,000 men strong. Of this number, 3,000 were Chin, 2,000 Kachin, 2,000 Karen and only 1,893 were Burman. The rest were Indian and Gurkhas. The Karen were trained especially as officers. Moreover, Chin, Kachin and Karen battalions in the British Burma Army were the forces which resisted and pushed back the invading Japanese army and the BIA, in order to defend the falling Empire. Thus, the BIA took revenge when they got the upper hand. Bo Tun Hla, a BIA officer, for instance, shot seventeen Karen elders at Papun. In Myangmya, 150 Karen were slaughtered, including a former cabinet minister, Saw Tha Pe, and his English wife and their children. The Japanese and the BIA ended the bloodbath after murdering more than 1,000 Karen civilians (cf. M. Smith 1991: 440; Gravers 1993: 45, note 46; Lintner 1989: 36; U Maung Maung 1989: 88–89).

This racial conflict between the Burman and the Karen was troublesome for Aung San who formulated from the very beginning a political ideology based on inclusive pluralism, for a 'non-racial, non-religious' secular society. He always wanted to maintain, as his daughter Aung San Suu Kyi observes, 'a good relationship between different races in Burma, knowing that it was essential for the unity of the nation'. She continues:

> In the 'Blue-Print for Burma' which he had drawn up for Suzuki in 1940, he had already stressed the need to 'bridge all gulfs now existing through British mechanisms between the major Burmese race and the hill tribes, Arakan and Shan States, and united all into one nation under the same treatment. The conflict between the Karens and the Burmans greatly troubled him throughout the latter part of 1943; he, Than Tun and Let Ya labored to bring about peace and understanding between the two races (1991: 20).

Although the racial conflict between the Karens and the Burmans was resolved in 1943, as Aung San Suu Kyi observes, the conflict between the two races returned as a full-scale war in 1949.

Meanwhile, in August 1943, the Japanese granted Burma the status of 'a sovereign independent nation and a co-equal member of the Greater East Asia Co-Prosperity Sphere' (ibid.). Even though the Japanese retained ultimate sovereignty, Dr. Ba Maw was installed as the head of state with the title of *Adipati* (Great lord) as well as Prime Minister, and Aung San became Minister of Defence. Although the establishment of an independent government was little more than nominal, the existence of the independent government gave Aung San the opportunity to strengthen his Burma National Army (BNA as the BIA was now called) to build up his political organization under the protection of his own defence forces. This would have an immediate effect on the Chin attitude towards the war.

Japan Ral: The Chin Resistance against Japanese Invasion

In early 1944, after the military administration was established all over Burma Proper, the Japanese launched a series of military offensives into Chinram, the strategic borderline between Burma and India.

By the time the Japanese began their military offensives, the Chin were well prepared to defend their land. Knowing about the atrocities committed by the BIA on the Karen people, and also the brutality of Japanese military rule, Chin soldiers of the British Burma Rifles returned to Chinram, together with a retreating British army, in order to defend their people. They feared that the BIA might take revenge against them because the Chin Hills Battalion had also been deployed to crush the Saza San rebellion. The British officers, particularly Colonel H. N. C. Stevenson, the Superintendent of the Chin Hills, used the mistrust of the Chin towards the Burman as an opportunity for the colonial war effort and recruited Chin soldiers as quickly as possible. He thus formed the Chin Levies in order to defend Chinram against the advancing Japanese army and the BIA.

Recruitment and training of the Chin Levies was under the administration of Colonel Stevenson and G. E. Naylor, with the objective of building resistance to protect Chin villages. Accordingly, two cadet Levies – A and B – were established. The B Levies were not professional soldiers. Their role was to defend their own villages effectively. This required that they be armed and that they be reliably on call at rotational intervals to ensure that each possessed a pool of runners from which inter-village messengers or porters could immediately be available at any hour of the day or night. The B Levies were rationed and paid only while on duty.

The A Levies were a full-time force and to a large extent recruited from discharged sepoys of the British Burma Rifles. Many of them were survivors of action in Burma where the Japanese and the BIA launched the invasion that wiped out their units in the British Burma Rifles. The A Levies fought side by side with the Chin Hills Battalion, which retreated together with the British Army to Imphal. The Chin Hills Battalion became the backbone of General Wingate's army, legendary in World War II and well-known in history as the *Chin-dit* Army.

According to Colonel Stevenson, commanding officer of the Chin Levies, his men killed '214 Japanese in the first eight nights of ambush'. The Levies captured maps, war diaries and considerable equipment. He also gave this tribute to the Chin Levies:

> The loyalty of the Chins to me and my small handful of brother officers passes belief. At a time when the whole might of the Commonwealth was collapsing about their heads, they stood firm by our side facing the mortars and machine guns of Japan with shotguns and ancient flintlocks ...
>
> Their resistance to the Japanese in the critical years after 1942 may well have decided the fate of India, for the infiltration tactics which led to the downfall of Burma were frustrated by their skill as jungle fighters (Stevenson 1963: iv).

Sir Reginald Hugh Dorman-Smith, GBC, Governor of Burma, wrote the following in the foreword to the book *The Economics of the Central Chin Tribes* by H. N. C Stevenson:

> Had the Chin let the Japs pass through on their conquering way into Manipur and Assam, the difficulties that would have befallen India are beyond computation. The Chins and Kachins as well as Nagas have come into the limelight as a result of the Japanese invasion of Burma. They have shown themselves to be sturdy guerrilla fighters as the Japanese have good reason to know ...
>
> Aided only by their mountainous environment and small irregular forces composed of the Local Frontier Forces Battalion (that is, the Chin Levies), itself largely Chins and discharged sepoys of the Burma Rifles, the people succeeded in spite of their paltry number and inadequate arms in throwing that enemy back from their borders. It is not easy to assess the service they have done to us by that lonely stand, but this we can say, that ... one thing is certain, and that is the empire as a whole and India in particular owe a very great debt to these sturdy hill men.[7]

The Chin Hills Battalion and the Chin Levies effectively stopped the BIA military administration east of the Chindwin River and successfully delayed the invading forces of the 33rd Japanese Army under Lt. General Yanagida at the battle of Tiddim. In the course of its advance down Tiddim road, '1,316 Japanese were killed – fresh corpses counted on the ground – another 533 wounded, and 53 were taken prisoners' by the Chin Hills Battalion and the British 17th Indian Division (cf. Vumson 1986: 173). This delay, according to Field Marshall Sir William Slim, was decisive for the Battle of Kohima-Imphal, where the Allies finally defeated the Japanese forces (cf. Slim, 1956).

THE CHIN CHRISTIAN MOVEMENT DURING WORLD WAR II

During the war, from April 1942 to February 1946, there was no foreign missionary present in Chinram. The Christian movement during that period was therefore carried on by the Chin themselves. As indicated above, the Japanese occupied a certain part of East Chinram, including Haka, Falam, and Tiddim. There was severe fighting in the Tiddim area in 1944 and 1945 when British troops fought back into Burma. Thus, the Chin people faced great hardships because most of the villages in the northern and central parts of East Chinram were destroyed in the struggle.

Although the churches faced great difficulties, the local leaders responded to the challenge faithfully, and the number of believers increased from 4,000 to 9,000 in the Haka area alone.[8] Chin scholar Za Tuah Ngur views 'the absence of foreign missionaries' during the war as 'one positive benefit of the war', not only because of church growth but also because it fostered 'the development of local leadership' in church activities (Za Tuah Ngur 1993: 71). Robert Johnson also registered significant developments during the war, as follows:

> The war years revealed the solid foundation laid by the American Baptist Mission in developing national leadership and self-support, and thus during the war years the Christians continued and even much increased their zeal. In fact, deprived of foreign money and leadership, and thrown in full reliance on the Spirit, they entered into a time of revival. This period came very near to a time of mass conversion (Johnson 1963: 391).

The major factor for church growth during the war was the contribution of the Chin Hills Battalion and the Chin Levies. As mentioned, the Chin soldiers in the British India Army who formed the First and Second Chin Hills Battalions had converted in large numbers ever since they returned from Europe efter World War I. Even so, since both Chin Hills Battalions were stationed in Burma, not in Chinram, they did not have an immediate impact on Chin society. However, during the war they retreated to Chinram and defended their homeland together with their own people. Their influence during the war years was enormous. They regularly held Sunday services under the supervision of their own army chaplains. This encouraged local church activities during the war years.

Another factor was what Za Tuah Ngur called 'the challenge to traditional religion' (1993: 71). As mentioned already, each village had a sacred place called the *Tual* ground where villagers made their annual sacrificial offerings to the village guardian god, *Khua-hrum*. As we have seen, the Chin considered these sacrificial places as both holy, and taboo, i.e. places where ordinary people were forbidden to go lest they desecrate them. During the war, however, all these *Tual* holy places were 'trampled by soldiers and became camps or battlefields' (ibid.). Even where the conflict spared it, the *Tual* became training grounds for the Chin Levies. Every morning, local lads recruited for the Chin Levies drilled on the *Tual*, which they used to regard as a holy place. Subsequently, the *Tual* was renamed with the addition of the adjective *Cer* as *Cer Tual*, meaning 'parade ground'.[9]

Indeed, it really was a time when the traditional religious concepts of the sacred and mundane started to lose their effectiveness and meaning. And it was, as Ngur concludes, 'a great challenge to their traditional beliefs. It was more powerful than one thousand sermons for the local people to witness that nothing happened to the soldiers who violated their taboos and profaned their holy places. As a consequence, many people were converted to Christianity' (ibid.: 72).

The third factor, according to Za Tuah Ngur, was the psychological insecurity of the people during the war, the psychological dimension of conversion. She writes:

> During the war years the people fled into the jungle or to more remote areas for safety. Psychologically they felt they were in need of someone to provide a refuge more than at any previous time. Their traditional beliefs and tribal gods could no longer assure them of security. When Christians witnessed, it seemed that there was a greater openness to the Gospel. Christianity offered inner security at a time when all else was uncertain, thus many people became Christians (ibid.: 71).

One remarkable event during the war, as Za Tuah Ngur observes, 'was the sending of Chin missionaries to the Matupi area in the southern part of the Chin State'. As indicated already, two indigenous missionaries, That Dun and Pa Hrek, were sent to Matupi to work among the Cho tribe of the Ngaleng group. They were able to begin their work during the war, because the war spared the southern and central parts of East Chinram, including the Matupi, Zotung, Zophei, Lautu and Mara areas. The Chin Christians regard the work of That Dun and Pa Hrek as the beginning of a genuine indigenous Chin mission because they were purely missionaries, not teachers-cum-preachers like Sakhong and Lian Kar. They were able to make converts and establish the church which grew to Association status soon afterwards.[10] Johnson describes this event as 'the beginning of mission work originated and sustained by the Chin themselves' (Johnson 1963: 392).

Different experiences during the war had different repercussions on the way that the Chin and the Burman nationalists moved towards the independence of Burma.

NOTES

1 R. A. Lorrain had published his book *Five Years in Unknown Jungle for God and Empire* already in 1912 so naturally it did not mention the event in 1938. Regrettably, I did not have a chance to visit the archive of the Lakher Pioneer Mission in Serkawr (Sherkor) when I was in Mizoram, India, in 1991.

2 B-201: 'A Brief Biography of Dr J. H. Cope', written by Chester U. Strait in 1938.

3 This figure did not include the schoolteachers, who always acted as Christian workers.

4 Symleah 1981; see also Coupland 1945: 164.

5 Gravers' own words: '1,600 Christian Karen and Chin *participated* in hunting down the rebel' (1996a: 303). The truth is that the Chin and Karen soldiers in the British India Army *participated* in that campaign, not as the Chin or the Karen national army, nor as a Chin or a Karen ethnicity, nor as Chin Christians; they were just deployed by the British government. It is importnt to differentiate between professional soldiers and a national army.

6 Their slogan at that time was Ingalate a.khet, Bama a-chiat (The British difficulty is Burman's opportunity).

7 See Dorman-Smith 1943, and Vum Kho Hau 1963.

8 Robert Johnson in Maung Shew Wa, *Burma Baptist Chronicle: Book II* (1963), p. 391. In contrast to Robert Johnson's account, Dr Strait reported in 1940 that there were some 6000 Christians in the Haka area alone.

9 Nowadays, the football ground is also called *Cer Tual* in Chin.

10 Although almost all of the students accepted Christian teaching at school, none of them were baptized because neither Sakhong nor Lian Kar were ordained preachers. In Baptist Churches only an ordained minister can perform a baptism.

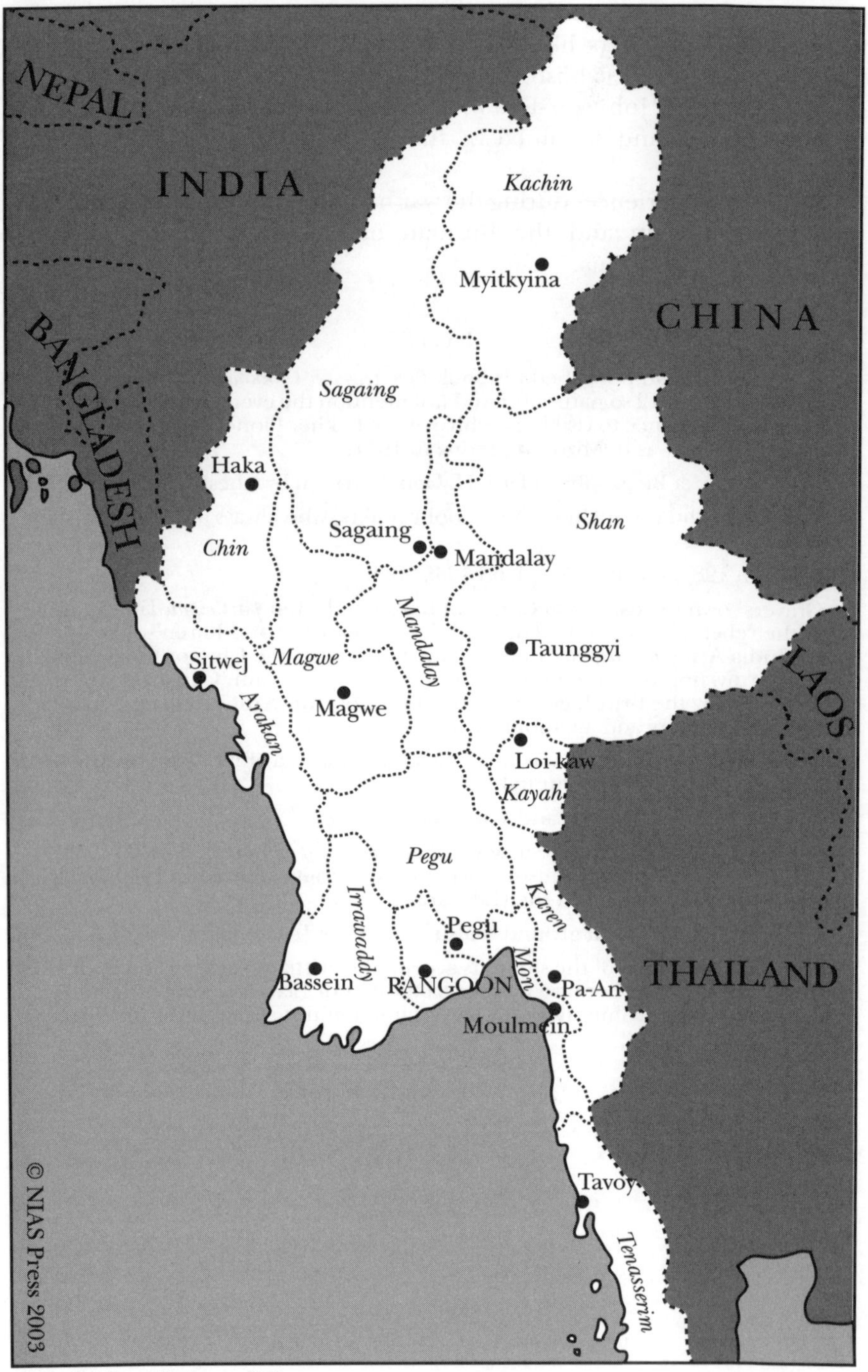

Map 2: The Union of Burma

NINE

The Joining of the Union of Burma

Soon after World War II, the British returned to Burma with new colonial schemes, including the 'Crown Colonial Scheme of 1941' which intended to create a separate British colonial province, the 'Chin-Lushai Country'. Before this project was implemented, however, the Conservative Party led by Sir Winston Churchill was defeated by Attlee's Labour Party in the first post-war general election. Eventually, Clement Attlee annulled the Conservative Party's 'Crown Colonial Scheme' when he came into power in 1945.

As far as Chin political development after World War II is concerned, the most dramatic turning point was the failure of the Crown Colonial Scheme. When H. N. C. Stevenson became the Director of Frontier Area Administration, he once more attempted to create a similar colonial province, which came to be known as the United Frontier Union. However, this attempt too was defeated. After the failure of such attempts, political developments in East Chinram led towards a dialogue with the Burmese independence movement. In this chapter, I am going to investigate how the Chin joined and supported independence for Burma in the final hours, when they realized that the British Labour government had abandoned them in the midst of uncertainty. Moreover, I shall pay particular attention to how Aung San's policy of secularism was reversed by U Nu's more confessional policy on religion.

POST-WAR SETTLEMENT AND THE *WHITE PAPER*

An Early Negotiation with the British

The Chin Levies and General Slim's 14th Army were able to hold the line against the Japanese, and by late 1944 General Wingate's *Chin-dit* Army was able to penetrate into and behind the Japanese front lines. In early 1945, the British Army launched a full counter-offensive into Burma. During this period, Aung San prepared to switch sides. He set up a secret 'Anti-Fascist

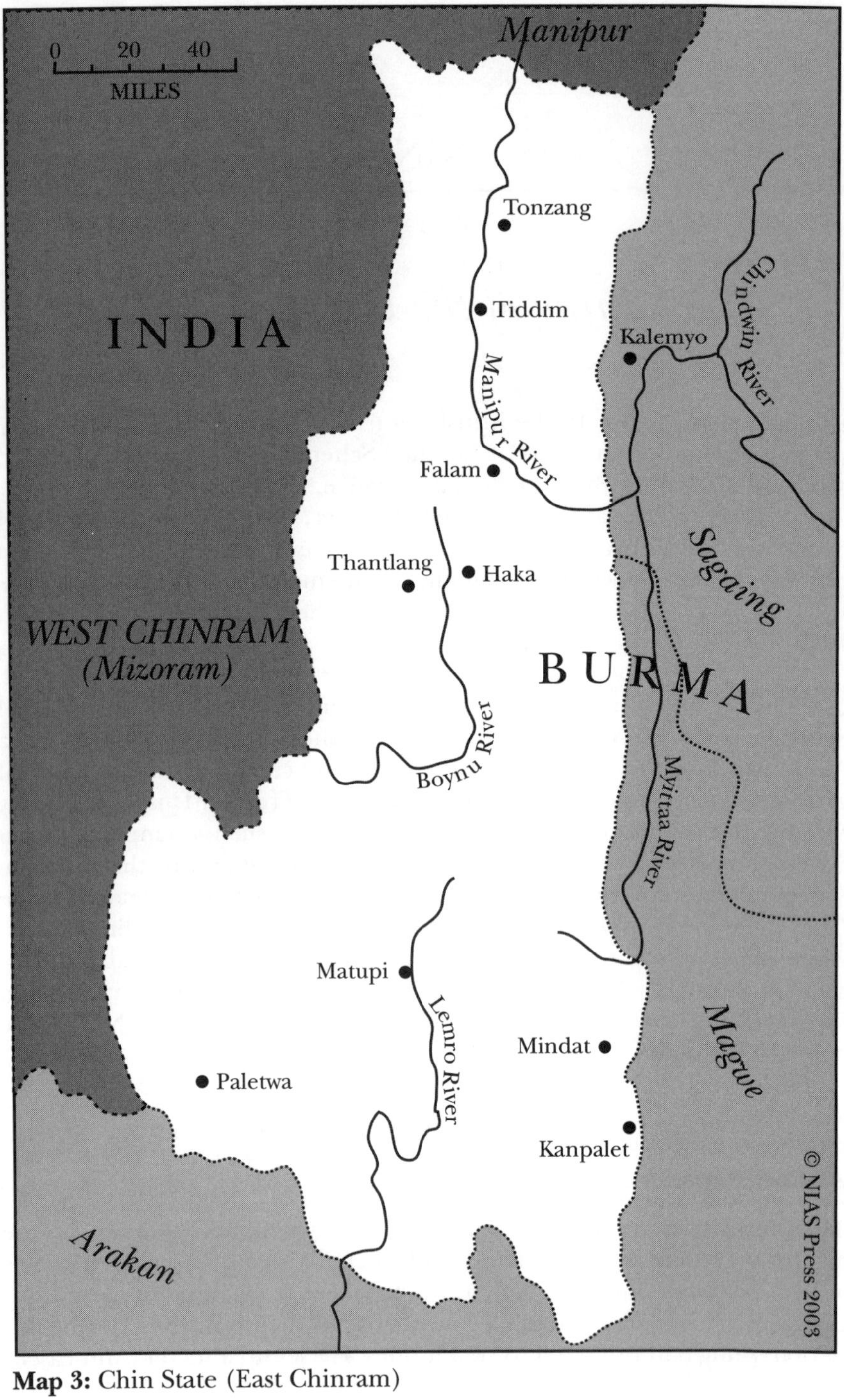

Map 3: Chin State (East Chinram)

Organization' (AFO), later transformed into the 'Anti-Fascist Peoples Freedom League' (AFPFL), whose mass base was provided by his Burma National Army. In March 1945, the BNA launched a rebellion against its allies.

When the BNA began its military offensive against the Japanese, General Slim's army had already recaptured almost two-thirds of Burma, including Mandalay and Upper Burma, from the hands of the Japanese. On 21 April 1945, General Slim met Aung San with the approval of Admiral Mountbatten, Supreme Commander of Southeast Asia. At that meeting, Aung San explained his position to General Slim: '[I]t was not that he disliked the British, but he did not want British or Japanese or any other foreigners to rule his country' (Slim 1958: 419–420). General Slim was greatly impressed by Aung San, as he mentioned several times in his book in such lines as, 'I admired his boldness', 'I liked his honesty', 'I was impressed', and so on. As a result of this meeting, the BNA was allowed to join the Allied troops in operations against the Japanese, and by June 1945 when a victory parade was held in Rangoon, the BNA soldiers 'participated alongside the units representing British Empire and Allied forces' (Aung San Suu Kyi 1991: 22).

The decision of the British to accept the offer of the AFO to fight the Japanese alongside the Allied forces, as Aung San Suu Kyi observes, 'had been due to the political acumen of Lord Mountbatten, Supreme Allied Commander, Southeast Asia'. Although Aung San and his BIA (renamed BNA) had been known to the British as the 'Burma Traitor Army', Mountbatten 'refused to accept this reading of the situation' because, as Tinker explains:

> He was very conscious that the task of driving Japanese out of South East Asia had only just begun. He needed a secure base for the hazardous assault upon Malaya. He could not risk a guerrilla rebellion in his rear. More profoundly, Mountbatten perceived that imperial high noon had passed away, with imperial sunset soon to follow. He recognized the urgency for the British to establish friendly relations with the younger generation of Asian nationalists (Tinker 1986: 461).

Mountbatten realized in advance the new post-war realities in Burma and Asia, and he also recognized the fact that post-war Britain would be too weak to impose its will any longer in areas where there was a substantial, organized opposition to British rule.

The Replacement of Dorman-Smith

When the British returned to Burma in the spring of 1945, they outlined their long-term plan for the future in a White Paper. This plan provided a three-year period of direct rule under the British governor, during which economic rehabilitation from the ravages of war was to be consolidated.

Then the Legislative Council of Ministerial Burma would be restored according to the 1935 Burma Act. Only after elections had been held under the existing 1935 Burma Act would the legislature be invited to frame a new constitution 'which would eventually provide the basis on which Burma would be granted dominion status' (Aung San Suu Kyi 1991: 22).

For Chinram and other areas which were excluded from Burma Proper, the White Paper provided for the maintenance of the prewar *status quo*. The Karenni State was still bound by the pre-colonial treaty as an independent nation. Since Chinram, the Kachin State and the Federated Shan States were excluded from the administration of Burma Proper, they would, according to the White Paper, have 'a special regime under the Governor'. The point is very clear: they would have remained under the direct rule of the Governor. As Aung San Suu Kyi correctly points out, Chinram and the other countries of non-Burman nationalities, or the so-called Excluded Areas or Frontier Areas would not have been included in Burma Proper 'unless the people of these areas specifically expressed their own desire to be amalgamated with the rest of Burma' (ibid.: 24).

Sir Reginald Dorman-Smith, the Governor of Burma, was particularly eager to emphasize the new imperial plans of the White Paper, which were 'designated to change the relationship between the imperial metropolis and its possession' of Burma. This new imperial plan emphasized partnership rather than dominance, but 'a partnership in which the former imperial power would retain a substantial degree of ultimate sovereignty and control over the political agenda' (Christie 1996: 152). In his view, the 'Burma battle for freedom was over' because he thought that a new partnership in a form of dominion status would have satisfied most of the Burman nationalists. Before the war, dominion status was beyond the dreams of many Burman nationalists. He therefore strongly emphasized 'reconstruction of Burma', rather than the 'battle for freedom'. He also persuaded the Burman nationalists to cooperate in the implementation of the White Paper. Thus, it was in effect from November 1945 to January 1947.

On this point, Dorman-Smith clearly misjudged Aung San and the new generation of Burman nationalists who had come of age during the war and led the independence movement with strong momentum under the banner of the Anti-Fascist People's Freedom League (AFPFL). Governor Dorman-Smith, unlike Mountbatten, was unable to see the real situation of post-war Burma. He was surrounded by 'civil servants who had lived through times when the young nationalists had challenged their authority and disrupted their administration' (Aung San Suu Kyi 1991: 23). These civil servants, both British and Burmese, were simply out of touch with the new reality. They were 'against any form of co-operation with the AFPFL, and during the period of military administration immediately following the British reoccupation of Burma, they wanted to declare the AFPFL illegal and arrest Aung San as a traitor' (ibid.). Moreover, when the civilian government was reinstalled and Dorman-Smith returned to Rangoon, he opted for the pre-war, old-time politicians as an alternative to the AFPFL. Not only did he form an advisory council of 'his cronies Sir John Wise, Sir Paw Tun, [and] Sir Htoon Aung Kyaw' – also appointed a 'Legislative Council' (which had no actual legislative power) of 35 politicians such as U Saw, U Chit Hlaing, Tun Ok and Ba Sein – those leaders who could be lured away from the AFPFL.

As noted already, the AFPFL had been created during the war by Aung San, contained a wide range of political organizations and had a nation-wide structure, a quasi-military formation and access to arms. Aung San, now the undisputed leader of the Burma independence movement, made clear to the returning Governor of Burma that 'the terms of the White Paper are totally unacceptable to the AFPFL'. For Aung San, the White Paper was nothing but an excuse to delay independence. He therefore declared, 'The battle for freedom has just begun'. Since Aung San was already determined to gain complete independence, for him, 'there was no question of dominion status' (ibid.).

Between the summer of 1945 and early autumn 1946, the British officers, including Governor Dorman-Smith, were to discover that Aung San and his AFPFL could make Burma ungovernable for the British. By that time, the Labour Government in Britain felt that Dorman-Smith had 'lost his grip'. Attlee therefore decided to replace Dorman-Smith with Major-General Hubert Rance, who had administered Burma during the temporary military administration in 1945. Rance arrived in Rangoon on 30 August 1946 and immediately informed London that 'the White Paper now is out of date'. He also cabled directly to Attlee: 'The AFPFL is the only horse to back' (M. Smith 1991: 77–78). Thus, together with the replacement of Governor Dorman-Smith, who closely associated himself with the Conservative Party, the Labour Government in Britain 'reversed its policy, and Burma's political agenda became largely a matter of bilateral negotiation between the British and Aung San's AFPFL' (Christie 1998: 155).

In the early stage of the post-war period, the colonial powers strongly highlighted the rights and interests of the Chin and other non-Burman nationalities from the so-called Frontier Areas. But in December 1946, the British invited only Aung San and his delegation from Burma Proper to London without a single representative from the Frontier Areas to discuss 'the steps that would be necessary to constitute Burma a sovereign independent nation' (Aung San Suu Kyi 1991: 23). Since Attlee's Labour Government had already prepared to grant Burma independence 'either within or without the Commonwealth', the London Talks were largely a formality, at most putting into more concrete form the principles to which they already had agreed. The only stumbling blocks were the issues relating to what they then called the Frontier Areas (U Maung Maung 1988: 253).

THE ISSUES OF NON-BURMAN NATIONALITIES

From Crown Colonial Scheme to the United Frontier Union

After they successfully expelled the Japanese from Chinram and subsequently from Burma, the Chin and other non-Burman nationalities such as the Kachin and Karen expected better treatment from the British government in the way of social, cultural, economical and, most importantly, political development projects for their homelands. They had, after all, served with great loyalty on the British side during the war. They were not prepared for the rapidly changing political situation after the Second World War, nor for independence of their respective homelands. While Indian and Burman nationalists had been fighting hard to gain freedom and become sovereign nation-states, the Chin, the Kachin and the Karen were still willing to remain under British rule.

As they expected, the British officers who had had a friendly relationship with the Chin people during the war, particularly Dorman-Smith and H. N. C. Stevenson, reminded the London government of its commitment to implement Robert Reid's Crown Colonial Scheme soon after the war. They updated the proposal, however, in accordance with the changing political situation during the war, which favoured the formation of not just a separate colonial province of Chinram but a province of the Commonwealth that would include East Chinram (the present Chin State in Burma), West Chinram (the present Mizoram State in India), the Chittagong Hill Tracts (in present Bangladesh), Arakan Hill Tracts, the Pakkoku, Nagaland, Manipur, North Cachar and Mikir areas, parts of the Chindwin District, the west bank of the Chindwin, that is, the Kale, Tamu and Kankaw Valleys, the hill areas of Sadiya and the hills of Tripura. Thus, in 1945, Robert Coupland, Professor in the History Department at Oxford University and an expert on constitutional law, was sent again to study the possible implementation of the resolution to make the Chinram (Chin-Lushai Country) into a province of the Commonwealth. He gave his affirmative report to the British government under the Crown Colonial Scheme (Coupland 1946: 201–205).

However, as already implied, these plans did not come to pass for the Chin, as the British Conservative Party then led by Prime Minister Winston Churchill lost in the succeeding British general election in 1945. After the war Great Britain was too weak to impose a political settlement of its own choice on the colonies. The British therefore voted for the Labour Party, which had long had the independence of India and Burma as part of its program. Since the Labour Party wanted to give away the entire subcontinent, the creation of a province of the Commonwealth between India and Burma was out of the question for them, and Attlee vetoed the plan soon after coming into power.

When H. N. C. Stevenson became the Director of the Frontier Areas Administration (FAA), however, he made another attempt to create a colonial province that came to be known as the 'United Frontier Union'. He developed a plan unifying all the Frontier Areas into one big union, which would have included East Chinram, the Kachin State in the north, the Federated Shan States and independent Karenni State in the east, parts of the Toungoo District and the Hills District of Salween, homeland of the Karen people, and Thaton and Moulmein, reaching as far south as Victoria Point at the southernmost tip of present Burma, including the entire mineral-rich Tenasserim Division where the Karen, Mon and others were living side by side (U Maung Maung 1989: 256). Stevenson's great scheme of unifying the Frontier Areas into the 'United Frontier Union' covered the homelands of the non-Burman nationalities.

The significance of this ambitious United Frontier Union plan was that it included both pre-colonial independent nationalities, such as the Chin, Kachin, Kaya and Shan, and the peoples already conquered by the Burman kings before the colonial period (e.g. Arakan, Mon, and even Karen). Yet Burman nationalists, including even moderate leaders like Aung San, could not accept the idea of the separation of Arakan and Mon from Burma Proper because the British had conquered them as part of the Burman Empire. The Burman King Alaungpaya had already conquered the Mon Kingdom in 1757, and Arakan was conquered by King Bodawpaya in 1784. Although the Burman and the Shan had many interrelationships and power struggles in their history, the British conquered at least the present Shan State if not the entire Shanland as an independent country, not as a domain of the Burman Empire.

The Karen situation was different and problematic. Although the Burman and the Karen had been living side by side but not together for centuries, no Burman king had declared a victory over the Karen as they had over the Mon and the Arakan. Unlike the Chin and Kachin, however, the Karen – especially those who lived in the delta areas – paid tribute to the successive Burman kings. Because of this, the Karen were legally the subjects of the Burman Kingdom, a fact admitted by the Karen leaders in the *Karen Memorial* presented to the British government as part of the campaign for Karen self-determination and the creation of an independent Karenistan in 1946. They admitted in the *Karen Memorial* that the Burmese kings and the Burmese people made Karen literally into slaves before the colonial period (cf. San C. Po, 1928; Tinker 1984: 492–497). Moreover, the Karen were included in the Legislative Council of Ministerial Burma which was created, according to the Burma Act of 1935, for Burma Proper or the former Burman Kingdom. That was the reason that Aung San, the Chief Minister of Ministerial Burma, could legitimately represent not only the Burman but also the Karen, Mon and Arakan as the subjects of Burma Proper, at the Panglong Conference in 1947.

With or without knowing such complicated historical issues of non-Burman nationalities, Stevenson tried to promote his great scheme to create the United Frontier Union, which actually began under Dorman-Smith and continued with Acting Governor Sir Henry Knight (before Hubert Rance arrived to Rangoon). Stevenson invited the leaders of the Frontier Areas to Panglong, the Shan State, in March 1946, which became known as the First Panglong Conference. While peoples from the Frontier Areas accepted Stevenson's plan with the greatest hope and enthusiasm, Burman nationalists like U Nu and U Saw – both attended the conference – criticized it as a 'policy of deliberate separation of the peoples, and of keeping the hill peoples in a stagnant primitive state' (Tinker 1984: 497). In spite of such an attack from Burman nationalists, the First Panglong Conference was a successful conference for both Stevenson and the peoples from the Frontier Areas. They decided to hold the Second Panglong Conference the following year.

Aung San's Inclusive Policy

As indicated above, the British policy of forming a United Frontier Union was strongly criticized by the Burman nationalists. At the London Talks in December 1946, the Burman delegates demanded that 'the amalgamation of the Frontier Areas and Ministerial Burma should take place at once, and that the Governor's responsibility for the Frontier Areas should end' (ibid.: 217). The London Talks were a bilateral negotiation between the British government and Aung San's AFPFL, without a single representative from non-Burman nationalities. Although there were at least three Karen members in the Constituent Assembly of the Interim Burmese government, none of them were included in the London Talks. Instead, Aung San included several councillors, civil servants and politicians in the delegation. He even included his main political rivals, such as U Saw and Ba Sein.

On the demand for amalgamation of the Frontier Areas with Ministerial Burma, the British delegates countered on the grounds that

> HMG for their part are bound by solemn undertakings to the people of those Areas to regard their wishes in this matter, and they have deep obligations to those peoples for the help that they gave during the war. According to the information available to HMG, the Frontier Areas are not yet ready or willing to amalgamate with Burma Proper (ibid.: 217).

During the talks, Attlee received a cable from the Shan *Sawbwa* (princes or chiefs), through the Frontier Areas Administration and governor, stating that 'Aung San and his delegation did not represent the Shan and the Frontier Areas' (Yawnghwe 1987: 99). Stevenson, the Director of Frontier Areas Administration, also cabled London:

> We understand that the Hon'ble U Aung San and the Burman Mission visiting London will seek the control of FA. If this is the case we wish to state emphatically that neither the Hon'ble Aung San nor his colleagues has any mandate to speak on behalf of FA (Tinker 1984: 277).

In short, Aung San and his delegation had no right to discuss the future of the Frontier Areas.

It is true that Aung San and his delegation neither represented nor had the right to discuss the future of the peoples from the so-called Frontier Areas, especially the Chin, Kachin, and Shan because they were independent peoples before the colonial period and conquered separately by the British as independent countries. Aung San could legitimately represent only Burma Proper, or the so-called Ministerial Burma, which belonged to the Old Kingdom of Burman before the colonial period. No Burman king had ever conquered Chinram. That was the reason why the British had applied two different administrative systems ever since the beginning of the colonial period, one for Burma Proper (Ministerial Burma) and one for the Excluded Areas (Frontier Areas). Thus, when Burma and India were to be given independence by the British, Chinram was not to be handed over to either India or Burma since it was not annexed by the British as a part of either country. It had the full right to be a sovereign independent state when the British withdrew its imperial administration from British India and Burma. In a nutshell, Aung San did not and could not represent the Chin and other nationalities from the so-called Frontier Areas without mandate from the peoples themselves.

During this critical period, Aung San showed his honesty and his ability of leadership, which eventually won him the trust of the non-Burman nationalities. He acknowledged the fact that the Chin and other non-Burman nationalities from the Frontier Areas had the right to regain their freedom, independence and sovereign nation-state status because they were not the subjects of the pre-colonial Burman kingdom. They had every right to self-determination, to make their own decisions about their future – whether to gain independence directly from Great Britain as sovereign nation-states without any attachment to Burma, or to join an Independent Burma, or even to remain as Provinces of the Commonwealth of Great Britain. Aung San therefore bravely and wisely put his signature on the historic agreement, which became known in as the Aung San–Attlee Agreement, signed on 27 January 1947. This document read:

> 8. Frontier Areas:
>
> (b) The leaders and the representatives of the peoples of the Frontier Areas shall be asked, either at the Panglong Conference to be held at the beginning of next month or at a special conference to

be convened for the purpose of expressing their views upon the form of association with the government of Burma which they consider acceptable during the transition period ...

(c) After the Panglong Conference, or the special conference, His Majesty's government and the government of Burma will agree upon the best method of advancing their common aims in accordance with the expressed views of the peoples of the Frontier Areas.

However, on the particular issue of non-Burman nationalities, two members of the Burman delegation refused to sign the Aung San–Attlee Agreement. One was U Saw, the former Prime Minister, and the other was Ba Sein, who had shared the leadership of the minority faction of Dobama Asi-Azone with Tun Ok after the 1938 split. In their view, the clause concerning the Frontier Area in the Aung San–Attlee Agreement entailed 'the implicit threat of dividing Burma into two parts' (U Maung Maung 1989: 255). Not only did they ignore the history of the non-Burman nationalities, such as the Chin, Kachin and Kaya, but also the will of the people from the so-called Frontier Areas. Upon their return to Rangoon, 'U Saw and Ba Sein joined Ba Maw and Paw Tun, another former Prime Minister, to form the National Opposition Front, claiming that Aung San had gone over to the imperialists for the sake of holding office' (Aung San Suu Kyi 1984: 46).

Aung San, however, was not unduly troubled by the accusations of his political opponents and plunged straight into negotiations with the pre-colonial independent nationalities, namely the Chin, Kachin and Shan. As mentioned above, the 'Aung San–Attlee Agreement' had left the future of the Frontier Areas to the decision of its peoples. The negotiations between Aung San, as the sole representative of an Interim Burmese government, and the Chin, Kachin and Shan, were held at the Panglong Conference (also known as the Second Panglong Conference), in February 1947. The Panglong Conference produced an agreement acknowledging that freedom would be 'more speedily achieved by the Shan, the Kachins and the Chins by their immediate cooperation with the interim Burmese government', and was the 'culmination of Aung San's mission to unite the diverse races of Burma which had begun with the *Thakins'* wartime efforts for racial harmony' (ibid.: 47).

The Panglong Agreement

During the London Talks, FFA Director Stevenson strongly advocated the creation of a United Frontier Union from a different channel. Since he successfully conducted the First Panglong Conference, he took another step and went to London to promote his plan. However, the atmosphere in London was quite different from that in Panglong. Although he received a great deal of support from old-hand bureaucrats like Political Undersecretary Henderson, who tried to push the plan, he did not receive enough support from the Labour Government and the Governor of Burma in Rangoon. Soon he found that his 'great plan' was 'tersely ordered to be set aside' by Prime Minister Attlee (U Maung Maung 1989: 256). It was the second time that Attlee had vetoed such a Crown Colonial Scheme for the Chin and non-Burman nationalities. As Attlee had frequently used his veto power against the rights and interests of those peoples, they no longer had a secure political position under the Labour Government, but found themselves increasingly marginalized.

When the prime minister vetoed Stevenson's plan for a United Frontier Union, Stevenson discovered that the main culprit was Governor Hubert Rance. Stevenson strongly criticized Rance, saying, 'The Governor had gone behind the back of the DFA (Director of the Frontier Areas Administration(in making arrangements to include the Shan State and possibly other Frontier Areas in Burma. (ibid.: 179). Hubert Rance retaliated with the support of the prime minister and secretary of state: 'Stevenson should retire and should not attend the vital Second Panglong Conference where future relations between the Burman and the FA peoples were to be decided in the presence of an HMG observer from London' (Tinker 1984: 303f). In this way, Stevenson was forced to resign just a week before the Second Panglong Conference, which he had personally organized and developed to assure the rights and interests of non-Burman nationalities, including the Chin.

Without knowing about such behind-the-scenes politics, the Chin prepared to attend the Second Panglong Conference in February 1947. Before they left for Panglong, they decided to have a Primary Conference of their own at Falam, which was held on 28 October 1946. The Falam Conference was attended by tribal chiefs, headmen, elders and representatives of various sub-tribes of the Tiddim, Falam and Haka. They discussed 'whether the Chin Hills should remain directly under the control of the Governor, or whether it should join Burma when the latter gets Dominion status'. The following resolutions were passed:

1. Chin Hills should remain under the direct control of the Governor.
2. They should join together and cooperate with their brother hill peoples, the Shan and Kachin (ibid.: 97f).

The second resolution, of course, concerned Stevenson's plan to create the United Frontier Union or to unify all the Frontier Areas. The Chin decided to support the plan. At the conference they also elected three leaders, Chief Hlur Hmung, Chief Thawng Za Khup and Chief Kio Mang, as the representatives of the Chin people from East Chinram.

When the Chin representatives reached Panglong, the political scene was not what they had expected. There were no more friendly faces. The absence of Stevenson was a big blow for them. He was the person whom they could trust without reservation. He knew their history, their culture and even spoke their language, because he had previously been the Commissioner of the Chin Hills District and the Commander of the Chin Levies during the war. Since Stevenson could speak Chin (Lai dialect), Chief Hlur Hmung of Lungbang (Falam) and Chief Kio Mang of Haka did not even bring their interpreters; only Chief Thawng Za Khup of Seizang (Tiddim) brought his interpreter, Vum Kho Hau, to Panglong. Because of the absence of Stevenson, Vum Kho Hau, who belonged to the Zomi tribe and spoke the Zo dialect, became the interpreter for the whole team although he could not speak the Lai dialect fluently, especially not the local Haka dialect.

When the Chin representatives arrived at Panglong on 7 February 1947, the Shan and Kachin representatives had already held their own discussions and signed a joint agreement between themselves on 6 February. It read:

> The freedom for the Shans and the Kachins would be achieved sooner through cooperation with the Burmans; as such, the two races would send in their respective representatives to take part in the Executive Council of the Burmese government during the transition period with such conditions as democratic rights and equal rights with the Burman, complete responsibility for Frontier Areas internal affairs, and in addition to have joint responsibility for external affairs, defense, railways, and customs departments in the Government of Burma, a separate Kachin State, and most importantly, the right of secession after independence.[1]

The Chin representatives joined the discussions with the Kachin and Shan soon after they arrived. At first they were stunned. Since the Kachin and Shan had already decided to join the Independent Burma, there was no more talk about a separate Frontier Union. They found themselves listening to Tin Aye, leader of a Shan radical group called the Shan People's Freedom League (SPFL), who strongly advocated that they join Independent Burma. They also felt that the British had abandoned them in the midst of uncertainty, and their political security was increasingly marginalized; this finally, at the last minute, led them to accommodation with the dominant Burman nationalist movement.

The Chin leaders, however, maintained the position that even if they joined Independent Burma, it would amount to a form of temporary cooperation under the condition of maintaining their pre-colonial independent status. They therefore said: '*Kan ram cu kan mah te in le kan phunglam ning te in uk kan duh*' (We want to rule our country by ourselves according to our own political systems).[2]

What the Chin leaders had tried to demand was a kind of commonwealth of independent nation-states with full autonomy and complete political powers of the legislative, judiciary and administrative authorities. The Chin word *ram* designates an independent country or a nation-state, not mere state that is a creature of a larger Union without political powers. On this particular point, the Chin were manipulated by both the British and the Burman and ended up without a separate nation-state in the Union of Burma. I shall come back to this later.

The second Chin demand called for the establishment of a Supreme Executive Council of the United Hills Peoples (SCOUHP), 'which shall have full powers of decision on all matters of policy between the Hill Peoples and the Government of Burma'.[3]

The Chin, as the Burman scholar Maung Maung wrote, 'were no more friendly to Burman than the rest' (U Maung Maung 1989: 280). Thus, to safeguard their own legitimate interests, they proposed a body that would be comprised of equal numbers of Chin, Kachin and Shan representatives. They also wanted the conference to confirm their 'right of secession from Burma at any time after attaining freedom,'[4] a matter which was already agreed upon for the Shan and Kachin on the previous day. The Chin representatives were also concerned about financial support from the Union, and the Chin–Kachin–Shan agreement thus made the following provision: 'Any deficiency in local finance to be made good from Burma revenues.'[5]

Aung San, Chief Minister of the interim Burmese government, arrived in Panglong on 8 February 1947 and persuaded the Chin, Kachin and Shan leaders to join Independent Burma. He promised 'the frontier peoples separate status with full autonomy within the Burma Union, active participation at the center in a kind of states Senate, protection of minority rights, and the privilege of secession' (Cady 1958: 539). He also promised 'to make the agreed terms into law so that they had the guarantee for the future, and told them to have no fear of the Burman' (U Maung Maung 1989: 282).

When the Chin delegation held a meeting with Aung San, they presented the demands mentioned above. On that occasion, Vum Kho Hau misinterpreted what Chief Hlur Hmung and Chief Kio Mang had said. He wrongly translated the Chin word *Phunglam* into the English word 'customary law'. The meaning of the word *Phunglam* is more than customary law; the correct translation must be 'ways of life', which includes traditional political systems, religious systems and almost every aspect of life. Customary law is only the judiciary aspect of the broader political system. It does not include legislative and administrative aspects of political power. Worst of all, he translated the Chin word *ram* to the English word 'district', not 'country' or 'nation-state'.[6]

Thus what Aung San understood, and what has become almost conventional truth in Burma, was that the Chin did not ask for a separate state but a district within the Union and their old customary law (U Maung Maung 1989: 280; Maung Maung 1959: 191), that is, only judicial power which did not include the essential political power of legislative and administrative authorities for the Chin. 'Aung San was somehow surprised', Chief Kio Mang later recalled, and 'he asked the Chin leaders not to hesitate to ask if there were some things he could do for them.'[7]

It was a surprise for the Chin leaders as well because they did not expect such sympathetic terms from the Burmans. They asked for schools, hospitals and roads, and Aung San promised everything they asked (cf. Vumson 1986: 192). Aung San and his advisor U Tin Htut decided to give East Chinram, when the Union Constitution was drafted, special administrative status because they thought that the Chin wanted to maintain their old customary law. Thus, the Chin ended up with a mere Special Division without statehood in the Union of Burma.

When Aung San had successfully persuaded the Chin, Kachin and Shan to join Independent Burma, the historic Panglong Agreement was signed on 12 February 1947. The Chin, Kachin and Shan did not surrender their rights of self-determination and sovereignty to the Burman in the Panglong Agreement; they signed it as a means to speed up their own freedom together with the Burman and other nationalities in Burma. The preamble of the Panglong Agreement declares:

> Believing that freedom will be more speedily achieved by the Shan, the Kachin, and the Chin by their immediate co-operation with the interim Burmese government.

The Panglong Agreement was a joint statement of the pre-colonial independent peoples – Chin, Kachin, Shan – and the Interim Burmese government led by Chief Minister Aung San, who came into power in August 1946 according to the Burma Act of 1935.[8]

The interim Burmese government was a government for the region formerly known as Burma Proper or Ministerial Burma, which included such non-Burman nationalities as the Arakan, Mon and Karen. The Arakan and Mon were included because they were occupied by the British not as independent peoples but as the subjects of the Burman king.[9]

The Karen were included in the Legislative Council of Ministerial Burma according to the 1935 Burma Act because the majority of Karen (more than two-thirds of the population) were living in delta areas side by side, but not together, with the Burmans.[10]

Since these peoples were included in the Legislative Council of Ministerial Burma, Aung San could represent them in Panglong as the head of their government. Thus, the Panglong Agreement was an agreement between peoples from pre-colonial independent countries, who had the right to regain their independence directly from Great Britain, and the right to form their own respective nation-states without any mutual attachment. It was an agreement between the peoples of post-colonial nation-states-to-be, each of which the Malaysian scholar Shamsul A. B. calls a 'nation-of-intent'.[11]

Since the Union of Burma gained independence in 1948, the date when the Panglong Agreement was signed is celebrated as Union Day. The observance of 12 February as Union Day means the mutual recognition of the Chin and other nationalities, including the Burmans, as 'different peoples historically and traditionally due to the differences in their languages as well as their cultural life'. It is also the implicit recognition of the distinct national identity of the Chin people, a people who had the right to gain their own independence and to establish their own nation-state separately. In other words, it is the recognition of the pre-colonial independent status of the Chin people as well as their post-colonial status of nation-state-to-be.

The Conditions for Joining the Union of Burma

According to the Aung San–Attlee Agreement at the London Talks, a Frontier Areas Committee of Enquiry (FACE) was formed to inquire about the frontier people's wishes through an additional and specific consultation. The British government appointed Col. D. R. Reese-William as chairman of the FACE. Since the committee conducted its inquiry after the signing of the Panglong Agreement, during March and April 1947, the evidence they heard was generally in favour of cooperation with Burma but under the conditions of 'equal rights with the Burman, full internal autonomy for the Hill Areas, and the right of secession from Burma at any time'.[12]

The Chin leaders also presented their own memorandum to the FACE in which the Chin's future relations with Burma were treated as follows:

(1) That the Chin shall have equal rights and privileges as the Burman,

(15) To secede, if they wish to do, as envisaged in the Panglong Agreement (Reese-William 1947: 71f).

But when the FACE asked the Chin delegation whether the Chin should become a part of 'Ministerial Burma or be joined in a Federation with Burma and the other Frontier Areas', the Chin leaders were unable to make a clear decision, because they did not know, as Chief Mang Ling said, 'the literal meaning of [the word] federation' (ibid.: 76). What they knew was that they wanted to rule their country according to their own political system and 'ways of life', with some financial assistance from the Union. Although they were unable to make a clear decision for a separate state within a federal system, the Chin 'did not want to be part of Ministerial Burma' either (U Maung Maung 1989: 291). Finally, Za Hre[13] the only Chin delegate with a college education, suggested to the FACE that 'we (Chin) should go into Ministerial Burma' (Reese-Williams1947: 76). In this way, the Chin ended up with a mere Chin Special Division and not a separate state within a Union.

The FACE concluded in its report to the Government that the majority, who supported cooperation with Burma, also demanded the 'right of secession by the States at any time'. The FACE report, particularly the part about the right of secession, was strongly criticized by such Burman nationalists as U Saw and Thakhin Ba Sein, who refused to sign the Aung San–Attlee Agreement. They accused Aung San of having given up Burman territory and argued that the Frontier Areas were just a consequence of the colonial policy of 'divide and rule', which, of course, was *not* true. Aung San dismissed this criticism as historically unfounded and politically unwise. And he said, 'The right of secession must be given, but it is our duty to work and show (our sincerity) so that they don't wish to leave' (cited in Tun Myint 1957: 10–11). Because he had promised the Chin, Kachin and Shan leaders at the Panglong Conference 'to make the agreed terms into law', the right of secession was legalized in the 1947 Union Constitution of Burma, Chapter X, Article 201, and 202:

> Chapter (X): The Right of Secession
>
> 201. Save as otherwise expressly provided in this Constitution or in any Act of Parliament made under section 199, *every state shall have the right to secede from the Union* in accordance with the condition hereinafter prescribed.
>
> 202. The *right of secession* shall not be exercised within ten years from the date on which this Constitution comes into operation.

Although the 'right of secession from the Union' became law in the Union Constitution, Burma did not become a genuine federal union. Aung San, who persuaded the Chin, Kachin, Shan and other non-Burman or nationalities to join Independent Burma, was assassinated by U Saw on 19 July 1947. He was succeeded by U Nu as leader of the AFPFL. When U Nu became leader, Burman politics pointed in a retro-historical direction, backwards toward the old kingdom of Burma. Backward striving policies did nothing to accommodate non-Myanmar nationalities who had agreed to join Independent Burma only for the sake of 'speeding up freedom'. As a leader of the AFPFL, the first thing U Nu did was to give an order to U Chan Htun to re-draft Aung San's version of the Union Constitution, which had already been approved by the AFPFL Convention in May 1947. U Chan Htun's version of the Union Constitution, not Aung San's, was therefore promulgated by the Constituent Assembly of the interim government of Burma in September 1947. The fate of the country and the people, especially the fate of the non-Burman nationalities, changed dramatically between July and September 1947. As a consequence, Burma did not become a genuine federal union, as U Chan Htun himself admitted to historian Hugh Tinker: 'Our country, though in theory federal, is in practice unitary' (Tinker 1957: 84).

THE END OF AUNG SAN'S VISION OF A SECULAR STATE

The Assassination of Aung San

The most tragic event in the history of independent Burma was the assassination of Aung San and almost all of the cabinet members of his Interim Burmese Government on 19 July 1947. Aung San had been the most charismatic political figure and the undisputed leader of the Independence Movement since 1938. He had negotiated successfully for the country's independence with Great Britain and also with the pre-colonial independent nationalities. He shaped a constitution for a Federal Union based on the principle of equality.

Together with Aung San, six of the most competent leaders, who had been preparing to play leading roles in Burma's transition from a British colony to an independent nation, sacrificed their lives when freedom was less than six months away. Three non-Burman nationalities were represented among the seven men who were assassinated. Soa Sam Htun was a *Sawbwa* or prince of the Mong Pawn from the Shan State, Mahn Ba Khaing was a prominent Karen leader and, Abdul Razat, originally from India, was a leader of the Muslims in Burma. They were all gunned down at a cabinet meeting held in Secretariat Hall, in the center of Rangoon.

U Saw, who refused to sign the Aung San–Attlee Agreement in 1947, and eight of his personal militiamen, who carried out the killing, were tried and found guilty of murder and sentenced to death.

From Aung San to U Nu

On the evening of the same day that Aung San and his cabinet members were assassinated, the leaders of the Anti-Fascist People's Freedom League (AFPFL) and the People's Volunteer Organization (PVO) held a joint meeting at the PVO office. The PVO, as mentioned already, was an association for wartime veterans, which, in effect, had became a militia force loyal to Aung San. At that meeting, the PVO leader, Thakin Tin, told U Nu: 'Trouble will befall the whole nation if U Saw is appointed in General Aung San's place. You must see the Governor so that you can fill the General's position' (cited by Kin Oung 1993: 62). The AFPFL's executive committee passed a resolution recommending U Nu for the post of Deputy Chairman of the Executive Council and Prime Minister of the Interim Burmese Government. U Nu hesitantly agreed on the condition that 'he would be allowed to resign six months after independence' (ibid. 62). As a result, the governor officially invited U Nu to become prime minister and form a government.

According to his own version, U Nu never wanted to become a politician. His life-long ambition was to become either what he called the 'Bernard Shaw of Burma', a playwright, or a Buddhist monk, or if possible, both (U Nu

1975). In fact, he did not stand for election in the general elections for the Constituent Assembly, held in April 1947. He went back to his native town, Wakhema, and resumed a writing career. Nevertheless, Aung San brought U Nu back into politics and appointed him speaker of the Constituent Assembly despite the fact that U Nu had not been elected to a seat in the assembly.[14]

When Aung San and his cabinet members were assassinated, U Nu was also one of the intended victims. U Saw ordered one of his best gunmen, Ba Nyunt, to kill U Nu at his office, as Ba Nyunt confessed during the trial (Kin Oung 1993: 12). Luckily, U Nu was not in his office but in town on an assignment during those bloody morning hours.

The two leaders began their friendship in the 1934–35 academic year when U Nu returned to Rangoon University to study law. He had received his BA degree already in 1929, and worked as superintendent of the National School in Pantanaw, where one of his colleagues was U Thant. When U Nu became Prime Minister of Burma, he appointed U Thant as ambassador to the United Nations, and the latter eventually became Secretary General of the United Nations. Aung San and U Nu quickly developed a friendship, and the 1936 students' strike cemented their relationship. In the traditional Burmese manner, Aung San affectionately called U Nu *Kogyi Nu* (my elder brother Nu), as U Nu was eight years his senior.

It seemed that U Nu was unprepared to hold political office, let alone become the leader of an entire nation, when he was appointed to take Aung San's place. While he tried hard to remain loyal to Aung San and his legacy, U Nu directly or indirectly betrayed Aung San's policy on at least two points. The first point, as mentioned already, was on the principle of the federal constitution, and the second point on religious policy. Although Aung San had opted for a 'secular state' with a strong emphasis on 'pluralism' and the 'policy of unity in diversity' (Silverstein in Lehman 1981: 51–58) in which all different religious and racial groups in the Union could live together peacefully and harmoniously, U Nu gradually reversed Aung San's secularism.

Hence, when U Nu became the leader of the AFPFL and prime minister of the Union of Burma, the 'politics of nationalism' in Burma did not accommodate non-Myanmar, non-Buddhist nationalities like the Chin and Kachin. Traditional 'politics of nationalism', which played a vital role for U Nu's political legitimacy in independent Burma, were not only associated with the old Buddhist Kingdom of Burma but rooted itself in Burman 'traditional nationalism', in which Buddhism was of the utmost importance and the major component. From the beginning, the religious-oriented, traditional nationalism of Burman excluded, as Donald Smith observes, 'various non-immigrant groups: Burma Muslims who had lived in the country for hundreds of years, Zerbadis (descendants of Muslim men and their Burmese wives), domiciled Hindus, Christians and the non-Buddhist ethnic minorities' (D. E. Smith 1965: 114).

Aung San clearly wanted a separation between religion and politics, as he declared at the AFPFL convention in January 1946: 'We must draw a clear line between politics and religion, because the two are not one and the same thing. If we mix religion with politics, then we offend the spirit of religion itself' (cited in Maung Maung 1962: 127). He appealed to the Sangha (the community of Buddhist monks) 'to purify Buddhism and broadcast it to all the world so that all mankind might be able to listen to its timeless message of Love and Brotherhood till eternity ... this is the highest politics which you can do for your country and people' (Aung San Suu Kyi 1991: 8). In that same speech, he condemned 'the exploitation, injustice, superstition, and priestcraft frequently associated with religion', while holding that Buddhism could become 'the greatest philosophy in the world' if its ritual could be eliminated (Maung Maung 1962: 124).

In May 1947, the AFPFL held its 'constituent assembly' at Jubilee Hall in Rangoon. Aung San presented a constitution for the future Union of Burma, which he himself drafted, to a 111-member committee of the 'constituent assembly'. During the discussion of the committee some of the more traditionalist leaders, especially *Deedok* U Ba Choe, pressed for the constitutional recognition of Buddhism as the state religion. They argued that '[i]ndependence would be empty, if it did not elevate Buddhism to that status in law, as it enjoyed that status in fact with over 80 per cent of the citizens professing it. All earlier kings proclaimed themselves to be guardians and upholders of religion, and led their armies to battle for religious causes' (Maung Maung 1959: 98). This kind of argument and proposal was 'vehemently rejected by Aung San, who was convinced that such a provision would seriously impair unity'.[15]

Such a clause in the constitution would inevitably be resented as an attempt to impose the religion of the majority on the non-Burman nationalities. Aung San declared that the 'state should be secular' (ibid.). His version of the draft Constitution approved by the AFPFL convention on 23 May 1947, contained the following provision:

> The Union shall observe neutrality in religion and religious matters. It may, however, extend material or other assistance to religious institutions.[16]

However, soon after Aung San was assassinated, U Nu instructed U Chan Htun, a devoted Buddhist nationalist who later became a secretary general of the Buddha Sasana Council, to redraft Aung San's version of the constitution which had already received approval by the AFPFL convention. Thus, two months after the assassination of Aung San, the Constituent Assembly adopted U Chan Htun's version of the Union Constitution of Burma, in which Section 21. declares:

> (1) The state recognizes the special position of Buddhism as the faith professed by the great majority of citizens of the Union.

(2) The state also recognizes Islam, Christianity, Hinduism and animism as some of the religions existing in the Union at the date of the coming into operation of this Constitution.

The changing of Aung San's version of the Union Constitution proved to be the end of his policy for a secular state and pluralism in Burma, which eventually led to the promulgation of Buddhism as a state religion in 1961. In fact, it was the beginning of religious fanaticism, which finally led Burma into racial conflicts and narrow-minded nationalism, and a five-decade-long civil war. This also had serious repercussions on the position of the Chin in the Union of Burma.

THE CHIN NATIONAL DAY: AN OPTION FOR DEMOCRACY

The 1947 Constitution of the Union of Burma provided for a bicameral legislature, and the National Assembly therefore had two chambers – the Chamber of Nationalities and the Chamber of Deputies. For the Chamber of Nationalities, rather like the House of Lords in the British Parliament, the hereditary chieftains of the nationalities, especially the Chin, Kachin and Shan, had the right to retain their traditional powers without popular vote or election whatsoever. The Shan *Sabwa* and Kachin *Duwa*, therefore, enjoyed their traditional political power and privileges in a new form in the Union of Burma as guaranteed by the Union Constitution. For the Chin *Ram-uk*, however, the Chamber of Nationalities did not become a guarantee for their traditional power but a battleground between traditional chieftainship and a democratic system.

According to the 1947 Constitution, eight parliamentary seats in the Chamber of Nationalities were reserved for six major tribes of the Chin, namely the Asho, C'ho, K'ho-mi, Laimi, Mizo (Lushai) and Zomi. The question was who would represent the Mizo tribe, which was already outside the power structure that could be exercised within the Constitution of the Union of Burma. The Asho tribe, which lives in the Chindwin and Irrawaddy valleys, was outside the administration of the 'Chin Special Division'. Moreover, who would represent the Laimi tribe, the Haka or the Falam? Among these groups two kinds of highly respected political systems had developed, *Ram-uk* and the *Tlaisun Democratic Council*. Which system would generate representatives for the Laimi tribe, the *Ram-uk* system of the Haka or the *Tlaisun Democratic Council* of the Falam? A problem for the C'ho and K'ho-mi tribes was that there was no system of hereditary chieftainship among them. Thus, no one was eligible to represent their whole tribes.

To solve their problem peacefully, all the Chin tribes held a conference at Falam in February 1948, where the Chin once developed a traditional democratic system themselves. The representatives from all parts of Chinram, even representatives of the Hualngo group of the Mizo tribe who live inside Burma, attended the conference and discussed the matter for three days. Most representatives felt that the traditional chiefs who had no modern education were no longer competent to cope with the rapidly changing political situation. Their performances during the Panglong Conference and the 'Frontier Areas Committee of Enquiry' were unsatisfactory for the majority, especially for the new generation of Chin Christians. Moreover, core Christian values such as individual freedom, equality and representative government – especially the Baptist polity which supported a representative, self-governing Association – became the ideal for most Chin Christians. Christians gradually dominated the Chin political scene after the Japanese invasion of Chinram and the creation of the Chin Levies.

On 20 February 1948, they made their decision democratically, by popular vote. An overwhelming majority voted down 'the traditional political system of chieftainship', not only for representatives of the Chamber of Nationalities at the Union Assembly but even for local level administration. They created six constituencies for the Chamber of Nationalities; four for the Northern Chin State – Haka, Falam, Thlantlang and Tiddim and four for the Southern Chin State – Matupi, Paletwa, Mindat and Kanpalet.

This was a historic day for the Chin people. It was the first time in their history that all tribes had gathered and made such an important decision together, a decision not only to abolish their own, many-centuries-old political system, but to establish a modern, democratic political system in Chinram instead. It was the first time that the Chin people had started to emerge from 'the world of passive existence as peoples' (*passiver Volkheit*) under the absolute power of traditional chieftainship into the world of 'self-determination' (Eric Hobsbawm in A. Smith and J. Hutchinson 1994: 177–184). Until that time, Chin society was more or less a tribal society – at least in terms of its political system. In tribal society, as Elie Kedourie explains:

> A tribesman's relation to his tribe is usually regulated in minute detail by custom which is followed unquestioningly and considered part of the natural or the divine order. Tribal custom is neither a decree of the General Will, nor an edict of legislative Reason. The tribesman is such by virtue of his birth, not by virtue of self-determination (Kedourie in A. Smith and J. Hutchinson 1994: 49–55).

On 20 February 1948, the Chin people abolished the power of traditional chieftainship, usually considered to be a legitimate order from *Khua-hrum*, and its instrument of tribal custom, and also adopted the 'virtue of self-determination' and the 'General Will' of the people, which is associated with a political system called 'representative government'. It was not just a landmark for the end of the traditional political system, but the end of the legitimacy of traditional Chin religion in political life. In other words, it was the end of the structural and functional pattern of traditional Chin religion, which had functioned as a unitary pattern of *god–chief–land*.

When the Chin abolished their traditional political system of chieftainship and feudalism of land ownership, their traditional worship of *Khua-hrum* became a defenceless religion. The end of the traditional political system and its effect on traditional religion can therefore be seen as a mutual correlation between religious and political elements in tribal society. As mentioned in Part One, religious, political, social, cultural and almost all aspects of life in tribal society were inseparably intertwined. Anything that affected one aspect of the society would affect the whole. Thus, when traditional religion lost its legitimacy in political life, and the political system lost its effectiveness in religious life, the society itself was thrown into crisis as the power of tribal identity in all its religious and political forms – the heart and soul of the old tradition – became meaningless. The whole structure of society needed to be changed, as Chin scholar and politician Lian Uk observed:

> Our forefathers ... worshipped the hills and dales in our territory thinking there were spirits in them. But the reason they worshipped these hills and dales is not only that they believed in the spirits in them, but also that they have in the remote regions of their consciousness that they are our protector from the invasion of the outside world, from the invasion of the alien people. It has been in some way these hills and dales of ours which have preserved our distinct national identity to reach this modern age as a people. But we should now adopt a political means with which we could develop our distinct national identity, as our hills and dales are no longer enough of a barrier to keep out the invasions of all kinds (Lian Uk 1997: 1).

The abolition of traditional chieftainship and the adoption of a democratic system by the people themselves was a 'formative moment' in Chin history. When the *chiefs* lost their powers, the *land* was gradually nationalized, and Christianity gradually replaced the worship of guardian god *khua-hrum*, the traditional functional pattern of *god–chief–land* was finally broken.

Especially from the religious and political points of view, 20 February 1948 marked not just the end of the tribal political system of chieftainship, but the beginning of the end of tribal identity in all its political and religious forms. It was the beginning of a new national identity, which holds and ties the whole Chin race together in the name 'Chin', and no longer in the tribal names – e.g. Asho, Chó, Khó-mi, Laimi, Mizo and Zomi. They all commemorate '20th February' as 'Chin National Day', not as a tribal day. Since they all celebrate 'Chin National Day' together, their tribal identities which were inseparably associated with their old traditional religious and political systems are no longer effective or necessary.

Since 1948, 20 February has been celebrated and observed as 'Chin National Day' not only by the Chin themselves but by successive governments of the Union of Burma as an official annual holiday within the Chin State of the Union. When the Union was under the parliamentary system, the president and the prime minister of the Union of Burma usually sent official greetings to the Chin people every Chin National Day. The same day has been recognized as State Inauguration Day among the Mizo tribe of Chin in India since 1986, as it was the chosen date on which the Mizoram State was proclaimed by the Indian prime minister.

The observance of 'Chin National Day' as an official holiday means a recognition of the distinct national identity of the Chin people by the successive governments of the Union of Burma and India. The Chin are recognized as people who have a clear, distinctive national identity, inhabiting a territory with its own population within a definite boundary. The Chin people in Burma and India are not be seen just as a minority group but a 'nation with our own distinctive culture and civilization, language, literature, names and nomenclature, sense of value and proportion, customary law and moral codes, aptitudes and ambitions; in short we have our own distinctive outlook on life. By all Canons of International Law, the Chin people in Burma, Bangladesh and India are a nation' (Lian Uk 1997: 22).

In retrospect, the observance and celebration of Chin National Day for five-decades has been important and meaningful for Chin people everywhere. It strengthens their sense of oneness in the name of Chin, which holds them and ties them together into an integral people. It also serves as a symbol of unity, as a springboard from which all affairs can be made to reflect the common cause, common interest and common goals of the entire Chin population. It is also a means to proclaim to the world that the Chin people share a distinct national identity and also a common ideal for peaceful co-existence among all races and religions, in Burma, in India and throughout the world. It also strengthens their inspiration for freedom and democracy, denouncing despotism in all its forms. It has become a Chin tradition and thus its celebration means much to every Chin. In short, Chin National Day is the most powerful political and national symbol, one that holds and ties all the Chin people together in the midst of three multi-racial and multi-religious countries – Burma, Bangladesh and India.

NOTES

1 'Shan–Kachin Agreement' signed on 6 February 1947; original document reprinted in Tinker 1984, pp. 424–425.

2 Interview with Kio Mang, ex-chief of the Haka who signed the historic Panglong Agreement on behalf of the Chin people on 12 February 1947. I conducted the interview as a publisher of Rangoon University Chin Students Magazine called *Chin Magazine (University of Rangoon)* during the academic year 1981–82.

3 'Chin–Kachin–Shan Agreement' signed on 7 February 1947; the full text of the document is reprinted in Tinker 1984, pp. 424f.

4 'Chin–Kachin–Shan Agreement', resolution no. 5, reprinted in Tinker 1984: pp. 425f.

5 'Chin–Kachin–Shan Agreement', resolution no. 7, reprinted in Tinker 1984: p. 426.

6 Perhaps Vum Kho Hau translated the Chin word *ram* into the English word 'district' because the northern part of East Chinram (present Northern Chin State) was at that time called the Chin Hills District. He might have picked up the familiar contemporary terminology when his tongue was tied with a foreign language. The same mistake was made by Chief Mang Ling during the interview with the 'Frontier Areas Committee of Enquiry'; see the report of 'FACE', p. 76.

7 Interview with Kio Mang in March 1982.

8 When Aung San accepted this post, he stated that 'the White Paper and Section 139 of the Government of Burma Act 1935 still existed legally' (see Maung Maung 1989: 241). But he accepted the post not as a mere Counsellor of the Executive Council but as Chief Minister of an Interim National Government under special arrangement with the new governor, Rance. He secured his position and furthered his power in the 1947 general elections for the Constituent Assembly, the body which drafted the constitution for independent Burma.

9 The Arakan Kingdom was conquered by the Burman King Bodawpaya in 1784 and the Mon Kingdom by King Alaung paya in 1757.

10 The Karen National Union (KNU) rejected the terms of the 1935 Burma Act in 1946 because they demanded independence for a separate homeland. Thus they boycotted general elections of the 1947 Constituent Assembly, but the Karen Youth Organization (KYO) entered the general elections.

11 My concept of 'nations-to-be' can be compared with Benedict Anderson's theory of 'imagined political community' and Shamsul's 'nations-of-intent'. See Anderson 1991 and Shamsul A. B. 1996.

12 See the resolutions of the Chin, Kachin and Shan leaders at the SCOUHP meeting on 23 March 1947 and the memorandum they presented to the FACE (FACE report 1947).

13 Son of Chief Thang Tin Lian. He was later known as Za Hre Lian and became Minister of Chin Affairs Council during U Nu's parliamentary regime.

14 The Constitution allowed anybody to hold political office but for no more than six months without being elected.

15 *The Nation*, 15 July 1961; its editorial 'Shades of Aung San' as U Nu's 'State Religion Bill' was being debated at the Union Assembly. Cited in Smith 1965: 230.

16 Aung San's version of the Union Constitution, approved by the AFPFL convention, was reprinted in 1989 by the United Nationalities League for Democracy (UNLD) which was formed by all the political parties of non-Burman nationalities in Burma during the 1988 democracy movement. Until that time it was virtually unknown in Burma, at least in Burmese. An English translation was included as an appendix in Maung Maung 1959: 247.

TEN

Christianity and Chin Identity

In previous chapters we have seen how the Chin, step by step, decided to join the Union of Burma. The Chin National Day of 1948 was the symbolic expression of the new social and political determination of the Chin people. As we have seen, Chin National Day served as the fulcrum of change for transition, transforming traditional Chin society into a new and modern one. Within the contemporary context of the Union of Burma, Chin identity continues to be articulated.

An additional dimension to the Chin National Day was also inspired by the ethos of the Chin Baptist Association, which gradually replaced traditional religion as the motivating religious force. In this final chapter, I shall summarize my findings of the transformation of Chin identity and highlight characteristics in the new Chin spirituality. This indigenous form of Christianity modifies the manner in which the Chin have adjusted to the new national framework of the Union of Burma.

THE CHIN CHRISTIAN MOVEMENTS ON THE EVE OF INDEPENDENCE

When U Nu became the leader of Burma and the country moved to establish Buddhism as the state religion,[1] Chin self-awareness and a common identity became the phenomena which then intertwined inseparably with Christianity. This Chin 'political identification with Christianity' was, as Lehman observes, important for the Chin, 'even for non-converts' (Lehman 1985: 5), when they joined the Union of Burma in which they faced a new environment of multi-racial and multi-religious pluralism.

The Return of Missionaries after the War

Soon after the war, the American Baptist missionaries quickly resumed their work in Chinram. The Nelsons returned to Tiddim with a new missionary couple, Elizabeth and Robert Johnson, in May 1946. The Johnsons subsequently moved to Haka in February 1947, where they were designated to work.

Robert Johnson was born on 1 May 1915 in Chicago, Illinois. His parents, Gustav Albert and Clara Josephine Johnson, emigrated from Jönköping, Sweden. He graduated from Wheaton College with a Bachelor of Arts in 1938 and from Eastern Theological Seminary with a Bachelor of Divinity in 1941. Elizabeth (Betty) Lue Johnson was born on 24 October 1916 in Decatur, Illinois. She received her education at Millikin University and got her Bachelor of Science degree in 1942. They were married on 21 May 1941 and appointed as missionaries to Chinram that same year. However, as mentioned above, the war disrupted their careers. Then they faithfully served among the Chin until all missionaries were expelled from Burma by General Ne Win's military dictatorship in 1966.

As they resumed work after World War II, the missionaries realized that the most pressing need for the Chin Christians in the postwar years was to train new leaders to meet the demand of the increasing number of converts. Thus, both Nelson and Johnson concentrated on teaching at the Bible schools and translating the scriptures and hymnbooks. This new development changed the role of missionaries in East Chinram as well; while the church leadership role was mostly taken over by the Chin, the missionaries played the intellectual role of lecturers at the Bible schools and as translators of the Bible.

When Rev. Nelson returned, he opened the Bible School at Tiddim in 1947; his heavy teaching load was shared by a local leader, T. Hau Go:

> They [the missionaries] have been greatly reinforced by the arrival of Hau Go, a Judson College student who completed his training at Madras Christian College during the war, taking his honors degree in Philosophy. Though offered a lucrative post in the Frontier service by the Burma Government, Hau Go has chosen to go into full time Christian work among his Chin people (Nelson 1948: FM-343).

At Tiddim Bible School, Nelson and Hau Go taught their students English and arithmetic in addition to regular Biblical courses, as Nelson wrote:

> Some of our students had little or no knowledge of English when they entered the school. Since there is very little Bible literature in Chin, we realized that after graduation they would have limited opportunity for continued study. Thus, we decided to teach English as well. Much emphasis was placed on the study of English the first two years. While progress varied with individual students, we feel that most students gained a working knowledge that will enable them to continue their studies by means of books and aids in the English language. Such books as an English Bible, Montgomery's translation of the New Testament, a concordance, a Topical Textbook, Halley's Bible Handbook, a small text for rural workers on diseases common to their villages, and treatment of those diseases ... were books given

to each student as a beginning of a small library, together with available books in the Chin language (Nelson 1950: FM–387).

The first batch of twenty students – eighteen men and two women – graduated from the Tiddim Bible School in 1950, and 'most of the men graduates were to become pastors in their areas, but one was scheduled to go down to the plains to become a chaplain to the Chin soldiers' at the Chin Hills Battalions, stationed in Pe-nwe-kung (Johnson 1988: 865). In 1948, Johnson also reopened the Bible School at Haka, which had been founded by Strait 1928 but discontinued in 1931. Along with Rev. and Mrs Johnson, Rev. Sang Ling, a graduate from Insein Divinity School, and Rev. Lal Hnin, who, like Hau Go, was trained in India at Cherrapunji Theological School during the war, also taught at the Haka Bible School. The first batch, thirty-nine students, were admitted, and they all graduated in 1951. Johnson correctly observes that the Bible Schools at Tiddim and Haka were 'the most single effective effort' (ibid.: 866), especially for the Nelsons who were unable to return to Chinram from their second furlough.

> The successful completion of the Bible School in Tiddim was one of the solid accomplishments of the Nelsons during their post-war term of service, one of the many things they did to strengthen the work of the Christian mission in the northern half of the Chin Hills. Their graduates, all of them from the Tiddim, Falam, and Kale Valley areas, went out to become loyal and effective workers for the Lord (ibid.).

In 1951 the Nelsons left Chinram for their second furlough. 'According to our present plan', he wrote, 'we expect to leave Burma about the middle of May. We are thinking longingly of a short visit to Sweden, the land of our forefathers, about midsummer' (Nelson 1950: FM–387). Their second furlough was the end of their missionary careers because the government of newly independent Burma refused to grant them re-entry visas.

The Reformation of the Chin Hills Baptist Association

When the Chin Hills Baptist Association was founded in 1907, the churches were weak and the number of Christians was small. Thus, in order to create an atmosphere of cheerful and encouraging good fellowship among new converts, they combined an annual church council meeting with a religious festival, called *civui*. As the number of Christians increased, however, it became increasingly difficult for all church members to come together, as it strained the ability of a village or small circle to host such large numbers of people. For instance, in 1938, when the CHBA *civui* was held in Tonzang, 900 persons attended for three days. It was then decided to hold the CHBA council meeting only once every three years, starting in 1938.

Since all the businesses of the CHBA, including financial matters, such as the salary of pastors and preachers, was decided at the triennial meeting, the CHBA *civui* became exceedingly important for the member churches. The system functioned in such a way that every three years, each church or circle of churches sent their preacher or other delegate, together with all their money – offerings and tithes – to the council meeting of the CHBA. There, the money from all the churches was counted. Other money, such as mission grants and gifts from abroad, was added. All this was totaled and announced to the assembly. Then there was a division of the funds to pay the salaries of pastors and other Christian workers, and for projects such as sending missionaries, and so on. In fact, it was a unitary, centralized system that needed to change (cf. Johnson 1988: 857–862).

In 1948, the CHBA was reformed at the triennial meeting held in Satawm, Falam township. Under the CHBA, three 'local associations' – the Falam Baptist Association, Haka Baptist Association and Tiddim Baptist Association – were formed. As part of the reform, a self-support governing system was recommended for the administration of the local associations. Thus, 'the old, unitary, control-at-the-top system' was abandoned, the treasury was divided and each local association was given the responsibility for the payment of salaries and pensions of the workers in their respective associations. However, 'to care for needed other projects, such as missionary work in the Matu area and the plains, for example, ten percent (10%) of all offerings would be used as a General Fund' (ibid.: 861). Moreover, in order to supervise local churches more closely, it was agreed to hold local association meetings during the intervening two years between the triennial meetings of CHBA *civui.* Although not officially renamed, 'All Chin Baptist Associations' was popularly used interchangeably with CHBA.[2]

In 1954, when the CHBA was reformed once again, the term 'Convention' was used for the central body of the Chin Baptist Churches, and the term 'Association' continued to be used for the local associations.

In 1948, the All Chin Baptist Associations claimed a membership of 180,000 baptized, and more than that number of unbaptized, which was more than one-third of the entire population of the Chin Hills District. Today, the number of local Baptist Associations has increased from three to twenty-five, and more than 80 per cent of the Chin people are Christians.

NEW PHENOMENA IN CHIN SOCIETY

Gradual Adjustment to the New Way of Life

Conversion to Christianity was not a single event but a long and gradual process of adjustment to the demands of the new faith, as opposed to the old ritual habits, and also to the necessities of life. The new life always began with the abandonment of all the ritual systems connected with their traditional religion of *Khua-hrum* worship. This included not only the sacrificial ceremonies but also many kinds of feasts and festivals which were regarded as practices of the 'old life', what the Chin called *Lai phung.* Moreover, all the converts were expected to stop the practice of polygamy and the consumption of intoxicants, *Zu le Zuhui.*

The missionaries also encouraged the early Chin Christians to separate the 'new', Christian ways of life, *Krifa Phung,* from the 'old', traditional Chin ways of life, *Lai Phung,* and they portrayed the 'old life' as sinful in nature and full of unhappiness and permanent unrest, while the 'new life' was brightly coloured and full of happiness and peace. When people converted to Christianity they located themselves in the 'new life' and conceived their identity in terms opposing the old *Lai Phung* to the new *Krifa Phung.* This was due to the fact that nineteenth century evangelicalism, represented by nearly all the Protestant missionaries, including the American Baptist Chin mission, 'placed emphasis upon Christianity as a way of life, a lifestyle'. For them, 'doctrine was important but meaningless if not associated with a transformed life'. Thus, as Frederick Downs observes, 'the idea of becoming a Christian meant adopting a new mode of life' (Downs 1992a: 146). In this way, conversion to Christianity and the adoption of new ways of life were inseparably linked to each other.

Within this long process of adjustment, at least three new forms of 'imagined' social world appeared – Sunday worship services, singing from the same hymnbooks and reading the Bible.[3] These three characteristics of the social world were related to their new ritual systems in Christianity and the expectation of salvation in Christ's life after death in *Mithi-khua.* The observation of the Sabbath, Sunday, was the first step that the Chin Christians took to 'imagine' themselves as 'living lives simultaneous with other lives in an homogenous time', measured not only by Biblical teaching about the Holy Sabbath (Genesis 2: 2, Exodus 20: 8), but also by modern clocks and calendars. It was novel, yet effective and meaningful for the Chin Christians to create an 'imaginary community' beyond their clan and tribal boundaries. As they observed Sunday, they also practised the same rituals of singing from the same hymn book and reading the same Holy Bible everywhere in Chinram. Whether they sang hymns or read the Bible, congregations at the Sunday service knew that they were singing and reading what many other believers were singing and reading at the same time with the same hope and faith in one God. These new parts of life gradually became the most powerful sources for the creation of an 'imagined community', and thereby the creation of a common national identity for all the Chin Christians (cf. B. Anderson 1991, especially Ch. 3).

New Rituals and Phenomena in Chin Society

The moment of conversion was not the end of the story but the beginning of a long process of adjustment to the demands of a new faith in relation to the ritual habits of traditional religion that it supplanted. In this, the adoption of a new faith and new identity was marked by some sort of ritual. The most observable phenomenon for the converts among the Chin was *inn-thianh,* 'cleansing the house'. The *inn-thianh* ritual was the first observable phenomenon of conversion and was usually followed by baptism.

The procedure of the *inn-thianh* ritual is almost the same everywhere in Chinram. Here, I shall quote from the eyewitness account of Dr East, reported in 1910. He described 'cleansing house' in the Tiddim area as follows:

> Mr. Cope and I were now in Tang Nu Kwa (Thangnuai village) by invitation. We were called there by people, as seven families have accepted 'the new way' and have asked us to help them break down the altars and all emblems of evil spirit worship, to remove all heads of animals sacrificed and all bamboo resting places for evil spirits. This is quite a job as there are many altars and many skulls, sometimes numbering fifty to a hundred, and as we have to take the initiative in order to embolden the people to touch things formerly held sacred and to help us in this most destructive breaking down of spiritualism and building up of Christianity. It took us from about 2 p.m. until the lengthening of the shadows.
>
> The last house we sanctified to God belonged to an old couple, and when all emblems had been taken out and readied to be burned, the old man asked, 'Will I dare to sleep in this house tonight since the spirits are not here to protect us and they have no resting place?' We then knelt and I placed my hands upon their heads and prayed that Jehovah God should now and thereafter dwell with them to protect them against sickness and all harms. Now they were ready to go back into the house and were not afraid, as God was greater than the evil sprits.

East concluded his letter to his wife:

> My dear, it is not an easy thing to break loose from an age-old custom and imprinted beliefs. Man's heart is hungry for atonement. Man's heart is seeking a resting place. Man's heart is only satisfied when it knows that it is protected by one stronger than himself. And so a tremendous struggle takes place in the hearts and minds of these people when they awaken from age-long ignorance and face God (East 1910: FM–186).

A similar account of the 'cleansing house' ritual was also reported by F. O. Nelson in 1949 as follows:

> Upon our arrival in the village we gathered at the home for the service. First of all the Christian elders from Tiddim asked a number of questions concerning their decision to follow Christ. Upon receiving suitable replies to all his questions, approval was pronounced. Then a number of the Christians proceeded to 'clean house'. The routine is almost always the same. Along the rafters near

> the door a number of skulls are found. These are the skulls of animals that have been sacrificed in days past to appease the evil spirits. The skulls are kept to guard the family from harm. Tied to the skulls are small bags of food for the spirits. The skulls and all are taken out of the house and burned. Oftentimes within the house will be found an 'altar', a fireplace that is used only during the time of sacrifice. This altar is also destroyed. Just outside the front gate still other skulls (which are the skulls of wild animals they killed in hunting expeditions) are buried in the earth. These are dug up and added to the bonfire. Quite often, as was the case yesterday, in the yard will be found a queerly shaped pole (*lawngpi-ttial*) somewhat resembling a totem pole. After a Chin family gives a certain sacrifice, one that is rather costly, they are entitled to erect such a pole in front of their house. It is a mark of honor among the Chins and is highly esteemed. I recall at a similar service a year or so ago, when a neighbor woman saw them cutting down this pole, she cried out incredulously, 'You are not taking away the pole, are you?' One of the Christians replied, jokingly, 'Yes, shall we come and take away yours, also?' The thought was so disturbing to the woman she was unable to answer. But when a Chin becomes a Christian, even these highly regarded objects are put away since they are related to their *khua-hrum* worship (Nelson 1949: FM–343).

Nelson was incorrect in saying 'the skulls are kept to guard the family from harm'. The skulls of sacrificial animals could not guard the family from harm – only the guardian god, *Khua-hrum,* and the household god, called *Chung-um,* could do that. The skulls were kept only as symbols of communion or the covenant that the family had made with its guardian god. Since they were the traditional symbols of that communion, the skulls were treated with homage and respect. As we recall, the Chin did not believe that the sacrificial animal skulls could kill people. The skulls had to be treated respectfully because they were the symbols of holy communion between man and god. But in the case of Thuam Hang, he challenged the symbols of the whole system of traditional religion and even attacked the symbol of holy communion. He expected to die because of his rejection of the protection from the guardian god, *Khua-hrum,* and his denial of the sacrificial ceremony to appease the evil spirit, *Khua-chia,* which caused all kinds of harm in life. Sometimes missionary texts are rather exaggerated and their accounts too romanticized. They therefore require careful reading.

When the Chin converted to Christianity, they wanted to abandon their traditional ways of life. This was encouraged by missionaries. Their traditional religion, after all, was inseparably linked with every aspect of life. As mentioned earlier, a house was not just a home for people, it was also an abode for the household god, called *Chung-um.* When they wanted to become Christians, the first thing the Chins needed to do was to let their guardian god leave the house. In order to do this, they needed to clean their house and destroy the altar called *Kho le Kheng*. That was the reason people used to say '*Kho le Kheng kan thlak cang!*' [We destroyed the *Kho le Keng* as a symbol of our conversion to Christianity] when they became Christians.

For many able-bodied people, destroying the *Kho le Kheng* was not enough. They even rebuilt their houses completely and usually changed the architectural style as well. Thus, most Christian houses are not in the traditional long-house style, but resemble the house that Dr East built in Haka. People also usually shifted away from their old house sites when they became Christians, for they knew that the sites themselves were associated too closely with their guardian gods. When a whole village became Christian, they even moved off their old village site, which was inseparably linked with the village guardian god, *Tual Khua-hrum*. In the new village, the church stood at the heart of the community instead of the *Tual* or *Tlenlai Lung* that had been the communal sacrificial stone or altar.

From the traditionalist point of view, as Nelson mentioned concerning the elderly Chin woman, this new phenomenon of the *inn-thianh* ritual was a painful memory of the destruction of centuries-old institutions. Today the Chin people view this from a different perspective. They now remember those stories as a symbolic departure from the 'old ways' and portray the epoch as an abrupt juncture, a fold in the course of social history marked by sharp changes in the ethos of everyday life. These new phenomena or rituals, such as *inn-thianh* and baptism, which are part of the conversion process, are the key social and ritual activities through which the transformation of identities and communities are accomplished.

Finding Common Ground and Common Knowledge

The views of the first missionary couple, Arthur and Laura Carson, on traditional Chin religion, culture and customs were completely negative. Thus, Arthur Carson would write bluntly, 'the Chin had no word for God, grace, mercy, forgiveness', or even 'love'. Laura Carson also wrote that 'sacrificing to evil spirits' was 'their only religion and system of medicine' (L. Carson 1927: 161). They passed, as Johnson points out, 'this attitude along to young Dr East' and other missionaries. The reason, according to Johnson, is that 'when Laura and Arthur Carson first came to the Chin Hills, they were so struck by the backwardness of the people that they overreacted' (Johnson 1988: 263).

East seemed to have changed his view after staying nine years in Chinram and travelling widely among many tribes. In 1909, he met 'a man over one hundred years old' at Roshi in the Falam area. He mentioned this old man in his letter, and that he would not be surprised 'if he was even older' than 100 years. This old man told him about the origin of the Chin and also the story of *Pa Lo Tuanbia*, which Dr East translated as 'The Story of Fatherless', and other myths and legends (East 1984: 144). Here is the story of *Pa Lo Tuanbia*, recorded by Pu Sakhong and confirmed by Pu Mang Lung of Aibur.

> Once upon a time, a young lady who was still a virgin was collecting firewood far away from her village. On her way home, she found a fruit called *Khuhlu* on the village sacrificial stone called *Tlenlailung*. Without thinking much about the consequences, she just picked up the fruit and ate it because she was so thirsty from working the whole day under the hot sun. However, the fruit was not just ordinary fruit, since it had been left there by *Khua-zing*, the Supreme God. Moreover, *Khua-zing* had mixed His own fluid with the juice of the fruit, for *Khuhlu* was a very juicy fruit. So the young lady conceived without knowing man and gave birth to a baby boy. Since she was unmarried, the baby was simply named Pa Lo, meaning 'Fatherless'. Pa Lo was extraordinarily strong and handsome. When he grew old enough, he asked his mother who his father was, where he lived, and so on. His mother told him that his father lived in heaven without mentioning His name. Their conversation was overheard by God from heaven. God had another son in heaven. So he sent his son in heaven to visit his younger brother on earth. Pa Lo's brother from heaven did not stay long on earth, but during his short stay he made all kinds of domestic animals out of clay, and then he turned them all to living animals. In this way, Pa Lo became the richest man in the world and performed all kinds of feasts and festivals such as *Bawite-bawi* and *Khuang-cawi* (Pu Sakhong 1971: 11).

East was also told the story of Theizam, similar to the biblical story of the Tower of Babel, and the story of Ngun Nu, similar to the biblical story of the Flood. Thus, he wrote on 4 February 1909:

> To me it seems that the easiest way of getting into their hearts is to meet them where we have common ground and common knowledge. I was led to believe that these people had no knowledge of God, no word for love, and no word for heaven. However, I could not accept that idea as I very thoroughly believe in racial unity and that 'God made all men out of one blood', and it is therefore unreasonable to believe that any part of that race should be absolutely destitute of all knowledge of the Creator. It is a certainty that the wild Chins believe in the God of Heaven as Creator. This knowledge is universal among them (East 1984: 144).

For the Chin, however, the most attractive common ground between their own traditional religion and Christianity consisted not only of stories such as the Tower of Babel (see pp. 13–14 above), the Flood, the Virgin Mary and Pa Lo, as East pointed out, but the theological similarity between the traditional Chin religious teaching of life after death in *Mithi-khua* and the biblical teaching of heaven or paradise. Among missionaries, only Strait applied this similar theological concept of eschatology in his ministry after East left the field. Strait did the first scholarly work on traditional Chin religion for his doctoral dissertation, *A History and Interpretation of Chin Sacrifice*, in 1933. His knowledge of traditional Chin religion helped in his ministry of preaching the Gospel and also when he translated the Bible and hymns. He could apply the idioms and concepts of the Supreme God and life after death in *Mithi-khua* to help the Chin find common ground.

The Chin saw no problem at all in reaching heaven after death, for their religion taught them that one must surely reach heaven if one died an ordinary death. The problems concerning heaven, which bothered the Chin most, were, 'Are we going to be rich or poor? What kind of house are we going to live in?' According to traditional Chin religious teaching, the house they built here on earth was the image of the house in their future life in *Mithi-khua*. If one could build a large good house here on earth, then one would surely have a large good house in the future life. But now the Chins were told by Christian missionaries that if they became Christians they would have a very big and good house glittering with silver and gold, a house prepared for them by God Himself. They did not need to build homes on earth for life in heaven, they were told, but they must abandon their own traditional way of life to become Christian! And they were told that if they did become Christians, their homes in heaven would be like the missionary house in Haka built by Dr East in 1907. Such concrete examples and promises were most attractive to the first generation of Chin Christians. Thus, for first and second generation Chin Christians, one of the most popular hymns translated by Dr Strait was:

> *Ni nak in a ceu khua a um ko,*
> *Zumnak in a hnu ah kan hmuh lai,*
> *Khi khin kanmah kanpa hngak len ko,*
> *Kannih umnak a ser lio dah ngai.*
>
> (There is a place brighter than the Sun, And we will reach it by faith, Our Father is preparing a house for us, That is glittering like silver and gold.)[4]

The next most attractive Christian teaching for the Chin was the substitute death of Christ on the cross for sinners. According to traditional Chin religious teaching, if one died an accidental or violent death, his soul must go to *Sarthi-khua*. No soul can be redeemed from *Sarthi-khua* except by a substitute violent death that is revenge for his or her death. To illustrate this point, I shall tell a story of an accident that took place in the early 1950s, in Leitak in the Zophei area.

Leitak, the principal village of the Zophei area, used to be ruled by a powerful chief, Mang Hnin of the Hlawn Ceu clan. When Mang Hnin died, he had only one son by his principal wife, *Nutak*. This son was to inherit his power and wealth although he had many more sons by his minor wives (*nuchun*). Unfortunately, a son of a former slave of Mang Hnin accidentally killed the heir of Mang Hnin's house, Lian Hei. According to the traditional Chin religious belief of *Laiphung*, Lian Hei would become the slave of his slayer, no matter who he was. For his mother, Pi Khuang Cin, it was completely unacceptable that her beloved son, the heir of Mang Hnin, would become the slave of her own former household slave. Thus, Pi Khuang Cin, Regent of Mang Hnin's house, gathered together all her relatives from far and near and ordered them to take revenge on her son's death. At that time the government strongly prohibited such action, but Pi Khuang Cin was determined to take revenge no matter what. She was ready to make war if the government of newly independent Burma intervened in any way.

On that occasion, Pu Sa Khawng, who was married to Lian Hei's elder sister, convinced Pi Khuang Cin that revenge had already been taken for Lian Hei, for he was a Christian. Lian Hei converted when he was in mission high school at Insein near Rangoon, where his uncle, Lt Tial Khuai, served in the Chin Hills Battalion of the British-turned-Burmese army. Pi Khuang Cin asked who had taken that revenge for his son, and she was then told about the substitute death of Jesus Christ on the cross. Pi Khuang Cin was convinced, because Jesus was crucified on the cross. She would not have been convinced if Jesus had died of natural causes, for she knew according to her religious teaching that only a violent death could substitute for her son's death. Thus, she was converted and baptized by Rev. Heng Cin, who was once expelled from Leitak by her late husband, Chief Mang Hnin.[5]

In this way, the theological similarity between traditional Chin religion and Christianity, especially the substitute death of Jesus, opened the doors for conversion, even for people who headed Chin religious and political institutions, like Pi Khuang Cin.

THE GOSPEL AND A NEW SOCIETY

In this section, I shall discuss how Christianity helped the Chin people to adjust to the change thrust upon them and create a new Christian society, and how Christianity provided a means of promoting self-awareness through its ideology and ecclesiastical structures.

The Feast as Worship

According to traditional Chin belief systems and ritual practices, there was no worship without feasts. Moreover, the feast was not just the sharing and partaking of food, but sharing the source of life (*zing-dangh*), and therefore a communion between god and man as well as between men. Since the Chin viewed the feast as sharing the source of life, there was no feast without animal sacrifice, known in Chin as *Sathi luan lo cun Do a si lo* (literally, 'there is no feast without pouring blood'). The concept of pouring blood is important because blood was the source of life, which must be shared as communion, not only between the family, clan and tribe, but also between them and their guardian god, *Khua-hrum.* In other words, pouring sacrificial animal blood was a symbol of communion between the guardian god *Khua-hrum* and man, and the sharing and partaking of the flesh and blood of sacrificial animals at the feast was a communion between men which united members of the community with each other and with their god, *Khua-hrum.* The idea of animal sacrifice in the worship of *Khua-hrum* could be compared with the idea of animal sacrifice in the Hebrew Bible. According to W. Robertson Smith,

> [t]he leading idea of animal sacrifices of the Semite was not that of a gift made over to the god, but an act of communion in which the god and his worshipers united by partaking together of the flesh and blood of a sacred victim. (W. R. Smith 1996)

When they became Christians, the Chin concept of communion as sharing the 'source of life' (*zing-dangh*) was gradually transformed into the concept of communion with the 'giver of life' (*Khua-zing*), that is, the Supreme God. Together with such a transition in belief, missionaries introduced a new ritual practice of 'Holy Communion' between the Supreme God and man, which was simply a symbol without animal sacrifice or feast. Since there was no sharing and partaking of a real meal at Christian Holy Communion, it was only a communion between man and the giver of life, the Supreme God *Khua-zing,* and not a communion between men. In the traditional Chin religious concept, there was no communion between men without sharing and partaking of 'the flesh and blood of a sacred victim' because the Chin considered it a source of life that must be shared between men so that the members of a family, clan and tribe who shared the same source of life could be united as one. By contrast, the Christian 'Holy Communion', which was usually held on the first Sunday of the month, did not bond people together as the same family or clan or tribe but rather as a community of believers. Moreover, the missionaries discouraged almost all traditional Chin feasts and festivals, which, of course, were related to traditional ritual practices and belief systems. To become Christian during the missionary era was to abandon the traditional Chin ways of life, *Lai-phung.*

However, the spiritual life of the Chin people could not survive without feasts and festivals, even in Christian churches. Traditional Chin religion received 'its own rhythm through the regular changes of the agrarian year', and people had to renew their spiritual life through the rhythm of seasonal changes which were usually marked by many feasts and festivals. Thus, the Chin church, as those in Africa, 'has adapted itself to this ancient rhythm but also has created new festivals and given the year a new meaning' (Sundkler 1980: 181). In this way, Christmas (*Krismas*), New Year (*Kumthar*) and Easter (*Tho*) became the most important social feasts and festivals for the new Christian Chin community. As already mentioned in previous chapters, they also created a new festival called *Civui*. Although Christmas, New Year and Easter are seasonal festivals, the Chin celebrate them rather like feasts, and a single family usually hosts the festival for the entire community.

Krismas, Kumthar and Tho

Although there was a transition from the Chin traditional ways of life to the new community of faith, the conversion to Christianity did not break down the ties of family, clan and tribe in Chin society overnight. Instead, a good relationship between members of the family, clan and tribe was somehow strengthened by the teaching of love in the Christian Gospel. The only thing that was needed for the new converts among the Chin was to create a new ethos of how to express such love and good feelings among themselves.

In their old culture, love and good feelings were expressed in feasts and festivals, which tied family, clan and tribe strongly together. Since Christian teaching encouraged good relationships and strong ties among people, feasts and festivals were also needed for newly converted Chin Christians so that they could express their love and ties with others and among members of the Church who shared the same faith in Christ the Savior. The only difference between the two – the old and the new – was that in the old culture, feasts and festivals were centered upon family, clan and tribe, emphasizing the difference between one family and another, one clan and another, one tribe and another, and so on. Hence, it reinforced clan and tribal identities, which in the long run resulted in deep separation between different tribes of the same people. In the new Christian society, however, feasts and festivals were centred upon a love that transcended the boundary of family, clan and tribe; they emphasized the Love of God and humanity, which everybody could share and enjoy freely. Thus, the celebration of Christmas, New Year and Easter Sunday in the new Chin society was not just the celebration of seasoning feasts and festivals but the transformation of Chin society from 'clan and tribal groups' to 'the religious community of the Church' with no boundaries. In this way, Chin society was gradually transformed from a 'clan and tribal oriented society' to 'the community of faith in Christ'.

Today, *Krismas, Kumthar* and *Tho* are celebrated like many other feasts and festivals in the old culture, as discussed in Part One. Normally, a family who has a harvest as good as expected declares that they will host *Krismas* or *Kumthar* or *Tho,* whichever is most appropriate for them. Most families prefer to host *Krismas* and *Kumthar* because *Tho* usually occurs during the farmers' busiest time, when they need to prepare their fields for sowing before the monsoon rains come in the middle of May. Even when the harvest is not as good as expected, a Christian family still offers a portion of money or food to the church and the church arranges to hold the feast as best as it can. *Krismas, Kumthar* and *Tho* are really community events for the new Chin society, and also a time for community fellowship and reunion. Even today, when things are changing rapidly, the Chin people still prefer to celebrate those feasts, especially *Krismas,* in their birthplace.

During the feast, the host family has to prepare meals for the entire community, but cooking and preparing food are traditionally the responsibility of the community as a whole, especially the youth. As long as the feast lasts, there are many activities in which almost all of the community or church members have to participate, such as song contests for solo and choir, dramas, speeches and sermons, etc. Villagers and the community commonly employ these events to recall the past and to share their faith and the rituals that define the new Chin society. In this way, those feasts and festivals gradually continue to help to create a new identity for a new society.

Civui and the Religious Festival

If *Krismas, Kumthar* and *Tho* are feasts and fellowship events for a single village or a single town, *Civui* is a festival for a larger community in which at least three villages participate. This is not just a religious festival but an expression of how the ecclesiastical structure of the Baptist church in Chinram created new festivals and has given the season a new meaning. Moreover, this is about how Christianity helped the Chin people to overcome clan and tribal barriers – which confined them separately in their history – and created a new tie and a new national identity for the whole Chin people. It is about how Christianity has helped Chin society to make the transition from a clan and tribal oriented society to a church-oriented community which transcends the old tribal identity.

As mentioned already, *Civui* means 'meeting', but it implies 'any Church administrative meeting which is larger than the local Church meeting' (Cung Lian Hup 1993: 105). Based on the ecclesiastical structure of the Baptist Church in Chinram, there are three levels of *Civui*: 1) Area Civui, 2) Association Civui, and 3) Convention Civui. Area Civui is held annually but Association and Convention Civuis are held triennially. (In some areas, especially in the Mizoram, *Civui* is known as *Khawm-pui,* meaning 'big gathering'.)

Civui is a combination of Church administrative meeting and religious festival for the new society in Chinram. For Area Civui, the Church business meeting usually starts on Thursday, but at the Association and Convention levels, where much more business is to be done, the meeting usually starts on Monday. In both cases, however, ordinary pilgrims (*Civui a zawh mi*) must arrive on Friday when the Church administrative business is over. In fact, Friday is the real beginning of *Civui*. Usually, there are at least seven worship services during the *Civui*: one on Friday night, three on Saturday and three on Sunday. The intervals between worship services are used for fellowship and games. During the services, there are many speeches, sermons, choirs and, at night, services and even dramas. The churches send their choir groups to highlight this special event, and the host village is responsible for providing room and board throughout the *Civui*. Nowadays, Area Civui is mainly held in rural areas, and Association and Convention Civuis are usually held in in towns such as Haka, Falam, Tiddim, etc. For the new Chin society, *Civui* has become a religious festival from which people can receive both spiritual and social benefits.

Civui is also a time when villagers and communities commonly recall their past, discuss their present and share visions for their future together in one hope and one faith. This is a time when people do not identify each other as members of family, clan or tribe but only as members of the community of faith. This is a time when people share their experiences, both past and present, and their emotions and faith. When people can freely share their experiences, emotions and faith, these are remembered as personal but also as collective memories or socially reconstructed narrative events. And when personal memory becomes common memory, this is called social history, a factor that holds 'history and identity together'. In this way, the Chin people do 'identity work' together, especially during such events as *Civui*.

Zubilee Civui: The Golden Jubilee at Haka in 1949

If *Civui* provides a time and place to create a common identity for the new Chin society, a 'formative moment' occurred at Haka in 1907 when the Chin Hills Baptist Association was formed. In 1949, a year after the Chin joined the Union of Burma, the 'confirmation moment' occurred when fifty years anniversary of Christianity in Chinram was celebrated with a great three-day jubilee meeting, known in Chin as *Zubilee Civui*.

The *Baptist News* reported that the official attendance of the Jubilee was 5,128, among them the 'visitors numbered almost 4,700', and 'some of them had walked five, six, up to nine days to attend the Jubilee'.

> As the jubilee got under way, a large cross dominated the entire scene, a mute reminder that the meetings were not primarily to honor the Carsons, first missionaries to the Chins, nor their Karen assistants, but to honor Jesus, the crucified and risen Savior, in whose name all mission work is done (Johnson 1949: FM–339).

Although the jubilee was held to glorify the name of Jesus as Lord (*Zisu Kri Kan Bawi A Si*, as the Jubilee signboard read), participants also remembered the first pioneer missionaries and the first Chin converts. During the jubilee many sermons and speeches were delivered and many hymns and choirs were presented, to praise the glory of God and to 'remind the Christians of the faithful labors of Rev. and Mrs Arthur Carson, Dr and Mrs East, Dr and Mrs Woodin, Rev. and Mrs J. Cope, and Rev. and Mrs Chester U. Strait' (Johnson 1949: FM-339).

The *Baptist News* highlighted the mass baptism during the jubilee; its headline read, '373 Baptized at Jubilee'.

> On Saturday afternoon thousands gathered around the tiny Haka Lake (*Ralkap Tili*) while eleven pastors baptized 373 new Christians. We heard the strong choir voices roll our familiar hymns over the water. As Franklin Nelson and I (Robert Johnson) watched the scene we thought back to the day, fifty years ago, when a lonely American and his wife arrived at this very spot, and there were no Christians at all and no baptisms. There were seven years of labor before the first Haka convert was made, but now the harvest has come (Johnson 1949: FM–339).

As Johnson reported in the *Baptist News*, the scenes of baptisms, worship services, sermons and choirs were momentous events to be remembered by new generations. More important for people at that time were the huge crowds who attended the jubilee. For them, the crowd itself was awesome because all came to worship together. These were people from all corners of Chinram, from different clans, tribes and language groups, who used to be rivals or even enemies. Such things rarely happened before. Now, different clans and tribes had come together to celebrate a religious festival. In the old culture, two tribes could never worship together except when they made a war alliance. Now they could worship together, even share food and partake of the source of life.

The sharing and partaking of food was very meaningful. If people regarded one another as friends, they provided food, especially during religious feasts and festivals. If they belonged to the same clan and tribe, they had to share and 'partake together of the flesh and blood of a sacred victim' during the religious festival as a communion. During the jubilee at Haka, Robert Johnson advised the Chin people – perhaps without realizing so much about the Chin custom – to break the tradition of providing food at religious festivals 'because of the unusually large number attending' (Johnson 1949: FM–339). But the church leaders at Haka refused to do so. They at least wanted to give away the meat for free because it was considered the source of blood and life, which must be shared as an act of communion among all the believers in one faith. Moreover, if they did not provide food, people from other tribes, especially those who were old rivals and enemies of the Haka, would think that they were unwelcome because of old rivalries. Now all were welcome and enjoyed the festival together. They participated in the new ritual of 'Holy Communion' and 'Baptism' on Sunday and also participated in the old communion of the source of life, as provided for them by the host church at Haka in the form of food, the mark of hospitality and love. This was the second 'formative moment', if not a 'confirmation moment', when traditional clan and tribal identities were in the process of breaking down, and a new identity was in the process of being established.

NOTES

1 U Nu promulgated Buddhism as the state religion of the Union of Burma in 1961.

2 This sometimes caused confusion for historians. Za Tuah Ngur, for instance, wrote that the 'All Chin Baptist Association was formed in 1906' (1993: 74).

3 I borrowed the theme of 'imagining social world' from Benedict Anderson's *Imagined Communities: Reflection on the Origins and Spread of Nationalism,* especially pp. 9–36.

4 This hymn is inscribed on the gravestone of Pu Hreng Kio of Aibur who became a Christian in early 1930s and passed away as a church leader of Aibur in 1961.

5 Interview with Pi Men Tang

ELEVEN

Concluding Remarks

In this study, I have argued that conversion to Christianity in Chin society was the product of a long-term process of change resulting from the pressures of external forces. Scholars from various disciplines have already affirmed the fact that there is a connection between religious change and social pressure in society. Social anthropologist Raymond Firth, for instance, writes, 'Religious change has often been a result of social pressures upon guardians of the doctrine and ritual.' He continues:

> Religion, like other social institutions, is continually in a process of change, due partly to external pressures of an economic or political kind, or to internal pressures of doctrinal debate or personal ambition (Firth 1996: 199).

Second, I have tried to analyse the paradoxical continuity between traditional Chin religion and Christianity. Doing this, I am mindful of the fact that if 'change is a lawful fact of social life', as scholars like Geertz have argued, how do we account for continuity? Scholars from various disciplines have tried to grapple with the problems of continuity within the process of change. In my hypothesis, continuity between traditional Chin religion and Christianity was possible because of the two religions' similar theological concepts and belief systems, especially the traditional Chin concept of a Supreme Being, *Khua-zing*, and the Christian understanding of a Supreme God, the creator of all things in this universe. As mentioned in Part One, the Chin had a clear concept of the Supreme God, *Khua-zing*, but they did not worship Him because He was good and did not require any sacrificial appeasement, as did *Khua-hrum* (guardian gods) and *Khua-chia* (evil spirits). Thus, traditional Chin religion was centred upon the worship of *Khua-hrum* and *Khua-chia*, not upon *Khua-zing*, the Supreme God.

When the Chin converted to Christianity, they inevitably passed through a long-term process of adjustment between the demands of a new faith and the ritual habits of their traditional religion. In this process of change and adjustment, the Chin people completely abandoned almost all their traditional ritual practices of sacrificial ceremonies and related value systems. This process of change made deep inroads into the original patterns and structures of traditional religion, and thereby society itself. The belief system, however, was transformed from *Khua-hrum*-centric to *Khua-zing*-centric within the same conceptual pattern of belief system. In other words, the worship of *Khua-hrum* was transformed into the worship of *Khua-zing*, or the worship of the lesser god to that of the Supreme God. This kind of 'conversion as transition in belief' can be compared with the theory of conversion of Robin Horton, who wrote about African religions in 1971.

According to Horton, traditional African religion incorporated a belief in a two-tiered structure. He explains:

> In the first tier we find the lesser spirits, which are in the main concerned with the affairs of the local community and its environment – i.e., with the microcosm. In the second tier we find a Supreme Being concerned with the world as a whole – i.e., with the macrocosm. Just as the microcosm is part of macrocosm, so the Supreme Being is defined as the ultimate controller and ground of the lesser spirits (Horton 1971: 101).

As explained in Part One, the lesser spirits of *Khua-hrum* worship were closely associated with exclusivism and isolationism, fundamental isolation from other societies where traditional instruments of integration, such as the family segment of wife-giver and wife-taker, clan, tribe, etc., could function properly. Thus, *Khua-hrum* worship could only be identified with family, clan and tribe, but never with the Chin people on the national level as a whole. The reason, as Horton explained in terms of micro- and macro-levels, was that

> [t]he essence of the pre-modern is that most events affecting the life of the individual occur within the microcosm of the local community, and that this microcosm is to a considerable extent insulated from the macrocosm of the wider world. Since most significant social interaction occurs within the local community, moral rules tend to apply within this community rather than universally – i.e., within the microcosm rather than within the macrocosm (ibid.).

Although the Chin had a clear concept of the *Khua-zing*, the notion of a Supreme Being is more highly developed in Christianity than in traditional Chin religion. The transition in belief and the shift of worship from *Khua-hrum* to *Khua-zing* offered the Chin a national identity, which they never had a chance to create under their traditional religious worship.

As we have seen, over the course of fifty years or so, Christianity became firmly rooted in Chinram and the church became the central pillar of a new Chin society, where people identified themselves and each other as members of a 'community of faith' rather than as members of a family, clan or tribe. Within the process of transition in beliefs from micro-level to macro-level, that is, *Khua-hrum* worship to *Khua-zing* worship, people's identification with each other was shifting as well from clan and tribal identities to a wider level or macro-level of Chin national identity or Chin-ness, just like in Horton's theory of conversion. All this took place within a broad framework of the community of faith. In this way, Christianity provided the means of overcoming clan and tribal identities, and at the same time helped to create a new society where people identified each other as brothers and sisters in one faith, or members of the community of faith.

Unlike other Christian communities in other parts of the world, the community of faith in Chin society was contained by the boundaries of Chinram, and therefore this community of faith was identical with 'Chin-ness' or a Chin national identity. Thus, Christianity and Chin-ness were inseparably intertwined in a new Chin society. The reason is, in my opinion, that Chinram was surrounded by peoples of other faiths – Burman Buddhists in the east, Bengali Muslims in the south, and Indian Hindus in the west – different races and different religions. As a result, the Chin had no immediate link or direct connection with peoples of their own new faith, other than of their own nationality, with whom they could identify as brothers and sisters.

Although isolationism, which provided the means of preserving Chin identity as a distinct people and nation before the colonial period, was broken down by the powerful outside forces of change represented by the British colonial power and its successors (independent Burma and India), and there were more and more social interactions between the Chin and other peoples, the process of de-tribalization did not become a process of de-humanization. Instead, the process of de-tribalization helped to strengthen the Chin's concept of national identity or Chin-ness in the midst of the multi-racial and multi-religious environments of both the colonial and post-colonial periods. This was mainly due to the fact that Christianity has positively provided the Chin people with a new world-view and value system, which apparently better equipped them to cope with the new challenges and change thrust upon them. In this way, Christianity provided the Chin people a means of preserving their national identity and promoting their interests in the face of powerful forces of change.

As Christianity and Chin-ness became inseparable phenomena in a new Chin society, Christianity or the church also played a very important role in the people's social and political lives, not just their religious lives, as they adapted to multi-ethnic/multi-religious environments which the Chin had never faced before. Chin self-awareness and common identity, especially after the colonial period, mirrored Chin political identification with Christianity. This political identification was very important for the Chin, 'even for non-converts', as Lehman observes.

> Political identification with Christianity, with the Church, gave the Chin a basis for treating the Burman on more or less equal footing. The Chin are traditionally animists, and as such are particularly looked down upon by traditional Burmans. To appear to the Burman as people with a literate and sophisticated cultural tradition, the Chin had, so to speak, two choices: to become Buddhists in the Burman mould, or to adhere to some other world religion. The first course was unacceptable to most Chin because it would have amounted to giving in to the Burman, so the Chin took the second course (1981: 5).

Lehman's observation is quite correct if looked at in terms of the new world-view and value system, which Christianity provided the Chin. This new perspective allowed the Chin to cope with the many problems of a wider world, which they had never had to face before, and it made it possible for them to stand on an equal footing with the Burman and other peoples. However, their 'political identification with Christianity and with the Church' was not the only reason that 'the Chin took the second course', that is, mass conversion to Christianity. Similar theological concepts between traditional Chin religion and Christianity, especially the concept of a Supreme God and the theology of eschatology, also played an important role, for it could channel transition and continuity between the two belief systems. Buddhism could not provide the concept of a Supreme God and a similar eschatological understanding of life after death, which were fundamental doctrines in both traditional Chin religion and Christianity.

While similarities in theology facilitated the transition between traditional Chin religion and Christianity and strengthened a broad Chin national identity, this broader identity has also elicited a serious challenge to the Chin in the political context of the Union of Burma. This was true when Burman politicians such as U Nu opted for a confessional policy on religion in the late 1940s and it remains true today. This suggests that a serious quest for inter-religious dialogue must be pursued by all, a dialogue, which is concerned with the basic issues of human rights as well as profound religious convictions, concepts and views of life.

APPENDIX I

Pioneers in Ethnic Studies

Early Ventures in the Study of Ethnicity

Ethnicity is a process or phenomenon, which existed long before any scholar paid attention to it. Unlike many other disciplines, the significance of the study of 'ethnicity' therefore is that 'the ethnic factor in many societies has been forced upon us more by events than by research' (Yinger 1994: 331). One of the first and most influential scholars of what we could call today 'ethnic consciousness' or 'ethnic identity' was the German nationalist and philosopher Johann Herder (1744–1803). He believes that 'humanity as a species is on a mission of higher development and achievement but that the agent of this development is not the species as a whole nor the individual but an intermediate level, the group' (Eller 1999: 50). According to Herder, 'the essential group is not just any type but a particular type, the *national group,* or *nationality*'. This is because each national group, he argues, is an 'organic unit', a 'national organism' with its 'own unique and natural qualities and genius, its own special culture and language, its own national soul' (ibid.: 51). Being that the 'natural and the national were synonymous in Herder's mind', national group as natural unit 'becomes a single being, an individuality, a personality, in which culture is the national personality, the group mind'. Herder therefore uses such terms as *Nationalgeist, Seele des Volks, Geist der Nation, Geist des Volks* to capture this national peculiarity that is, to him, 'inexpressible' (ibid.: 51).

Max Weber (1864–1920), the founder of modern sociology, was probably among the first scholars to employ terms such as 'ethnic groups', 'nations', and 'nationalism' in the modern concept of our daily usage. While he emphasizes cultural and political factors as key influences on economic development and individual behaviour, he also maintains that 'race does not give rise to an identity or a group until it is recognized *subjectively* as common and salient' and until it is employed as 'the basis for joint (mainly political) action' (Weber 1968: 389; A. Smith and J. Hutchinson 1996: 35–40). Here Weber defined race as 'common inherited and inheritable traits that actually derive from common descent', and from this understanding of descent he offers his definition of ethnic groups as follows:

> Those human groups that entertain a subjective belief in their common descent because of similarities of physical type or of customs or both, or because of memories of colonization and migration; conversely, it does not matter whether or not an objective blood relationship exists (Weber 1968: 389).

In short, for Weber, descent relations are neither a necessary nor a sufficient condition for identity and group formation, until and unless they are employed subjectively by the group themselves. Similar to Weber's subjective definition, DeVos also defines it as 'subjective symbolic or emblematic use of any aspect of culture [by a group], in order to differentiate themselves from other groups' (DeVos 1975: 5).

In addition to European Romantic Nationalists such as Herder and early modern sociologists such as Weber, modern anthropologists also contributed a great deal of the study in early ethnology, which eventually became the foundations for a Social Science of Ethnicity in our time. The great early ethnographers such as Malinowski, Evans-Pritchard, Mead and others 'tended to give us images of discrete peoples living in traditional cultures in general isolation from other groups and from outside influences, especially Western influence' (Eller 1999: 65). Certainly, colonialism of *Pax Britannica* or *Pax Teutonica* made much of this anthropological investigation possible and important, and anthropology therefore has an undeniable debt to colonialism for some of the interests, problems and methods it came to depend upon. In fact, many anthropologists were in direct or indirect service of colonial administrations, and often colonialism and the changes it wreaked on traditional societies appeared at the forefront of the literature.

Burma under British colonial rule produced many such outstanding works. One of the most influential is John S. Furnivall's *Colonial Policy and Practice: A Comparative Study of Burma and Netherlands India,* written soon after World War II at the request of the British government of Burma in their search for solutions to the socio-economic and political problems that arose after the war. It was first published in book form in 1948. He analysed the origins and structures of the colonial governments in Burma and the Dutch East Indies. In Burma, he argues, colonial rule destroyed the social structure, which was based 'on personal authority, custom and religion' (Furnival 1948: 92). Furthermore, colonial rule in Burma aided in the rise of a polarized society, which led to racial labour divisions and the growth of crime, the latter especially evident in Lower Burma. Thus, the representative government introduced by the British colonial power under such arrangements as the 1921 Act of Home Rule and the 1935 Burma Act failed to integrate the society. Instead, it set the stage for the activation, aggravation and escalation of 'sectional friction' (ibid.: 487). The book analysed the pre-colonial economy of Burma and the devastating impact that the capitalistic West had on Burmese traditional ways of life, as had his 1931 publication *An Introduction to the Political Economy of Burma*.

Furnivall described Burma as a 'plural society' – one that consisted of 'several groups living side by side but separately, not united for common welfare or for any other common end, but divided from one another by the common desire for individual profit' (1948: 487). For Furnivall, 'plural society' has two domains of meaning, political and economic. Politically, a plural society comprises separate racial sections; there is a mix of different groups, each with its own culture, language, religion, etc., but more crucial than their co-presence is their social segregation. 'They mix but do not combine' (1948: 304). They live side by side as citizens of the same polity but do not form a society in any significant way; they constitute a sort of caste system. Although Furnivall recognized the plural nature of Burmese society, he did not mention anything about the pre-colonial independent status of the Chin people and other ethnic national groups in what was then British Burma. However, Furnivall's works and his analysis of the 'plural society' in Burma are now recognized as the major pioneering contribution to the study of 'ethnicity' by such well-known scholars in the field as John Hutchinson, Anthony D. Smith, John Armstrong, Jack D. Eller and others.[1]

Another major contribution in the study of early ethnography and relations between ethnic groups and cultures appeared in 1954 when Edmund Leach, a former British administrator in Burma, published his now classical book *Political Systems of Highland Burma: A Study of Kachin Social Structure*. Edmund Leach convincingly argues that ethnic groups cannot be understood merely in terms of culture or a social system. He rejected the traditional anthropological notion of 'equilibrium in the society' by saying 'the ordinary ethnographic conventions as to what constitutes *a* culture or *a* tribe are hopelessly inappropriate' (Leach 1954: 281). He therefore suggests a more appropriate way of unit analysis in the ethnographic context, which in his case was the entire Kachin State, then called the Kachin Hills Area, with its many cultures, languages and named collectivities; and he reminds us that such collectivities are not 'social isolates' but elements in a larger and more inclusive social system (ibid.: 291).

Leach's second thesis, in which he argued against '*stable* equilibrium', states that the collectivities in society which compose the system, and indeed the system itself, should not be seen as *stable* through time because the very form of the system may change and the groups of the community or village may undergo social structural transformation from one political system to another, or from one belief system to another, as time proceeds. Since the social reality of political structures – in his case, the Kachin political structures of *Gumsa* and *Gumlao* – are essentially unstable, he maintains that 'they only become intelligible in terms of the contrast provided by the polar types' (ibid: 9), that is, the feudal hierarchy type of political system called *Gumsa,* and the democratic independent village-state system[2] (which Leach considered anarchistic and egalitarian) called *Gumlao.* At another level, the polar types of political systems can be seen in the contrast between the Kachin *Gumlao* system and the Shan autocracy called *Sawbua.* In this way, he proposes a new theory of what he calls the 'structural description of a social system' which he said would 'provide us with an idealized model which states the correct status relations existing between groups within the total system and between the social persons who make up particular groups' (ibid: 10). His theory is now commonly known as 'structural-functionalism' (Erickson 1999: 103). It is applied mainly in the analysis of socio-cultural change and conflict.

Since Leach views culture and society as open, fluid and even invented, he concludes his argument by saying that cultural categories and cultural identity are not objective. They are not tangible things, but subjective and symbolic. The identity or boundary of social collectivity in his view therefore 'is not necessarily ascertainable in the realm of empirical facts; it is a question, in part at any rate, of the attitudes and ideas of particular individuals at a particular time' (Leach 1954: 288).

Two Schools of Thought

In 1963, Clifford Geertz published his major work *Old Societies and New States: The Quest for Modernity in Asia and Africa,* which eventually became a pioneer work in the then developing school of thought on the study of ethnicity called 'primordialism'. In the very same year, Glazer and Moynihan published their book *Beyond the Melting Pot: The Negroes, Puerto Ricans, Jews, Italians, and Irish of New York City,* which was significantly different from Geertz's book, and eventually became the landmark for another school of thought called 'circumstantialism' or 'instrumentalism'. Since then the two schools have produced several interesting studies of the political role of ethnic ties and movements in various part of the world.

In *Old Societies and New States,* with a number of scholars contributing, by far the most influential chapter is 'The Integrative Revolution: Primordial Sentiments and Civil Politics in the New States', written by Geertz himself. In this epoch-making paper, Geertz included Burma as one example of ethnic diversity and the plural nature of newly independent states in Asia. The term 'primordial' was first applied in 1957 by Edward Shils, who was influenced by his reading in the sociology of religion. Shils sought to distinguish certain kinds of social bonds – personal, primordial, sacred and civil ties – and to show how even in modern, civil societies other types of social bonding persisted (cf. Hutchinson and Smith 1996: 41–42). This basic idea was taken up by Geertz, who effectively applied the term and the concept to illuminate problems of the new states in Asia and Africa stemming from 'the primordial diversity of the enclosed societies within the states' (Geertz 1973: 259–260). This diversity often complicated the creation of a statewide society or 'civil order'. He went on to explain his notion of 'primordial attachment' as follows:

> By a primordial attachment is meant one that stems from the 'given' – or, more precisely, as culture is inevitably involved in such matters, the assumed 'given' – of social existence: immediate contiguity and kin connection mainly, but beyond them the givenness that stems from being born into a particular religious community, speaking a particular language, or even a dialect, and following particular social practices. These congruities of blood, speech, custom, and so on, are seen to have an ineffable, and at times overpowering, coerciveness in and of themselves. One is bound to one's kinsman, one's neighbor, one's fellow believer, *ipso facto*; as the result not merely of personal affection, practical necessity, common interest, or incurred obligation, but at least in great part by virtue of some unaccountable absolute import attributed to the very tie itself. The general strength of such primordial bonds, and the types of them that are important, differ from person to person, from society to society, and from time to time. But for virtually every person, in every society, at almost all times, some attachments seem to follow more from a sense of natural – some would say spiritual – affinity than from social interaction (ibid.: 260).

Geertz continued by saying that such primordial elements as blood ties, race, language, region, religion and custom 'are usually involved concurrently, sometimes at cross-purposes with one another' (ibid: 261). In a nutshell, Geertz's primordial interpretation of ethnicity is something given and persistent in social affairs, a 'primordial' and even sacred tie between the members of the group. Ethnic attachments are among several that appear 'natural' for individuals, being rooted in the non-rational foundations of personality. For Fisher and van den Berghe, who follow the primordial school of thought, these 'attachments are grounded in man's biological and genetic make-up, and are based upon genealogical descent, or at any rate a belief in such common ancestry' (A. Smith 1983: xxix).

Glazer and Moynihan's book and its 'instrumentalist school of thought' present the exact opposite, contending that such attachments are liable to considerable change over time, that people may come to view their linguistic, religious, kinship and territorial attachments quite differently in highly modern societies, with some attachments falling by the wayside. Even the belief in common descent, though powerful and widespread, may not encompass all culture-defined groups which lay claim to special rights on account of their affinities and features. In their view, 'ethnic groups are formed and mobilized by specific social conditions and for special purposes. These conditions may include minority status, discrimination, stratification, segregation and others; the purpose may include equality, inclusion, social justice, economic opportunity, cultural preservation, and so forth' (Eller 1999: 72). They therefore argue that 'we cannot completely understand ethnicity and ethnic conflict by looking at the group in isolation or at the cultural content of the group' (ibid.: 72). Rather, as Eller observes them, 'We must add to this analysis the relation of the group to its context, to the things outside it, especially the dominant society and its categories, structures, and stratification. The ethnic group and its behavior are not *priori* facts but social products of collectivities, cultures and circumstances' (ibid.: 72).

In 1969 the Norwegian anthropologist Fredrik Barth published his work *Ethnic Groups and Boundaries,* which eventually became the 'bible for the instrumentalist or circumstantialist school of thought', as Jack Eller put it. Barth maintains that ethnic groups can be understood 'as a form of social organization, a particular species of social categorization that classifies a person in terms of his basic, most general identity, presumptively determined by his origin and background' (Barth 1969: 13). However, Barth argues that 'social identity is not entirely ascribed, that people can in fact change their identity and their social/ethnic affiliation in many cases', and that the continuity of ethnic groups depends neither on biological nor cultural continuity, but 'the continuity of ethnic groups ... depends on the maintenance of a boundary. The cultural features that signal the boundary may change, and the cultural characteristics of members may likewise be transformed, indeed, even the organizational form of the group may change, yet the fact of continuing dichotomization between members and outsiders *always remains*' (ibid.: 14). In short, for Barth, it is 'the ethnic boundary that defines the group, not the cultural stuff that [it] encloses' (ibid.: 15). In his view, this boundary is a process of functions within what he called the 'structuring of interaction'. He argued that ethnic categories 'provide an organizational vessel' within which people can ascribe characteristics to individuals and judge the performance of individuals by categorical standards. Each ethnic category comprises a specific field of meaning and value, consisting of mutually exclusive rules and criteria of evaluation. Ethnic identities, therefore, 'function as categories of inclusion/exclusion and interaction', in other words, ethnicity situated in the realm of social interaction between the groups. In this way, Barth concludes his theory by emphasizing that ethnic groups are socially constructed (ibid: 14–15).

According to Paul Brass, however, it is possible to reconcile the two viewpoints, at least to the extent of recognizing their utility for different types of ethnic community; and he argues that 'one possible route towards reconciling the perspectives of primordialists and instrumentalists may lie in simply recognizing that cultural groups differ in the strength and richness of their cultural traditions and even more importantly in the strength of traditional institutions and social structure' (Brass 1985: 5).

NOTES

1 The term village-state is used by Frederick S. Downs in his study of the Christian movement among ethnic groups in the Northeast India. See Frederick Downs 1983 and 1992 (a).

2 See Hutchinson and A. Smith 1996; A. Smith 1986; Armstrong 1982; Eller 1999. In Burma, John Furnivall was mostly remembered as a colonial administrator, professor at Rangoon University and the founder of the 'Burma Research Society' in 1910 and the 'Burma Book Club' in 1924. In the 1950s he was the advisor to the government of newly independent Burma. When he died, *The Guardian* paid homage to him on 6 April 1960 as the 'Grand Old Man of Burmese Scholarship'.

APPENDIX II

Previous Literature in Chin Studies

The Chins did not learn the art of writing until British administrators and American missionaries reduced their language to written form. Thus, the earliest written works about the Chins were produced almost exclusively by Westerners, mainly British military and administrative officers who had some connection with the Chins after the British East India Company occupied Chittagong, the country neighbouring the Chins, in 1760. During the first hundred years of their relationship, they produced no substantial study on the Chins; only after the British authorities appointed Colonel Tom Hubert Lewin as Superintendent of the newly created Chittagong Hill Tracts District in 1859 did the earliest historical and anthropological studies on the Chin appear.

Thomas Lewin was the first scholar in the field. He produced at least three works on the Chin: *The Hill Tracts of Chittagong and the Dwellers Therein* (1869), *Wild Races of South-Eastern India* (1870) and *A Fly on the Wheel or How I Helped to Govern India* (1912). Another of his pioneer works, *Lushai Primer,* became an indispensable handbook for British officers and missionaries in the early days. In addition to his classical historical, anthropological and linguistic studies of the Chin, his greatest contribution to Chin Studies was a map of pre-colonial Chinram, which he drew based on his own field research. He describes Chinram as an independent country ruled by independent tribal chiefs. Tom Lewin particularly explored the Western Chinram, then called the Arakan Hill Tracts, Chittagong Hill Tracts and Lushai Hills. He learned the Lushai (Mizo) dialect of the Chin language, dined with the Lushai chiefs, danced with the villagers and helped them in disease and distress. At the beginning, Lewin faced strong opposition, but in the end he became so dear to the Chins that they accepted him as their great chief and lovingly called him 'Thang Liana' (cf. Chatterjee 1990: 564–573).

Another British civil servant who contributed an early study on the Chin before Chinram became colonialized was C. A. Soppitt, who later became Assistant Commissioner of Burma and Sub-Divisional Officer in the North Cacher Hills. In his *A Short Account of the Kuki-Lushai Tribes on the North-*

Eastern Frontier with an Outline Grammar of Rangkhol-Lushai Language and A Comparison of Lushai with other Dialects (1893), Soppitt gave a detailed account of tribal developments among the Kuki-Lushai groups of the Chin from a linguistic perspective. His book was reprinted in 1978 by the Tribal Research Institute in Aizawl, Mizoram State, under the title of *Kuki-Lushai Tribes.* Lewin's and Soppitt's works provide us with the earliest substantial published information about the Chin.

In Burma, Father Sangermano briefly mentioned the Chin in his now classical work *The Burmese Empire* written in 1783, though only published a hundred years after his death, in 1893. Arthur Phayer, in his 1883 work, *History of Burma* also mentioned early Chin history in the Chindwin Valley.

A second type of writing comes from military officers who gave detailed accounts of military operations in what was then referred to as Chin-Lushai Country. These include R. G. Woodthorpe's *The Lushai Expedition, 1871–72* (1872), A. S. Reid's *Chin-Lushai Land: Including a description of the various Expeditions into the Chin-Lushai Hills and the Final Annexation of the Country with Maps and Illustrations* (1893), A. G. E. Newland's *The Images of War* (1894), Sir Charles Crosthwaite's *The Pacification of Burma* (1912) and L. W. Shakespear's *History of Assam Rifles* (1929). Apparently, this type of writing reflected the self-consciousness of the mighty army of the British Empire. The books were written from the British conqueror's perspective. From the Chin point of view, this kind of writing reflects occupation, exploitation, persecution and marginalization – in short, the reality of the historical experiences of the Chin people for many decades.

A third type of writing, on Chin-ness, or the ethnic origin of the Chin, should be regarded as paramount. It investigates the origin of the Chin and the 'primordial identity' based on such concepts as 'collective memories', 'belief and value systems', 'symbols and myths' – especially the 'myth of common descent', the analysis of common elements in their cultures and social structures, and a 'sense of solidarity' based on blood ties. Most of this literature was written with the aim of creating a single administrative unit for the entire Chin people under a single law and concomitant regulations. However, most such writings were kept by the British colonial power and its successors – the Indian and Burmese governments – as 'strictly confidential documents' and inaccessible to the public, researchers or scholars. Only recently, the Tribal Research Institute in Aizawl, sponsored by the government of the Mizoram State, reprinted a number of these early works, thus making them available to scholars. It also published, for the first time, confidential government reports.[1]

One of these, *Foreign Department Report on Chin Lushai Hills, September, 1892,* describes the intense efforts made by the British government to bring the entire Chinram – now split between three countries, India, Burma and Bangladesh – into a single administrative unit. Follow-up efforts resulted in two important books.

The first was Bertram S. Carey's and H. N. Tuck's *The Chin Hills: A History of the People, Our Dealing with Them, Their Customs and Manners, and a Gazetteer of their Country,* Volumes 1 & 2, (1896, reprinted in 1976 and 1983); the other was *The Chin Hills Regulation* (1896), drafted by B. S. Carey himself. The purpose of these books was to provide the blueprint for the administration of newly occupied Chinram into a single administrative unit. John Shakespear's *The Lushai Kuki Clans,* Parts 1 & 2 (1912) was also written in line with this purpose. Although Chinram was divided into three administrative units in the end, the studies nevertheless made important contributions to our understanding of the essential unity in ethnic origin of the Chin people. To these books should be added William Shaw's *Notes on Thado Kukis* (1929) and G. A. Grierson's monumental work, *Linguistic Survey of India, III, 3* (1904), both of which make the same point from different perspectives.

A fourth group of writings, subsequent to the establishment of British administration, concerns such subjects as the government's dealing with the Chin and the impact of Western culture upon the people. Sir Robert Reid's *The Lushai Hills* (1942) deals mainly with changes in the political life of the Chin people in today's Mizoram State, then called the Lushai Hills. Two works, which contain deeper anthropological insights, are N. E. Parry's *The Lakhers* (1932) and A. G. McCall's *Lushai Chrysalis* (1949). Unlike early military officers and administrators such as B. S. Carey and Captain Dury, Parry and McCall, Political Officer and Superintendent of the Chin Hills District respectively, were concerned about the traumatic cultural changes thrust upon Chin society as a result of the impact of Western culture through the colonial power and Christian missions. While admitting that the advent of British colonial power brought large-scale changes in Chin society, they illustrated that the most active, dynamic and sustained instruments of change were the Christian missions. As McCall put it, the role of the government was simply 'the provision of law and order, and a modicum of utility services' (McCall 1949: 199). In fact they tended to see the British government as the upholder and preserver of Chin customs against the 'full-scale assault' on those customs launched by the missionaries, whom, Parry believed, were denationalizing the people. Parry indeed sought to preserve the pristine nature of Chin culture and its traditional religion. While sharing the same concern, McCall believed that some changes were not only inevitable but desirable. While he admitted that changes were inescapable, he pleaded that they should be indigenously ignited and properly guided, and not forced upon the people by 'over-zealous' outsiders. His book was an attempt to explain government policies in this respect, and to provide guidelines for moral and economic development of the people (cf. Kipgen 1996: 8–9).

N. E. Parry's book *The Lakher* can also be categorized as representing the fifth group of writings, which particularly deal with traditional Chin religion and culture. I believe Parry's book is the most comprehensive and reliable work on the subject in this category of writing. Just one year after Parry published his book, American Baptist missionary C. U. Strait also presented his doctoral dissertation *A History of Interpretation of Chin Sacrifice* at the Berkeley Baptist Divinity School in 1933. As a missionary to Chinram, Strait had been able to conduct field study among the Chin. But his missionary bias, too obvious throughout his presentation, made him unable to see the deep meaning of traditional Chin religion. Moreover, he misinterpreted many Chin religious concepts and terminology. For instance, he described *Ding-thlu, Lai-rel* and *Do-Dang*, as different deities (Cf. Strait 1933, especially chapter 3). As I explained in Part One of this study, *Ding-thlu* and *Lai-rel* are not names of deities but attributes of the Supreme God, called *Khua-zing*. *Do-Dang* or *Do-Dang-tu* is not the name of a deity either, but the title of the chief who made the sacrificial animal offering to the guardian god called *Khua-hrum*, on behalf of the community. However, Strait's work should be recognized as *the* pioneer study on traditional Chin religion and its related culture. Among Chin scholars, Sing Kho Khai in his B.R.E thesis *The Theological Concept of Zo in the Chin Tradition and Culture*, presented at the Burma Institute of Theology in 1984, has produced rather a comprehensive work. He was the first scholar to systematically analyse the theological significance of Chin tradition and coined the term '*Tual* community' to describe traditional Chin society (cf. Sing Kho Khai 1984, especially chapter VI).

In 1943, H. N. C. Stevenson published *The Economics of the Central Chin Tribes*, in which he analysed traditional Chin social systems from a functional perspective, and applied economic theory to investigate Chin culture.[2]

Although his main emphasis was on economic production, focusing on agriculture, animal husbandry, forest products, land tenure and trade systems, he also analysed the interrelationship between the economic, political and religious aspects of life in Chin society under the heading 'distribution and consumption'. He correctly pointed out that the economic choice in Chin traditional society was 'conditioned by the customs which regulate the amounts payable as a marriage price, the nature of resources used at each stage in the Feast of Merit, and the traditional sacrifices for various types of illness', but acknowledged at the same time that this 'influence of custom on economic choice is not absolute' (Stevenson 1943: 6–7). Since his functional analysis of Chin society centred upon the economic system of 'agriculture activities', the religion or religious ritual, which, in his view, had 'little or nothing of [a] function in tribal life', took second or even the third place in his analytical structures. His denial of religion as the central component of Chin society made him fail to see the interconnection between various kinds of tax systems – especially religious taxation – and different political systems in Chin social structures. He certainly mentioned the religious taxes, called *Sakhua* and *Khuaman*, but without explaining their differences. *Sakhua* was a religious tax collected by the chief or headman under a political system called *Khuabawi*, where the *Tual Khua-hrum* worship took the central place in religious life. *Khuaman* was a tax regimen for a political system called *Ram-uk*, where *Tlang Khua-hrum* worship was practised. His negligence of religious roots in Chin social structure also weakened his analyses of Chin traditional political systems. Though he was the first scholar to deploy the term 'Tlaisun Democratic Council' for the Chin's indigenous form of democratic administration, he failed to categorize three kinds of traditional Chin political systems, namely the *Khuabawi* system, *Ram-uk* system, and Tlaisun Democratic Council. This kind of ignorance might be the reason why the British colonial power abolished the traditional Chin political systems of *Ram-uk* and the Tlaisun Democratic Council in the first place.

In 1963, F. K. Lehman published his book *The Structure of Chin Society*, which provides fresh information about Chin socio-cultural life. Lehman analyses the structure of Chin society based on his anthropological field study in the Chin State of Burma in 1957–58. He discusses Chin society and culture in terms of its adaptation to local resources and its response to Burman civilization, with chapters on Chin history, agricultural and economic systems, land tenure and inheritance, Northern Chin social system, Southern Chin social system, the conceptual structure of traditional Chin religion, Chin attitudes and psychological orientations and, social and cultural changes. The significance of Lehman's work lies in the fact that it was carried out just before almost the entire populations of the Chin State became Christian. His book, therefore, deals only with pre-Christian traditional Chin society. Social and cultural change for him is defined in terms of economic resource accommodation with the Burman civilization, not a deeper level of social and cultural change from the traditional ways of life – i.e. *Lai Phung* in Chin, to the Christian way of life, *Krifa Phung*. Secondly, Lehman conducted his field research just before Burma closed her doors to international researchers. In 1962, the Burmese Army led by General Ne Win took over the offices of state from a democratically elected government and practised self-imposed isolationism for almost three decades. Thus, after Lehman, no scholar from Western countries has been able to carry out field study in the Chin State of Burma until the present day.

The study of the impact of Christianity upon Chin society has mainly been undertaken by individuals concerned with both Western missionaries and Chin Christians. The perspective from which such writings are usually composed is one of missionary expansion. The earliest works of this type were Laura Carson's *Pioneer Trails, Trials and Triumphs* (1927) and Dr East's *Memoirs* of his days in East Chinram (now part of the present Chin State in Burma) as a missionary, circulated widely in America in 1914 but not published in Burma until 1983, and then under the somewhat misleading title of *Burma Manuscript.* Both Laura Carson's and Dr East's purpose was to inform Western readers and 'to encourage the supporters of missionaries in the West to support them even more generously', as Frederick Downs put it (Downs 1994: 24). Their intention was not to promote self-understanding among Chin Christians, nor did they write for the Chin audience. The Chin are mentioned, of course, but only as the objects of missionary work. In order to draw the sympathy of their readers, they described the hardship of their work by portraying Chin social life as darkness 'beyond description' and the Chin society as 'beyond the pale of civilization' (Carson 1927: 145). Their descriptions of the moral and religious life of the Chin are therefore unreliable because their purpose in writing was not to help their readers understand traditional Chin culture objectively, but to solicit their support in bringing about changes 'for the better' in that culture.

Unfortunately, the principle of writing adopted by Laura Carson and Dr East at the beginning of the twentieth century has still been applied by many scholars, such as Frederick Dickinson, Robert Johnson and even to some extent by Cung Lian Hup. Frederick Dickinson, who wrote under the pseudonym Maung Shwe Wa, still used the paradigm of traditional mission history in his *Burma Baptist Chronicle,* published in 1963, just three years before all foreign missionaries were expelled from Burma. The basic assumption of traditional mission history was that what mattered most was 'what [the] missionary did and thought' (Downs 1994: 25). Frederick Dickinson therefore recorded the establishment of Christian churches in Burma as the energetic work of Western missionaries, not as an integral part of the socio-cultural history of the local people. Hence, conversion to Christianity was viewed as 'conquering the native people', not as their response to the Gospel. In this way, the Chin response to the Christian Gospel was regarded merely as the object of the mighty missionary expansion from the West. Thus, Frederick Dickinson would describe the Chin as 'wild or uncivilized', a people who should be conquered in the name of Christianity and related Western civilization and 'development'.

Robert Johnson also wrote his massive two-volume work *The American Baptist Chin Mission* from this perspective in 1988, though the book was an improvement over what had gone before by being more sensitive to the importance of Chin culture. As a former missionary to East Chinram, Robert Johnson has the great advantage of knowing the languages – Chin and English – and the nature of the mission fields. Moreover, he has had full access to the Official Archives of the American Baptist Churches at Valley Forge, Pennsylvania, and used tremendous amounts of materials from it. The main objective of Johnson's book, stated in his Preface, was not to contribute to the interaction between Christianity and Chin culture, nor to Chin self-understanding of their identity as Christians. It was not even a critical study on the American Baptist Chin Mission. The purpose was mainly to provide 'a storehouse of information about this particular facet of foreign mission' (Johnson 1988: 8a). He thus compiled 'many dates, list of pastors, villages, students, convention leaders', and many other miscellaneous things and events, such as hymn numbers sung at special services and convention meetings, a wide variety of information about the American Baptist missionaries in Chinram and even how they educated their children, 'believing that all of this will help to make missionary life real and vibrant' (1988: 8a).

More recently, Cung Lian Hup presented his doctoral dissertation 'Innocent Pioneers and Their Triumphs in A Foreign Land' at the Lutheran School of Theology in Chicago in 1993. Though he is a Chin from the Thlantlang area, Cung Lian Hup seems still influenced by the traditional missiological perspective of mission expansion. He did not analyse his material from the local Chin perspective, nor did he view the Christian movement in Chinram as an integral part of Chin socio-cultural history. Moreover, he heavily relied on Johnson's book and failed to visit Rochester or Valley Forge where the original materials are officially preserved. Za Tuah Ngur also drew heavily on Johnson's book when she presented her master's thesis 'Towards An Understanding of Christian Mission from a Chin Perspective', at the Theology Department of Birmingham University, UK, in 1993. Za Tuah Ngur not only applied but strongly recommended the traditional perspective of mission expansion, or what she called the 'all conquering approach to mission' (1993: 6). Thang Za Vung's master's thesis 'Christian Mission in the Northern Chin Hills', was presented at the History Department of Mandalay University in Burma, in 1975. His materials come mostly from the Archive of the Burma Baptist Convention in Rangoon; he also used some local sources. Pun Suan Pau applied Donald A. McGarvan's theory of 'church growth' when he presented his doctoral dissertation *Growth of Baptist Churches in Chin State: The Chins for Christ in One Century Experience,* at Union Theological Seminary, Manila, in 1998. He discusses in detail rapid church growth after the missionary era in the Chin State, and argues that revival movements were the main factors behind church growth.

In his *Christianity in North East India: Historical Perspectives* (1983), *History of Christianity in India: North East India in the Nineteenth and Twentieth Centuries* (1992) and *Essays on Christianity in North-East India* (1994), Frederick S. Downs effectively applied a socio-cultural perspective when he analysed Christian movements in northeast India, which includes Western Chinram or the Mizoram State in India. Frederick Downs's theory has been adopted and effectively applied in the study of Chin Christian movements by his student, Mangkhosat Kipgen, in his 1996 doctoral dissertation *Christianity and Mizo Culture.* Mangkhosat Kipgen, along with Downs, argues that Christianity in West Chinram of Mizoram can best be 'understood when the revivals there are understood as instruments of indigenization' (Kipgen 1996: 13). Although Kipgen deeply investigated the cultural roots of the revival movements in West Chinram, he paid no attention to the interaction between the Christian movement and the Chin nationalist movement.

While generally agreeing with Frederick Downs and Mangkhosat Kipgen, especially with the theory that they have adopted, my target was to investigate the development at a deeper level of cultural roots, i.e. the gradual shift of conceptual structures and belief systems of traditional Chin religion and the interaction between the Chin Christian movements and socio-political developments, especially Chin nationalist movements, which immensely influenced both the indigenization of Christianity and church growth and Chin political identification with Christianity in both India and Burma.

NOTES

1 In Burma these kinds of documents are still inaccessible except for the scholars who are patronized, controlled and even indoctrinated by the military government.

2 Stevenson was a typical British colonial administrator. He was a trained anthropologist, like John Furnivall and Edmund Leach, and he was a student of the legendary Malinowski. He served in various posts of the colonial administration, such as a military officer, and he was commanding officer of the Chin Levies, the Superintendent of the Chin Hills District, and Director of Frontier Areas Administration of Burma; and he played an important role during the transition period of Burmese independence.

References

A. ARCHIVES

The Archive of Tribal Research Institute, Aizawl, Mizoram State, India

- Records on British administration in Chinram during the colonial period, especially the period between the 1890s and 1910s.
- Early studies on traditional Chin society, culture, religion and history, including original copies of T. H. Lewin's books.

The National Archives, Rangoon, Burma

- Records on British administration in East Chinram (Chin Hills) during the colonial period, especially after the Anglo-Chin War (1917–1919).

  (Previously preserved in DC Office in Falam, Chin State. Moved to Rangoon by order of General Ne Win in the early 1980s.)

American Baptist Historical Society, Rochester, New York, USA

American Baptist Foreign Missions Overseas Letters, Rochester, New York, USA

Microfilm Records File on Missionaries Who Served in East Chinram in 1899–1966, Rochester, New York, USA

(a) Rev. Arthur E. & Laura Carson	FM-9	1887–1900*
	FM-182	1901–1919
(b) Mrs Laura Carson	FM-234	1920–1924
(c) Dr. Erik H. East	FM-186	1901–1915
(d) Rev. Joseph H. Cope	FM-184	1908–1919
	FM-236	1920–1929
	FM-282	1930–1934
	FM-283	1935–1938

* Including their years in Thayetmyo, Mission to Asho tribe of Chin.

(e) Dr. John G. Woodin	FM-327	1910–1915
(f) Rev. Chester U. Strait	FM-259	1925–1929
	FM-303	1930–1939
	FM-341	1940–
(g) Rev. Franklin Nelson	FM-298	1939
	FM-343	1940–1949
	FM-387	1950–1955
(h) Rev. Robert Johnson	FM-339	1941–1949
	FM-384	1950–1959

Archives of the American Baptist Churches in the USA, Valley Forge, Pennsylvania, USA

- Original correspondences, including reports, minutes and letters.
- Files of individual missionaries who worked overseas in the service of the American Baptist Missionary Union. The biographical files on missionaries who served in East Chinram are as follows:

(a) Rev. Arthur Carson	B. 111
(b) Mrs Laura Carson	B. 171
(c) Dr Erik H. East	B. 248
(d) Rev. Joseph H. Cope	B. 201
(e) Dr John G. Woodin	B. 277
(f) Rev. Chester U.Strait	B. 251
(g) Rev. Franklin Nelson	B. 484
(h) Rev. Robert Johnson	B. 476

B. PRIVATE COLLECTION

Pu Sakhong's diaries, especially from his years as missionary-cum-teacher in Matupi in 1930s.

Sakhong, Pu (1969) 'Kan Pupa Thawn-keh-nak Kong' [The origin and genealogy of our forefathers]. Aibur: Manuscript.

—— (1971) 'Kan Pupa Phunglam le Biak-nak Kong' [The religion and traditions of our forefathers]. Aibur: Manuscript.

C. BILIOGRAPHY

Anderson, Benedict (1991) *Imagined Communities: Reflections on the Origin and Spread of Nationalism.* London: Verso (2nd ed.).

Anderson, Courtney (1987) *To the Golden Shore: The Life of Adoniaram Judson.* Valley Forge: Judson Press.

Appleby, Joyce (1998) 'The Power of History'. *The American Historical Review,* No. 1. February 1998, pp. 1–14

——, Lynn Hunt and Margaret Jacob (1994) *Telling the Truth about History.* London and New York: W.W. Norton & Company.

Armstrong, John (1982) *Nations before Nationalism.* London and New York: Columbia University Press.

Aung San Suu Kyi (1984) *Aung San of Burma: A Biography Portrait by His Daughter* Edinburgh: University of Queensland Press.

—— (1991) *Freedom from Fear.* London: Penguin Books.

Aung-Thwin, Michael (1985) *Pagan: The Origins of Modern Burma.* Honolulu: University of Hawai'i Press.

Barth, Frederick (ed.) (1969) *Ethnic Groups and Boundaries.* Boston: Little Brown.

Bealey, Frank (1999) *The Blackwell Dictionary of Political Science.* Malden, Massachusetts: Blackwell Publishers.

Bechert, Heinz (1984) 'To be a Burmese is to be a Buddhist: Buddhism in Burma'. In H. Bechert and R. Gombrich (eds), *The World of Buddhism.* London: Thames and Hudson, pp. 147–158.

Bigandit, Paul A. (1967 [1887]) *An Outline of the History of the Catholic Burmese Missions from the Year 1720 to 1887* (2nd ed.). Rangoon: Hanthawaddy Press.

Bosch, David J. (1993) *Transforming Mission: Paradigm Shifts in Theology of Mission.* Maryknoll, New York: Orbis Books.

Brass, Paul (ed.) (1985) *Ethnic Groups and the State.* Beckenham: Groom Helm.

Brown, G. (1933 [1904]) *Burma Gazetteer, Upper Chindwin.* Rangoon: Government Printing.

Butwell, Richard (1969) *U Nu of Burma.* Stanford: Standford University Press.

Cady, John F. (1958) *A History of Modern Burma.* New York: Cornell University Press.

Caldwell, John (ed.) (1961) *A Survey of World Missions.* Chicago: Moody Press.

Carey, B. S. and H.N. Tuck (1976 [1896]) *The Chin Hills: A History of People, Our dealing with them, their Custom and Manners, and Gazetteer of their Country.* Vol. 1 & 2, Rangoon: Government Printing. Reprinted, Aizawl: Tribal Institute.

Carr, Edward H (1963 [1961]) *What Is History?* New York: Alfred A. Knopf.

Carson, Laura S. (1927) *Pioneer Trails, Trials and Triumph.* New York: Baptist Board Publication.

Ceu Mang (1981) 'Zophei Thawh Keh Nak' [The origin of Zophei]. *Lai Hold Rel Tu.* Vol. 3, No. 4, November 1981, pp. 12–19

Chatterjee, Suhas (ed.) (1990) *Mizoram Encyclopaedia,* Three Volumes. Bombay, Delhi, Calcutta: Jaico Publishing House

Chawn Kio (1993) 'Laimi Thawh Keh Nak' [The Origin of the Laimi Tribe]. In *Krifa Buk Bau;* Vol.1, No. 3, February 1993, pp. 12–21.

Christensen, Torben (1977) *The Church in a Changing Society.* Uppsala: Uppsala University.

Christie, Clive J. (1996) *Modern History of Sourth East Asia.* London: Tauris Academic Studies.

Coupland, Robert (1945) *The Future of India.* Oxford: Oxford University Press.

Crider, Donald M. (1963) 'The Works Among Kachins'. In Sowards, G. and E. Sowards (eds), *Burma Baptist Chronicle: Book Two.* Rangoon: Burma Baptist Convention, pp. 367–382.

Crosthwaite, Sir Charles (1912) *The Pacification of Burma.* London: 1912. Reprinted London: Frank Cass & Co. Ltd., 1968.

Cung Lian Hup (1993) 'Innocent Pioneers and Their Triumphs in a Foreign Land'. Unpublished D.Th. dissertation, Lutheran School of Theology, Chicago.

Dawt Sung (1976) *Chin Lu-myo-mya ih Ne-cheh Sanh-cin-ze Tamaing* [A history of the Chin struggle against British imperialism]. M.A. thesis, published in Burmese, Mandalay University.

Dorman-Smith, Sir Reginald Hugh (1943) 'Foreword', in H. N. C. Stevenson, *The Economics of the Central Chin Tribes.* Bombay: The Times of India Press.

Downs, Frederick S. (1983) *Christianity in North East India: Historical Perspective.* Delhi: Indus Publishing Co.

—— (1992a). *History of Christianity in India: North East India in the Nineteenth and Twentieth Centuries.* Vol. V Part 5, Bangalore: The Church History Association of India.

—— (1992b). 'Identity: The Integrative Perspective'. *Asian Journal of Theology,* Volume 6, Number 2, October 1992, pp. 238–249.

—— (1994) *Essays on Christianity in North-East India.* New Delhi: Indus Publishing Co.

East, Erik Hjalmar (1984 [1914]) *Burma Manuscripts.* Falam: Zomi Baptist Convention.

Eliade, Mercea (1963) *Pattern of Comparative Religion.* New York: Meridian Books.

Eliade, Mercea (ed.) (1987) *Encyclopedia of Religion.* Vols. 5, 6, 8, New York: Macmillan Publishing Co.

Eller, Jack D. (1999) *From Culture to Ethnicity to Conflict: An Anthropological Perspective on International Ethnic Conflict.* Ann Arbor: The University of Michigan Press.

Elly, E. B. (1978 [1893]) *Military Report on the Chin-Lushai Country.* Aizawl: Tribal Research Institute.

Enriquez, C. M. (1924) *Races of Burma.* Rangoon: Government Printing.

Erickson, Paul A. (1999) *A History of Anthropological Theory.* Toronto: Broadview Press.

Firth, Raymond (1996) *Religion: A Humanist Interpretation.* London: Routledge.

FACE (Frontier Areas Committee of Enquiry Report) (1948) *Frontier Areas Committee of Enquiry, 1947.* Rangoon: Government Printing

(Foreign Department Report) (1980) *Foreign Department Report on Chin-Lushai Hill: September,* 1892. Aizawl: Tribal Research Institute.

Furnivall, John S. (1929) *Christianity and Buddhism in Burma.* Rangoon: Peoples' Literature Committee and House, Rangoon Diocesan Council.

—— (1948) *Colonial Policy and Practice: A Comparative Study of Burma and Netherlands Indies.* Rangoon: Government Press.

Gangte, T. S. (1993) *Kuki of Manipur.* Delhi: Gyan Publishing House.

Geertz, Clifford (1973) *The Interpretation of Cultures.* London: Fontana Press.

—— (1996) 'Premordial Ties'. In John Hutchinson and Anthony Smith, (eds), *Ethnicity.* Oxford and New York: Oxford University Press.

—— (ed.) (1963) *Old Societies and New State: The Quest for Modernity in Asia and* Africa. New York: Free Press.

Gellner, Ernest (1983) *Nations and Nationalism.* Oxford: Blackwell.

Gin Za Thang (1973) *Zomi Innkuan Laibu* [A history of Zomi tribe] Rangoon: Published by the author.

Glazer, Nathan and Daniel P. Moynihan (eds), (1963) *Beyond the Melting Pot: The Negroes, Puerto Ricans, Jews, Italians, and Irish of New York City.* Cambridge: MIT Press and Harvard University Press.

Goukhenpau, T. and S. Vungkhom (Signatories) (1960) *Re-Unification of the Chin People.* Imphal, Manipur: Azad Printing. (Memorandum Submitted by the Paite National Council to Prime Minister of India for Re-unification of Chin People of India and Burma under One Country.)

Gravers, Mikael (1993) *Nationalism as Political Paranoia in Burma.* Richmond: Curzon Press, Nordic Institute of Asian Studies (NIAS) series 11. Reprinted and substantially revised and expanded in 1999.

—— (1996a). 'Questioning Autocratic in Burma: Buddhism between Traditionalism and Modernism'. In David Westerlund (ed.) *Questioning the Secular State: The Worldwide Resurgence of Religion in Politics.* London: Hurst & Company, pp. 297–322.

—— (1996b).'The Karen Making of a Nation'. In Stein Tønnesson and Hans Antlöv (eds), *Asian Forms of Nation.* Richmond: Curzon Press, Nordic Institute of Asian Studies, Studies in Asian Topics, No. 23, pp. 237–269.

Grierson, G. A. (1967 [1904]) *Linguistic Survey of India, Vol. III: Tibeto-Burman Family,* Part III. Delhi: Motilal Banarsdass.

Hall, D.G.E. (1968 [1954]) *A History of South East Asia.* London: St Martin's Press.

Hallencreutz, Carl F. (1988) 'Policy of Religion: The New Framework'. In Carl F. Hallencreutz and Ambrose Moyo (eds), *Church and State in Zimbabwe.* Gweru, Zimbabwe: Mambo Press, pp. 1–25.

Hallencreutz, Carl F. and David Westerlund (1996) 'Anti-Secularist Policies of Religion'. In David Westerlund (ed.), *Questioning the Secular State: The Worldwide Resurgence of Religion in Politics.* London: Hurst & Company, pp. 1–23.

Harvey, G.E. (1967 [1924]) *History of Burma.* London: Frank Cass & Co. Ltd.

Head, R. E. (1910) *Haka Chin Customs.* Rangoon: Government Printing.

Herbert, Patricia (1982) *The Hsaya San Rebellion (1930–1931) reappraise.* Melbourne University Press.

Hobsbawm, Eric (1994) 'The Rise of Ethno-Linguistic Nationalism', in John Hutchinson and Anthony Smith (eds), *Nationalism.* Oxford: Oxford University Press, pp. 177–184.

—— (1997) *On History.* London: Abacus Books.

Hockings, P. et al. (eds) (1992) *Encycopedia of World Culture,* Volume III, South Asia. Boston: G. K. Hall.

Horton, Robin (1971) 'African Conversion', in *Africa,* Vol.XLV, No. 2, 85–108.

—— (1975) 'On the Rationality of Conversion', Part 1, in *Africa* Vol. XLV, No. 3, pp. 219–35.

Houtman, Gustaaf (1999) *Mental Culture in Burmese Crisis Politics: Aung San Suu Kyi and the National League for Democracy.* London: Institute for the Study of Languages and Cultures of Asia and Africa. Monograph series No. 33.

Hovemyr, Anders P. (1989) *In Search of the Karen King.* Uppsala: Studia Missionalia Upsaliensia.

Hre Lian Kio (1975) 'Animism in Chin Hills'. Unpublished B.R.E. thesis, Burma Divinity School.

Htin Aung, Maung (1962) *Folk Elements in Burmese Buddhism.* London: Oxford University Press.

Hunt, Lynn (ed.) (1989) *The New Cultural History.* Berkeley, Los Angeles and London: University of California Press.

Hutchinson, John and Anthony A. Smith (eds) (1994) *Nationalism.* Oxford and New York: Oxford University Press

—— (1996) *Ethnicity.* Oxford and New York: Oxford University Press

Hutchison, William R. (1982) 'A Moral Equivalent for Imperialism: Americans and the Promotion of Christian Civilization, 1880–1910'. In Torben Christensen and William Hutchison (ed.), *Missionary Ideologies in the Imperialist Era: 1880–1920.* Aarhus: Forlaget Aros, pp. 167–178.

—— (1987) *Errand to the World: American Protestant Missionary Thought and Foreign Missions.* Chicago & London: The University of Chicago Press.

Hutton, J. H. (1929) *Manipur Hills.* New Delhi: Government Printing.

Igger, George G. (1984) *New Directions in European Historiography.* Hanover, New Hampshire: Wesleyan University Press.

—— (1997) *Historiography in the Twentieth Century: From Scientific Objectivity to the Postmodern Challenge.* Hanover and London: Wesleyan University Press.

Johnson, Robert G. (1949) 'Haka Zubilee'. *The Baptist News,* April 19, 1949.

—— (1963) 'The Church in the Chin Hills'. In Maung Shew Wa (ed.), *Burma Baptist Chronicle.* Rangoon: Rangoon University Press, 383–397.

—— (1988) *History of American Baptist Chin Mission.* Vols 1–2. Valley Forge: Published by the author.

Kedourie, Elie (1994) 'Nationalism and Self-determination'. In John Hutchinson and Anthony Smith (eds), *Nationalism.* Oxford and New York: Oxford University Press, pp. 49–55.

Khin Yi (1988) *The Dobama Movement in Burma, 1930–1938.* Ithaca, New York: Cornell University Press.

Kin Oung (1993) *Who Killed Aung San?* Bangkok: White Lotus.

Kipgen, Mangkhosak (1996) *Christianity and Mizo Culture.* Jorhat, Assam and Aizawl: Mizo Theological Conference.

Kyaw Yin (1969) *Boyoke Aung San i Maint-khun-mya* [Collections of Aung San's speeches]Rangoon: Sapebih-hman Press.

Laisum, David C. (1994) 'Naming God in Burma Today'. Unpublished D. Mins. dissertation, University of Chicago.

Lal Thang Lian (1976) 'History of Mizo in Burma'. Unpublished M.A. thesis, Mandalay University.

Lattourette, K. S. (1975) *A History of Christianity,* Vol. 2. San Francisco: Harper-Collins Publishers.

Leach, E. (1964 [1954]) *Political Systems of Highland Burma: A Study of Kachin Social Structure.* London: University of London, 1964. (First published by G. Bell & Son Ltd.)

Lehman, F.K. (1963) *The Structure of Chin Society.* Urbana: Illinois University Press.

—— (1967) 'Ethnic Categories in Burma and Theory of Social Systems'. In P. Kunstadter (ed.), *Southeast Asian Tribes, Minorities and Nations.* Princeton: Princeton University Press.

—— (1987) 'Burmese Religion'. In M. Eliade, *et al* (eds), *The Encyclopedia of Religion,* Vol. 2. New York: Macmillan.

—— (1992) 'Chin', in *Encycopedia of World Culture,* Volume III, South Asia, ed. by P. Hockings. Boston: G.K. Hall.

—— (1999) 'On the name of Chin, Lai and Zo'. *Chin National Journal,* No. 7, March 1999.

—— (ed.) (1981) *Military Rule in Burma Since 1962.* Singapore: Maruzer Asia.

Lewin, T. H. (1869) *The Hills Tracts of Chittagong and the Dwellers Therein; with composition vocabularies of the Hill Dialects.* Calcutta: Calcutta Central Press Company.

—— (1977 [1912]) *A Fly on the Wheel or How I Helped to Govern India.* Calcutta. Repr. Aizawl: Tribal Research Institute.

—— (1978 [1870]) *Wild Races of South-Eastern India.* Calcutta. Repr. Aizawl: Tribal Research Institute.

Lian Uk (1968) 'Chin Customary Law'. Unpublished LL.B. thesis, Rangoon University.

—— (1995) 'Crusade Message for the Chin People'. *Burma Debate,* Vol. 1. No. 3. January 1995, 15–19.

—— (1998a). 'Centennial Year of the Chin Hills Regulation'. *Chin* Journal, No. 6, February 1998, 201–205.

—— (1998b). 'A Message on the Golden Jubilee of National Chin Day'. *Chin Journal,* No. 6, February 1998, 185–187.

Lincoln, Bruce (ed.) (1985) *Religion, Revolution, and Rebellion.* London: Macmillan Press.

Lintner, Bertil (1989) *Outrage: Burma's Struggle for Democracy.* Hongkong: Review Publishing Limited.

—— (1994) *Burma in Revolt.* Oxford: Westview Press.

Lowis, C. C. (1910) *The Tribes of Burma.* Rangoon: Government Printing.

Luce, G.H. (1959a) 'Chin Hills Linguistic Tour' (December 1954). *Journal of Burma Research Society.* Vol. XLII, No. 1, June 1959. Rangoon: Rangoon University Press, pp. 19–31.

—— (1959b) 'Old Kyaukse and the Coming of Burman'. *Journal of Burma Research Society,* Vol. XLII, No. 1, June 1959. Rangoon: Rangoon University Press, pp. 75–109.

Malcolm, (Rev.) Howard (1839) *Travels in South-East Asia,* 2 vols. London: Tilt.

Mang Tling, John (1988) 'Birth of the Union of Burma followed by Independence'. *Chin Journal.* No. 6, February 1998, 188–190.

Martin, Luther (ed.) (1993) *Religious Transformations and Socio-Political Change.* Berlin: Mouton de Gruyter.

Mason, Francis (1860) *The Karen Apostle: Memoir of Ko-Tha-Byu, the First Karen Convert, with Notices Concerning His Nation.* London: The Religious Tract Society.

Maung Maung (1959) *Burma's Constitution.* The Hague: Martinus Nijhoff.

—— (ed.) (1962) *Aung San of Burma.* The Hague: Martinus Nijhoff, for Yale University Southeast Asia Studies.

Maung Maung, U (1980) *From Sangha to Laity: Nationalist Movements in Burma,* 1920–1940. New Delhi: Manohar.

—— (1989) *Burmese Nationalist Movement (1940–1948).* Edinburgh: Kiscadale Publication.

McCall, A. G. (1949) *Lushai Chrysalis.* London: Luzac &Co.

McCullouch, Major M. (1955) *The Valley of Manipur.* New Delhi: Government Printing.

McLeish, Alexander (1929) *Christian Progress in Burma.* New York: World Dominion Press.

Miller, David (1995) *On Nationality.* Oxford: Clarendon Press.

Naylor, L.P. (1937) *A Practical Handbook of Chin Language.* Rangoon: Government Printing.

Newland, A. G. E. (1894) *The Image of War: Service on the Chin Hills.* Calcutta: Thacker & Sprink.

Ni Ni Myint (1983) *Burma's Struggle Against British Imperialism, 1885–1895*. Rangoon: Rangoon University Press

Nu, U. (1975) *Saturday's Son.* New Haven, Connecticut and London: Yale University Press.

Parry, N.E. (1976 [1932]) *The Lakhers.* London. Repr. Aizawl: Tribal Research Institute.

Phayre, A.P. (1967 [1883]) *History of Burma.* London. Repr. New York: Augustus M. Kelly.

Po, San C. (1928) *Burma and the Karen.* London: Elliot Stock

Pum Suan Pau (1998) *Growth of Baptist Churches in Chin State: The Chins for* Christ in One Century Experience. Manila: Union Theological Seminary.

Rambo, Lewis (1993) *Understanding Religious Conversion.* New Haven and London: Yale University Press.

Reid, Robert(1976 [1942]) *A Note on the Future of the Present and Partially Excluded Areas.* Repr. Aizawl: Tribal Research Institute.

Reid, S.A. (1983 [1893]) *Chin-Lushai Land.* Calcutta: Government Printing. Repr. Aizawl: Tribal Research Institute.

Sakhong, Lian H. (1988a) 'A History of Baptist Mission in Chin State of Burma'. Unpublished B.D. thesis: Burma Institute of Theology, Rangoon.

—— (1988b) 'A Historical Study of Chin Traditional Religion'. Unpublished M.A. thesis, Rangoon University.

—— (1983) 'Chin A-myo-tha Neh' [Chin National Day]. *Chin Magazine,* Rangoon University. Vol. 5, No. 2, May 1983. Published by Chins' Student Literature and Cultural Committee of Rangoon University. In Burmese.

—— (2000) *Religion and Politics among the Chin People in Burma, 1896–1949.* Doctoral Dissertation, Uppsala University: *Studia Missionalia Upsaliensia LXXX.*

—— (ed.) (1999) *Nyin-Chian-Swa A-tu Let-twi Ne-thaing-ze Wa-dah* [Peaceful co-existence: towards Federal Union of Burma.]. Chiangmai: Published by the National Reconciliation Program, the Democratic Forces of the Union of Burma.

Sakhong, Za Peng (1983) 'Lai History'. Unpublished M.A. thesis, Mandalay University.

Sangermano, Vincenzo (1995 [1893]) *The Burmese Empire: A Hundred Years Ago as Described by Father Sangermano* (Introduction and Notes by John Jardine). Westminster: Archibald Constable and Co. Reprinted, Bangkok: White Orchid Press.

Scherer, James A. (1987) *Gospel, Church, & Kingdom: Comparative Studies in World Mission Theology.* Minneapolis: Augsburg Publishing House.

Schecter, Jerold. (1967) *The New Face of Buddha: The Fusion of Religion and Politics in Contemporary Buddhism.* London: Victor Gollance.

Shakespear, J. (1912) *The Lushei Kuki Clans, Part 1 & 11.* London: Macmillan and Co., Limited.

—— (1955) 'The Kuki Lushei Clan'. *Journal of Royal Anthropological Institute,* Vol. 2, No. 1.

Shakespear, L.W. (1977 [1929]) *A History of Assam Rifles.* Reprinted Aizawl: Tribal Research Institute.

Shamsul A. B. (1996) 'Nations-of-Intent in Malaysia'. In Stein Tønnesson and Hans Antlöv (eds.) *Asian Forms of Nation.* Richmond: Curzon Press, Nordic Institute of Asian Studies, Studies in Asian Topics, No. 23, pp. 323–347.

Shaw, William (1929) *Notes on Thado-Kukis.* Edited with introduction, notes, appendices and index by J. H. Hutton. Calcutta: The Asiatic Society of Bengal on behalf of the Government of Assam.

Shwe Lu Maung (1989) *Burma: Nationalism and Ideology.* Dhaka: Dhaka University Press.

Shwe Wa, Maung (1963) *Burma Baptist Chronicle,* Book 1. Rangoon: Rangoon University Press for the Burma Baptist Convention.

Siang Uk (1984) 'Ni Tial Khua-vang' [The *Ni Tial* shaman]. *Lai Holh Rel Tu Magazine* , Vol. 4, No. 2; January 1984, pp. 17–22.

Silverstein, Josef (1993) *The Political Legacy of Aung San.* Ithaca and New York: Cornell University Press.

Sing Kho Khai (1984) 'The Theological Concept of Zo in Chin Tradition and Culture'. Unpublished B.R.E thesis, Burma Institute of Theology.

Slim, Sir William (1958) *Defeat into Victory.* London: Lansborough Publication.

Smith, Anthony (1983) *Theories of Nationalism.* New York: Holmes & Meier Publishers.

—— (1986) *The Ethnic Origins of Nations.* Oxford: Blackwell.

—— (1991) *National Identity.* Harmondsworth, Penguin.

Smith, Bardwell (ed.) (1978) *Religion and Legitimation of Power in Thailand, Laos, and Burma.* Cornell University Press.

Smith, Donald E. (1965) *Religion and Politics in Burma.* Princeton: Princeton University Press.

Smith, Martin (1991) *Burma: Insurgency and Politics of Ethnicity.* London: Zed Books Ltd.

Smith, William Robertson (1996 [1894]) 'Sacrifice: Preliminary Survey'. In *Lectures on the Religion of the Semites: The Fundamental Institutions* (2d ed. London: Black), reprinted in C. E. Carter and C. L. Meyers (eds), *Community, Identity, and Ideology: Social Science Approaches to the Hebrew Bible.* Winona Lake, Indiana: Eisenbrauns, pp. 43–64.

Snodgrass, Major J. J. (1997 [1827]) *The Burmese War, 1824–1826.* London: John Murray & Co. Repr. Bangkok: Ava House.

Soppit, C.A. (1978 [1893]) *A History of Kuki-Lushai Tribes.* Shillong: Government Printing. Repr. Aizawl: Tribal Society.

Sowards, E and U Zan (1963) 'Baptist Work Among the Karen' in G. Sowards and E. Sowards (eds), *Burma Baptist Chronicle: Book Two.* Rangoon: Burma Baptist Convention, 1963, pp. 304–325.

Sowards, G. & E. (eds.) (1963) *Burma Baptist Chronicle: Book Two.* Rangoon: Burma Baptist Convention.

Spiro, Melford E. (1980) *Buddhism and Society: A Great Tradition and Its Burmese Vicissitudes.* Berkeley, California: University of California Press.

Stevenson, H. N. C. (1943) *The Economics of the Central Chin Tribes.* Bombay: The Times of India Press.

—— (1944) *The Hill Peoples of Burma.* Burma Pamphlets No. 6. London: Government Printing.

—— (1963) 'Foreword'. In Vum Kho Hau, *Profile of Burma Frontier Man.* Bundung: Published by the author.

Strait, Chester U. (1933) 'A History and Interpretation of Chin Sacrifices'. Unpublished D.Theol. dissertation at Berkeley Baptist Divinity School.

Sundkler, Bengt (1980) *Bara Bukoba: Church and Community in Tanzania.* London: C. Hurst & Company.

Symleah, D. R. (1981) *Crown Colony Scheme,* reprinted as *Proceeding of North East India in a series of History Association,* 2nd Session, 1981.

Tegenfeldt, H. G. (1974) *A Century of Growth: The Kachin Baptist Church of Burma.* South Pasadena, California: William Carey Library.

Than Tun (1988) *Essays on the History and Buddhism of Burma.* Whiting Bay, Scotland: Kiscadale Publications.

Thang Za Vung (1975) 'Christian Missions in Northern Chin Hills'. Unpublished M.A.thesis, Mandalay University.

Tinker, Hugh (1967) *The Union of Burma: A Study of the First Years of Independence.* London and New York: Oxford University Press, 4th Edition.

—— (1986) 'Burma's Struggle for Independence: The Transfer of Power Re-examined'. *Modern Asian Studies,* Vol. 20, No. 3.

—— (ed.) (1984) *Burma: Struggle for Independence (1944–1948),* 2 volumes. London: Her Majesty's Stationery Office.

Torbet, Robert G. (1987) *A History of the Baptists.* Valley Forge: Judson Press, 3rd. edition.

Trager, Hallen C. (1966) *Burma Through Alien Eyes: The Missionary Views of the Burmese in the Nineteenth Century.* London: Asia Publishing House.

Tun Myint (1957) *Shan Pyi Khoi-thoih-ye Piah-ta-na* [The Shan state secession issue]. Rangoon: (In Burmese.)

Vaipum, Max (1980) 'Metaphysics, Parapsychology, and Supernatural Things in Chinland'. Unpublished B.D. thesis, Burma Institute of Theology.

Vum Kho Hau (1963) *Profile of Burma Frontier Man.* Bundung: Published by the author.

Vumson, Suantak (1986) *Zo History.* Aizawl: published by the author.

Wayland, (Rev.) Francis (1853) *Memoir of the Life and Labors of the Rev. Adonairam Judson, D.D.* Boston: Phillips, Sampson, and Company.

Weber, Max (1968) *Economy and Society, Vol.1.* Ed. Guenther Roth and Claus Wittich. New York: Bedminster Press.

—— (1996) 'The origin of ethnic groups' in John Hutchinson and Anthony D. Smith, ed., *Ethnicity.* Oxford, New York: Oxford University Press, 1996, pp. 35–40.

Woodman, Dorothy (1962) *The Making of Burma.* London: The Cresset Press.

Woodthorpe, R. G. (1978 [1873]) *The Lushai Expedition.* London. Repr. Aizawl: Tribal Research Institute.

White, Geoffrey M. (1995) *Identity Through History.* Cambridge: Cambridge University Press.

Yawnghwe, C. T. (1987) *The Shan of Burma.* Singapore: Institute of South-East Asia Studies.

Za Tuah Ngur (1993) 'Towards an Understanding of Christian Mission from a Chin Perspective'. Unpublished Master Thesis: University of Birmingham.

Zawla (1976) *Mizo Pi Pu te leh an Thlateh Cancin* [The ancestors of Mizo and their history] Aizawl: published by the author.

Index